Low-dimensional semiconductor structures

Low-dimensional semiconductor structures provides a seamless, atoms-to-devices introduction to the latest quantum heterostructures. It covers their fabrication, their electronic, optical, and transport properties, their role in exploring new physical phenomena, and their utilization in devices.

The authors begin with a detailed description of the epitaxial growth of semiconductors. They then deal with the physical behaviour of electrons and phonons in low-dimensional structures. A discussion of localization effects and quantum transport phenomena is followed by coverage of the optical properties of quantum wells. They then go on to discuss non-linear optics in quantum heterostructures. The final chapters deal with semiconductor lasers, mesoscopic devices, and high-speed heterostructure devices.

The book contains many exercises and comprehensive references. It is suitable as a textbook for graduate-level courses in electrical engineering and applied physics. It will also be of interest to engineers involved in the development of new semiconductor devices.

Keith Barnham received his PhD from the University of Birmingham. He is a Professor of Physics at Imperial College and the author of over 150 technical papers.

Dimitri Vvedensky received his PhD from the Massachusetts Institute of Technology. He is a Professor of Theoretical Condensed Matter Physics at Imperial College and the author of over 200 technical papers.

LOW-DIMENSIONAL
semiconductor structures

Fundamentals and device applications

Edited by

Keith Barnham and
Dimitri Vvedensky

CAMBRIDGE
UNIVERSITY PRESS

PHYS

PUBLISHED BY THE PRESS SYNDICATE OF THE UNIVERSITY OF CAMBRIDGE
The Pitt Building, Trumpington Street, Cambridge, United Kingdom

CAMBRIDGE UNIVERSITY PRESS
The Edinburgh Building, Cambridge CB2 2RU, UK
40 West 20th Street, New York, NY 10011-4211, USA
10 Stamford Road, Oakleigh, VIC 3166, Australia
Ruiz de Alarcón 13, 28014 Madrid, Spain
Dock House, The Waterfront, Cape Town 8001, South Africa

http://www.cambridge.org

First published 2001

Printed in the United Kingdom at the University Press, Cambridge

Typeface *Times*   System *3B2*

A catalogue record for this book is available from the British Library

Library of Congress Cataloguing in Publication data

Low-dimensional semiconductor structures: fundamentals and device applications /
edited by Keith Barnham and Dimitri Vvedensky.
    p. cm
  Includes bibliographical references and index.
  ISBN 0 521 59103 1
    1. Low-dimensional semiconductors. 2. Crystal growth. 3. Quantum Hall effect. I.
Barnham, Keith, 1943–II. Vvedensky, Dimitri D. (Dimitri Dimitrievich)

QC611.6.S9 L67 2000
537.6'226–dc21   99-04544:

ISBN 0 521 59103 1 hardback

Contents

2   Electrons in Quantum Semiconductor Structures: An Introduction
E. A. Johnson

3   Electrons in Quantum Semiconductors Structures: More Advanced Systems and Methods
E. A. Johnson

6 Electronic States and Optical Properties of Quantum Wells
J. Nelson

7   Non-Linear Optics in Low-dimensional Semiconductors
C. C. Phillips

8   Semiconductor Lasers
A. Khan, P. N. Stavrinou and G. Parry

9   Mesoscopic Devices
T. J. Thornton

Contributors

Professor M. P. Blencowe
Department of Physics and Astronomy
Dartmouth College
Hanover
NH 03755-3528
USA

Dr J. J. Harris
Centre for Electronic Materials
  and Devices
The Blackett Laboratory
Imperial College
London SW7 2BZ
UK

Dr E. A. Johnson
The Blackett Laboratory
Imperial College
London SW7 2BZ
UK

Dr A. Khan
Centre for Electronic Materials
  and Devices
The Blackett Laboratory
Imperial College
London SW7 2BZ
UK

Professor A. MacKinnon
The Blackett Laboratory
Imperial College
London SW7 2BZ
UK

Dr J. Nelson
The Blackett Laboratory
Imperial College
London SW7 2BZ
UK

Professor G. Parry
Centre for Electronic Materials
  and Devices
The Blackett Laboratory
Imperial College
London SW7 2BZ
UK

Professor C. C. Phillips
The Blackett Laboratory
Imperial College
London SW7 2BZ
UK

Dr P. N. Stravrinou
Centre for Electronic Materials
  and Devices
The Blackett Laboratory
Imperial College
London SW7 2BZ
UK

Professor T. J. Thornton
Department of Electrical Engineering
Arizona State University
Tempe
Arizona 85287-5707
USA

Professor D. D. Vvedensky
The Blackett Laboratory
Imperial College
London SW7 2BZ
UK

Preface

Everyone who studies, develops or utilizes modern semiconductor materials must be aware of the importance of low-dimensional structures in optical and electronic devices, crystal growth, semiconductor theory and experiment, and semiconductor material science and chemistry. Virtually every major university in the world has one or several growth facilities dedicated to basic studies of growth processes, the fabrication of heterostructures for device applications, or exploration studies of the properties of new structures. Such facilities reside in physics, chemistry, materials science departments or in electrical, chemical, or mechanical engineering departments. These factors underscore both the inherent interdisciplinarity of modern semiconductor science and technology as well as the need for basic textbooks which are appropriate for students across these disciplines.

This book is aimed at the graduate-level student who has completed a first degree in physics, material science, chemistry, or one of the major engineering disciplines. It is based on an advanced Masters-level course which has been given at Imperial College for some years. Like the Masters course itself, the interdisciplinary nature of the subject is reflected in the choice of authors from different departments, colleges and from the University of London Interdisciplinary Research Centre for Semiconductor Materials (now the Centre for Electronic Materials and Devices). Many of the exercises which follow each chapter, and which we regard as an integral part of this book, have been tested by our students either as assignments during the course or as adaptations of examination questions. Detailed solutions to some of these exercises may be found at the end of this book.

Low-dimensional structures entered modern semiconductor technology from a number of different directions. The development of techniques for the controlled growth of semiconductor interfaces was of evident importance as an enabling methodology to attain reduced dimensionality and, hence, Chapter 1 describes our current understanding of the fundamentals of growth processes. However, in parallel with these advances, device developments were also pointing the way to reduced dimensionality. In the 1970s, the field-effect transistor took over from the bipolar junction transistor as the fundamental logic element in semiconductor chips. The current in a field-effect transistor is carried by electrons in an inversion layer, which formed the first practical example of a two-dimensional electron gas, as discussed in Chapters 2 and 10. When the electron confinement within an inversion layer becomes comparable with the de Broglie wavelength of the electrons and a perpendicular magnetic field is applied to confine electrons further in the two orthogonal directions, new and unexpected quantum physics emerge. This led to the discovery of the quantum Hall effect and the award of the Nobel Prize to Klaus von Klitzing and, later, to the discovery of the fractional quantum Hall effect and the award of the Nobel Prize to Horst Störmer, David Tsui, and Robert Laughlin. Quantum transport in low-dimensional systems is discussed in Chapter 5.

In parallel with these developments, the need for lower-threshold current in III–V lasers led to thinner active regions and the double-heterostructure device evolved into the quantum well laser (Chapter 8). Quantum confinement effects also led to the exploitation of different wavelengths within quantum wells. Their optical properties (Chapters 6) and, in particular, their non-linear optical properties (Chapter 7), turn out to be important for the logic elements of potentially much faster optical computers. Finally, the need for higher-speed operation led to fabrication technologies capable of processing smaller and smaller electronic devices. This enabled quantum confinement to be imposed in the other two dimensions directly by the architecture of the heterostructure, leading to a whole host of novel devices discussed in Chapters 3 and 9.

The completion of this textbook has required the assistance of a large number of people, including many of our colleagues, who have been extremely generous in providing original (and sometimes irreplaceable) diagrams. However, we would, in particular, like to thank Neal Powell for his excellent work in producing many of the figures.

Keith Barnham   *September 2000*
Dimitri Vvedensky   *London*

1 Epitaxial Growth of Semiconductors

D. D. Vvedensky

1.1 Introduction

Epitaxial growth is a process during which a crystal is formed on an underlying crystalline surface as the result of deposition of new material onto that surface. The study of this process dates back over 150 years, but it was not until the work of Louis Royer in the 1920s that the systematics of epitaxial growth began to be revealed (Royer, 1928). Royer carried out an extensive study of the growth of ionic crystals on one another and on mica, mainly from aqueous solution and, using optical microscopy, summarized his observations with a set of rules based on crystal structure. These rules led Royer to coin the term 'epitaxy', which is a combination of the Greek words *epi*, meaning 'upon', and *taxis*, meaning 'order', to convey the notion of growing a new crystal whose orientation is determined by a crystalline substrate and to distinguish epitaxial growth from polycrystalline and amorphous growth. A review of the history of epitaxial growth has been given by Pashley (1956).

The modern era of the epitaxial growth of semiconductors began with the work of Henry Theurer at Bell Telephone Laboratories in Murray Hill, New Jersey (Theurer, 1961). Motivated by the need to reduce the base resistance of discrete bi-polar transistors, Theurer demonstrated that thin epitaxial silicon layers could be grown on a silicon substrate. The idea that epitaxial structures could lead to new electronic and optical phenomena was founded on a suggestion in the late 1960s by Leo Esaki and Raphael Tsu (1970), then working at the IBM Research Laboratories in Yorktown Heights, New York. They proposed that structures composed of layered regions of semiconductors with different band gaps would have a spatially varying potential energy surface that would confine carriers to the narrower band-gap material. If there were few enough adjacent layers of this material, then the carriers could be confined within regions comparable to their de Broglie wavelength – the natural length scale that governs their quantum mechanical behaviour. For this reason, these narrow regions are now called 'quantum wells'. Electrons (and holes) in quantum wells were predicted to exhibit remarkable optical and transport properties that could be controlled by varying the width of the wells and their barriers.

At the time that Esaki and Tsu made their proposal, the available technology could not produce materials of sufficient quality to verify the predicted effects.

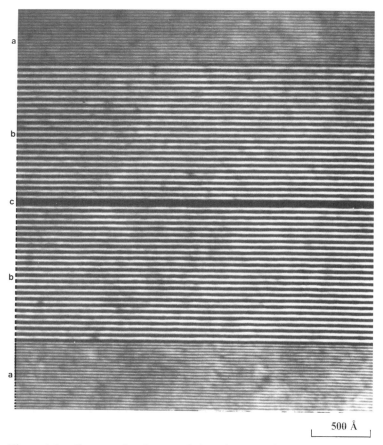

500 Å

Figure 1.1.   Cross-sectional transmission electron micrograph of a GaAs quantum well and GaAs and AlAs superlattices in an all-superlattice laser diode. At (a), (b) and (c) the layer thickness in monolayers (2.8 Å) and period number are $(4+4) \times 663$, $(8+8) \times 23$ and $(20) \times 1$. The dark regions correspond to the GaAs and the lighter regions to AlAs. (Courtesy J. Gowers and T. Foxon, Philips Research Laboratories, Redhill, UK)

However, with many major subsequent developments, epitaxial growth techniques have matured to the point where atomic-scale control of planar interfaces has become a matter of routine (Fig. 1.1). Quantum wells are now utilized in practical devices such as lasers (Chapter 8), which are used in compact disk players, and high-electron-mobility transistors (Chapter 10), which find application in satellite television receivers and mobile telephones. The control over interface definition and doping profiles has also made epitaxial structures a popular and fruitful testing ground for many fundamental ideas in condensed matter physics, such as the Wigner electron crystal, and has led to the discovery of new physical phenomena, such as the quantum Hall and fractional quantum Hall effects (Chapter 5).

This chapter will describe epitaxial growth techniques and their impact on the fabrication of quantum wells and other carrier-confining semiconductor materials, collectively referred to as 'quantum heterostructures'. Because the motion of carriers within these structures is restricted in one or more directions, the effective carrier dimensionality is reduced (Chapter 2), so the name '*low-dimensional* quantum heterostructures' is also used. We begin by reviewing the experimental and theoretical work that has shaped our understanding of the epitaxial growth process. The growth of quantum wells is then described briefly, but since the current interest in these and other planar heterostructures centres largely around utilizing their properties (which is covered at length in several chapters of this book), rather than their growth, most of our discussion will be devoted to heterostructures that confine carriers to one or zero spatial dimensions – called quantum 'wires' and quantum 'dots', respectively. Quantum mechanics predicts that the additional carrier confinement endows these structures with electronic properties which are superior even to those of quantum wells. But producing quantum wires and dots of sufficient quality to manifest all of the benefits of carrier confinement requires considerably more control over the growth process than that needed for quantum wells. We will review the strategies which have been developed to produce these lower-dimensional quantum heterostructures, highlight their successes, and discuss their future prospects.

A number of textbooks (Stringfellow, 1989; Tsao, 1993; Yang *et al.*, 1993; Barabási and Stanley, 1995; Markov, 1995; Herman and Sitter, 1996; Saito, 1996; Pimpinelli and Villain, 1998) have appeared in recent years that cover various aspects of epitaxial phenomena. These should be consulted for more extensive discussions than those provided here.

1.2   Epitaxial Growth Techniques

Advances in epitaxial growth techniques have played a pivotal role in the realization of ever more ambitiously designed quantum heterostructures. The experimental methodologies that have been developed to achieve the required control over composition, doping, and interface definition are described in this section. A summary is provided in Table 1.1 and a schematic depiction of the main methods is shown in Fig. 1.2.

1.2.1   Molecular-beam Epitaxy

The simplest way conceptually of fabricating semiconductor heterostructures is with a process known as molecular-beam epitaxy (MBE). This technique has its origins in a series of experiments, based on silicon, carried out by Bruce Joyce and his colleagues at the Allen Clark Research Centre of what was then the Plessey Company in Caswell, England, in the mid-1960s (Joyce and Bradley, 1966;

Table 1.1. A glossary of epitaxial growth techniques.

Method	Materials	Sources	Delivery
MBE	III–V, IV	Elemental	Molecular
MOMBE	III–V	Metal-organic (III), Elemental (V)	Molecular
CBE	III–V	Metal-organic (III), Hydride Organo-substituted Hydride (V)	Molecular
GSMBE	IV	Hydride, Chloride, Chloro-hydride	Molecular
MOVPE	III–V	Metal-organic (III), Hydride, Organo-substituted Hydride (V)	Hydrodynamic
VPE	IV	Hydride, Chloride, Chloro-hydride	Hydrodynamic

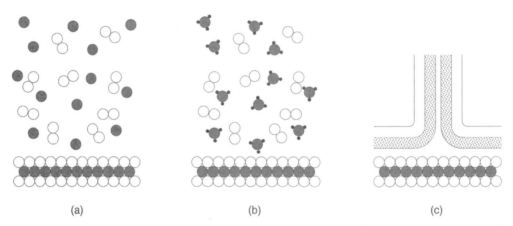

(a)   (b)   (c)

Figure 1.2.   Schematic illustration of each of the main methods of deposition for epitaxial growth of a compound semiconductor. (a) Evaporation of solid sources for molecular-beam epitaxy, (b) utilization of gas sources for metalorganic molecular-beam epitaxy, and (c) the hydrodynamic delivery of gas sources in metalorganic vapour phase epitaxy.

Booker and Joyce, 1966; Joyce *et al.*, 1967). Major developments, particularly in the application to III–V compound semiconductors, took place at Bell Telephone Laboratories in Murray Hill, New Jersey, some three to four years later, inspired by Al Cho and John Arthur. A historical review based on many of the seminal papers has been assembled by Cho (1994).

MBE is essentially a two-step process carried out in an ultra-high vacuum (UHV) environment. In the first step, atoms or simple homoatomic molecules which are the constituents of the growing material (e.g. atomic Ga and either As_2 or As_4 for GaAs, atomic Si for Si) are evaporated from solid sources in heated cells, known as *Knudsen cells* (Exercises 1–4), collimated into beams and directed toward a heated substrate which is typically a few centimetres in size (Fig. 1.3). The particles within these beams neither react nor collide with one another, i.e. the deposition onto the substrate is *ballistic* and the particles are said to undergo

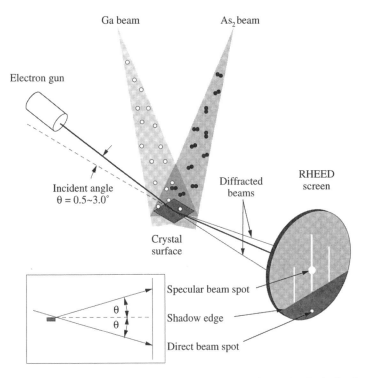

Figure 1.3.   The arrangement of the substrate, the RHEED (reflection high-energy electron diffraction; see Section 1.4.1) measurement apparatus, consisting of an electron gun and a collector screen, and the deposition of material within the UHV environment of an MBE growth chamber (after Shitara, 1992).

molecular flow – hence the name *molecular-beam* epitaxy. The substrate is often rotated for more uniform deposition rates across the substrate.

The second step of MBE is the migration of the deposited species on the surface prior to their incorporation into the growing material. This determines the profile, or *morphology*, of the film and its effectiveness depends on a number of factors, including the deposition rates of the constituent species, the surface temperature, the surface material, and its crystallographic orientation, to name just a few. The dependence of the morphology on the deposition rate of new material means that MBE (and other epitaxial growth techniques) are inherently *nonequilibrium*, or *driven*, processes. This provides an important distinction from crystal growth from a solution, where the supply of material to the growing crystal takes places by diffusion through the surrounding solution, and is therefore a *near-equilibrium* process (van der Eerden, 1993). Growth near equilibrium is governed almost exclusively by *thermodynamics*. For epitaxial growth, thermodynamics still provides the overall driving force for the morphological evolution of the surface, but the extent to which equilibrium is attained even locally is mediated by *kinetics*, i.e. the rates of processes that determine how a system evolves (Madhukar, 1983).

A major strength of MBE is that the UHV environment enables the application of *in situ* surface analytical techniques to characterize the evolution of the growing material at various levels of resolution – from microns to the arrangements of atoms. Particular techniques and the information they provide will be discussed in Section 1.4.

1.2.2   Vapour-phase Epitaxy

An alternative to deposition by molecular beams is the hydrodynamic transport of material to the substrate from gas sources (Stringfellow, 1989). In this scenario, which is called *vapour-phase epitaxy* (VPE), the constituents of the growing surface are delivered within heteroatomic molecules called *precursors*. For group IV materials, the precursors are hydrides, chlorides, or chloro-hydrides. The growth of III–V materials uses precursors for the group III species which contain carbon, i.e. they are metalorganic molecules, and the group V elements are supplied as hydrides, though, for reasons of safety, organo-substituted hydrides are often used. For these materials, therefore, this technique is referred to as *metal-organic* vapour-phase epitaxy (MOVPE).

The pressures inside a vapour-phase reactor can vary from 10^{-2} torr up to atmospheric, so the flow of the gas is viscous and the chemicals reach the substrate by diffusion through a boundary layer. Thus, the delivery of material to the growing film encompasses gas phase and surface chemical reactions, as well as mass transport within the injected fluid as it flows through the reactor, the latter being highly dependent on the system pressure and the reactor design (Jensen, 1993).

The use of gas sources has several attractive features for the epitaxial growth of semiconductor heterostructures. They can be used at room temperature, thus causing less contamination than higher-temperature sources and, with a very simple reactor design, can give a more uniform flux than that of a molecular beam, so that the substrate does not need to be rotated. An operational advantage over MBE is that, because there is no depletion, the growth chamber does not need to be opened and exposed to air to replenish the source material. However, this is compensated somewhat by the availability of multi-slice methods for both gas and solid sources, which enable the simultaneous growth of several substrates. Finally, an important practical disadvantage of vapour-phase techniques is that the gas sources can be highly toxic.

The comparative advantages of MBE and (MO)VPE are the subject of frequent and often lively debate, due in part to their strengths and weaknesses being in many respects complementary. MBE is carried out at relatively slow growth rates in a controlled UHV environment for which *in situ* diagnostic techniques for the growing film are readily available. On the other hand, (MO)VPE relies on the hydrodynamic delivery of material to the substrate, so only optical techniques are appropriate for *in situ* diagnostics. And yet, despite these differences, the choice between MBE and (MO)VPE is often dictated solely by materials issues. For example, the production of aluminum-containing heterostructures for optical applications, e.g. AlGaAs lasers, is well-suited to MBE but not to MOVPE, since the aluminum precursors used in MOVPE typically contain oxygen which, if incorporated into the material, acts as a non-radiative recombination centre that severely degrades the performance of the laser. Alternatively, materials containing phosphorus are usually produced with MOVPE because the residual phosphorus from the solid source used for MBE condenses as white phosphorus, which is highly pyrophoric.

A significant recent application of MOVPE is the development of light-emitting diodes (LEDs) and laser diodes (LDs) based on group-III nitrides (Nakamura and Fasol, 1997; Nakamura, 1998). These diodes are considerably brighter and last much longer than their counterparts based on other materials and have the important advantage of providing red, green, and blue LEDs within a single materials system (InGaN). Although the initial demonstration of a blue GaN LED, and certainly its almost immediate (two years!) commercialization, were due to a judicious use of MOVPE, there is now considerable effort aimed at producing such LEDs and LDs in the more controlled setting of MBE. The delivery of nitrogen to the substrate in this case is either in atomic form, produced by the dissociation of N_2 in a plasma, or as ammonia (NH_3), which undergoes thermal dissociation on the surface.

1.2.3   Molecular-beam Epitaxy with Heteroatomic Precursors

The desire to harness the complementary strengths of both MBE and (MO)VPE is embodied in a family of techniques based on the replacement of one or more of the elemental sources with heteroatomic (gas or liquid) sources whose delivery to the substrate is molecular (Abernathy, 1995; Foord *et al.*, 1997). There are several realizations of this approach (Table 1.1). In metal-organic molecular-beam epitaxy (MOMBE), a term used mainly for III–V materials, the group III source is a metal-organic molecule of the type used in MOVPE, but the group V source remains elemental. In chemical-beam epitaxy (CBE), both the group III and group V sources are heteroatomic, typically the same as those used in MOVPE. The term 'gas-source molecular-beam epitaxy' (GSMBE) is usually reserved for the growth of group IV materials, where the gaseous source material is silane (SiH_4) or disilane (Si_2H_6) for Si and germane (GeH_4) for Ge.

The main features of these methods are as follows. The absence of chemical reactions prior to arrival onto the substrate facilitates precise control over the thickness and composition of the growing film. The use of gas (or liquid) sources alleviates the need for opening the UHV system to air to replenish the Knudsen cells. Additionally, since growth occurs at low pressures, the *in situ* UHV diagnostic techniques used for MBE are also available here. But the use of heteroatomic precursors requires a more complex equipment design and there are new factors introduced into the growth kinetics because of surface chemical reactions, the resulting intermediates, and possible site specificity of these reactions. Nevertheless, systematic comparisons between heteroatomic and elemental source molecular-beam techniques have begun to reveal some of the atomistic origins of the differences between these two types of sources (Okuno *et al.*, 1990; Shitara *et al.*, 1993).

1.3   Epitaxial Growth Modes

Numerous experiments (Kern *et al.*, 1979, Venables *et al.*, 1984) have revealed that, for small amounts of one material deposited onto the surface of another (possibly different) material, the epitaxial growth morphology is one of three distinct types. By convention (Bauer, 1958; Le Lay and Kern, 1978), these are referred to as: *Frank–van der Merwe* morphology, with flat single crystal films consisting of successive complete layers; *Volmer–Weber* morphology, with three-dimensional (3D) islands that leave part of the substrate exposed; and *Stranski–Krastanov* morphology, with 3D islands atop a thin flat 'wetting' film that completely covers the substrate. These morphologies are illustrated in Fig. 1.4.

For lattice-matched systems, the Frank–van der Merwe and Volmer–Weber morphologies can be understood from thermodynamic wetting arguments based on interfacial free energies (Bauer, 1958). We denote the free energy of the epilayer/vacuum interface by γ_e, that of the epilayer/substrate interface by γ_i, and that of the substrate/vacuum interface by γ_s. The Frank–van der Merwe growth mode is favoured if

$$\gamma_e + \gamma_i < \gamma_s \qquad (1.1)$$

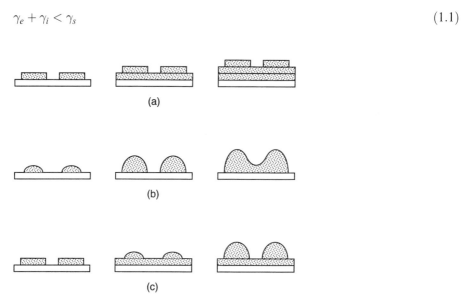

Figure 1.4.   Schematic evolution of the (a) Frank–van der Merwe, (b) Volmer–Weber, and (c) Stranski–Krastanov heteroepitaxial growth morphologies.

In this case, as the epilayers are formed, the free energy *decreases* initially before attaining a steady-state value for thicker films. Alternatively, if

$$\gamma_e + \gamma_i > \gamma_s \tag{1.2}$$

then Volmer–Weber growth is favoured. Here, the free energy *increases* if epilayers are formed on the substrate, rendering a uniform layer thermodynamically unstable against a break-up into regions where the substrate is covered and those where it is uncovered.

The Stranski–Krastanov morphology is observed in systems where there is appreciable lattice mismatch between the epilayer and the substrate. This growth mode is thought to be related to the accommodation of the resulting misfit strain, which changes the balance between the surface and interfacial free energies as the strain energy increases with the film thickness. Thus, although the growth of 'wetting' layers is favoured initially, the build-up of strain energy eventually makes subsequent layer growth unfavourable. Beyond this point, the deposition of additional material leads to the appearance of 3D islands within which strain is relaxed through the formation of misfit dislocations. However, there is another scenario within the Stranski–Krastanov morphology: the formation of islands *without* dislocations – called *coherent* islands (Fig. 1.5) – atop one or more wetting layers (Eaglesham and Cerullo, 1990; Madhukar and Rajkumar, 1990). This phenomenon, which has been observed for a number of systems (Petroff and

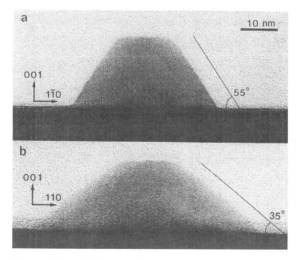

Figure 1.5.   High-resolution cross-section micrograph of an uncapped InP island on GaInP grown by MOVPE at 580 °C along the (a) [1$\bar{1}$0] and (b) [110] directions (Georgsson *et al.*, 1995). Note that the islands are elongated along [110] and that the planes of atoms are appreciably curved toward the centre of the island near the substrate caused by the compressive strain, but there is no evidence of any dislocations. (Courtesy W. Seifert)

DenBaars, 1994; Seifert *et al.*, 1996) and has the potential for many applications, will be discussed in Section 1.8.

Although wetting arguments based on interfacial free energies provide a useful classification scheme for the *equilibrium* morphology of thin films, the inherently kinetic nature of epitaxial growth means that many fundamental issues are left open. Foremost among these is the competition between different strain relaxation mechanisms. Consider, for example, the growth of InAs on the three low-index surfaces of GaAs. The Stranski–Krastanov morphology is observed only on the (001) surface; on the other two orientations strain relaxation involves misfit dislocation formation and a two-dimensional (2D) growth mode (Belk *et al.*, 1997; Yamaguchi *et al.*, 1997). For the growth of Ge and SiGe alloys on Si(001), strain relaxation can occur by several mechanisms of dislocation formation whose relative efficacies are determined by the morphology of the epilayer which, in turn, depends on the magnitude of the strain (Tersoff and LeGoues, 1994; LeGoues, 1996). These observations suggest that heteroepitaxial phenomena – particularly those involving materials with appreciable lattice mismatch – occupy a far richer and much more complex arena than arguments based solely on thermodynamics would suggest.

1.4   *In Situ* Observation of Growth Kinetics and Surface Morphology

An important operational advantage of the UHV environment of MBE is the wealth of analytic techniques available for examining the morphology of the growing surface *in situ*. The most prevalent of these utilize either diffraction or real-space imaging. Diffraction techniques (Yang *et al.*, 1993) include reflection high-energy electron diffraction (RHEED), low-energy electron diffraction, helium-atom scattering and grazing-incidence X-ray diffraction. Real-space imaging centres largely around the scanning tunneling microscope (STM) and the atomic-force microscope (AFM), but also includes techniques based on electron microscopy (Yagi, 1993; Bauer, 1994, 1996). Notable advances have also been made in the application of optical techniques (Auciello and Krauss, 1995), particularly to (MO)VPE, where the higher-pressure, reactive environment renders most other *in situ* techniques unsuitable. The *ex situ* analysis of epitaxial morphology is most often carried out with cross-sectional transmission electron microscopy (Hirsch *et al.* 1977), as shown in Figs. 1.1 and 1.5. This provides a method for imaging strain fields, dislocations, the crystallographic orientation of the substrate and epilayers, and the coherency and abruptness of heterogeneous interfaces. In this section, we will describe the most common methods used for the *in situ* observation and characterization of epitaxial phenomena: RHEED, the STM and the AFM.

1.4.1 Reflection High-energy Electron Diffraction

Surface electron diffraction is a standard method for examining the structure of surfaces both in equilibrium and in the presence of a deposition flux (Larsen and Dobson, 1988). A RHEED measurement is carried out by directing a high-energy (10–20 keV) beam of electrons at a glancing angle ($\approx 0.5°$–$3.0°$) toward the surface (Fig. 1.3). The electrons penetrate a few layers into the surface and those which emerge are recorded on a phosphorescent screen. There are three principal reasons why RHEED is so suitable as a diagnostic tool for MBE: (i) it is a simple measurement to set up, requiring only an electron gun and a recording screen, (ii) it is geometrically compatible with the molecular beams emanating from the Knudsen cells and so does not interfere with the growth process and, thus, (iii) it can be carried out *in situ* under *normal* growth conditions. The primary disadvantage of RHEED is that the 'images' of the surface are diffraction patterns. These are difficult to interpret quantitatively in real-space terms because the strong interaction between the electrons and the surface causes the incident electrons to be scattered several times before emerging from the crystal. This 'multiple scattering' means that, unlike kinematic diffraction patterns, RHEED patterns cannot be 'inverted' by performing a Fourier transform.

RHEED provides several types of information about a surface, including its crystallographic symmetry (from the symmetry of the diffraction pattern), the extent of long-range order (from the sharpness of the pattern), and whether growth is proceeding in a 2D or a 3D mode (Fig. 1.4). But one of the most common applications of RHEED is based on measuring the intensity of the *specular* beam (equal incident and reflected angles). A typical example, taken during the growth of GaAs(001), is shown in Fig. 1.6. Most apparent in this trace are the oscillations and their decaying envelope. The oscillations are due

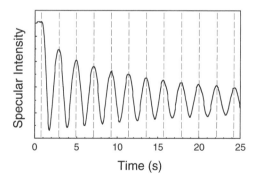

Figure 1.6.   Specular RHEED intensity oscillations from a singular GaAs(001) (a misorientation of $0\pm0.05°$) at a temperature near $575\,°C$. The incident azimuth of the electron beam is [010], the incident polar angle is $1°$, and the electron beam energy is $14\,keV$. The broken lines indicate the points where the specular intensity is a local maximum; these correspond to the deposition of additional Ga-As bilayers (after Shitara, 1992).

to the repeated formation of bi-atomic Ga-As layers (Fig. 1.4(a)) and provided the first direct evidence of layer-by-layer growth in this system (Neave *et al.*, 1983; Van Hove *et al.*, 1983). The decaying envelope results from the fact that this layer-by-layer growth is *imperfect*, i.e. subsequent layers begin to form before the preceding layers are complete (Section 1.6.3).

The period of the oscillations in Fig. 1.6 indicates that the time required to form a complete bi-layer is of the order of seconds. Since the molecular beams can be turned on and off mechanically with a shutter, the amount of material deposited can be controlled to within a fraction of a layer. Thus, a prescribed number of layers of one material (e.g. GaAs) can be deposited onto a surface, followed by a prescribed number of layers of a second material (e.g. AlAs). This process can be repeated to form a quantum well *superlattice* (Fig. 1.1). The electronic properties of such superlattices can be engineered by varying the lateral size of the quantum well and the depth of the carrier-confining potential by controlling the amount and types of materials deposited (Chapter 6).

1.4.2   Scanning Tunnelling Microscopy

The scanning tunnelling microscope, invented in 1982 by Gerd Binnig and Heinrich Rohrer (Binnig and Rohrer, 1987) at the IBM Research Laboratories in Rüschlikon, Switzerland, uses an atomically sharp tip placed sufficiently close (a few ångstroms) to a surface to produce an electron tunnelling current between the tip and the surface. By measuring this current as the tip scans the surface, 'images' of the surface are obtained which, under favourable circumstances, have a lateral resolution of ~ 1 Å and a vertical resolution of ~ 0.1 Å.

The basic principle of the STM can be understood with a model introduced some years ago by Tersoff and Hamann (1983). The tip is represented by a spherical potential well within which the Schrödinger equation is solved. By retaining only the spherically symmetric solutions, a simple expression is obtained for the tunnelling current I at low bias voltage V: $I \sim eV \varrho(\mathbf{r}, E_\mathrm{F})$, where $\varrho(\mathbf{r}, E_\mathrm{F})$ is the local density of states at the Fermi energy, E_F, of the surface at the position $\mathbf{r}$ of the tip. Thus, scans taken at constant current measure contours of constant Fermi-level charge density at the surface. Although this expression ignores the properties of the tip, which can substantially modify the tunnelling current, it does show that the STM is sensitive to *charge densities*, rather than simply *atomic positions*.

The STM revolutionized the field of surface science and has seen applications that extend far beyond traditional boundaries of condensed matter physics. Its impact on fundamental studies of epitaxial growth has also been immediate and far-reaching, but the inherently kinetic nature of growth does introduce some technical complications which are absent in studies of equilibrium surfaces. If an STM is placed in a conventional growth chamber, the tip shadows the incoming molecular beam. Thus, the imaging of growing surfaces has had to rely on one of two indirect strategies. The most common is to image a surface that has been quenched after a prescribed period of growth, thereby providing a 'snapshot' of the surface. Recently, however, it has become possible to arrange scan and growth rates within specially designed growth chambers to image the *same* region of a surface *during* growth (Voigtländer and Zinner, 1993; Pearson *et al.*, 1996). Though technically more demanding, this approach is the more desirable in principle because particular kinetic processes can be tracked and no quenching is required, thus providing a more faithful record of surface evolution. But, because of the very slow growth rates used in current implementations of this '*in vivo*' method, the growing surface is closer to equilibrium than for more typical growth rates and, moreover, is exposed for relatively long times to the ambient impurities which are always present in any growth chamber. These factors can affect the growth in several ways, so care must be taken when interpreting such images to ensure that they reflect the intrinsic growth characteristics of the material.

STM images of the (001) surfaces of Si and GaAs are shown in Fig. 1.7. These images reveal an important feature that is typical of semiconductor surfaces (and surfaces of many other materials). The creation of a surface produces broken, or dangling, bonds which leave the surface in an unstable high-energy state. The formation of new bonds to lower the surface free energy results in displacements of surface atoms from their bulk-terminated positions. We will distinguish between two types of such atomic rearrangement: *relaxation* and *reconstruction*. A relaxation preserves the symmetry and periodicity of the bulk unit cell. Expressed in units of the 2D primitive lattice vectors, such a structure is said to be (1×1). This is the typical case for non-polar semiconductor surfaces. A reconstruction involves more complex atomic displacements which modify the size and symmetry of the unit cell, leading generically to an $(n \times m)$ structure, i.e. a unit cell whose dimensions extend n and m times along the 2D primitive lattice vectors of the unreconstructed surface. In the images shown in Fig. 1.7(a), adjacent atoms on the surface of Si(001) form *dimers*, which produce a doubling of the unit cell along the axis of these dimers, i.e. a (2×1) reconstruction (Fig. 1.8(a)). The reconstruction on the GaAs(001) surface, shown in Fig. 1.7(b), is formed by a layer of dimerized As atoms, with alternate pairs of dimers missing, thus generating a (2×4) unit cell (Fig. 1.8(b)).

1.4.3   Atomic Force Microscopy

When the tip of an STM is brought close to a surface, the atoms near the apex of the tip exert a force on that surface which is of the same order of magnitude as the interatomic forces within the surface. This effect is the principle behind the atomic force microscope (Binnig *et al.*, 1986). An STM tip, mounted on a flexible beam, is brought just above a surface. The force between the surface and the tip causes a small deflection of the beam. The surface is then scanned while a constant force is

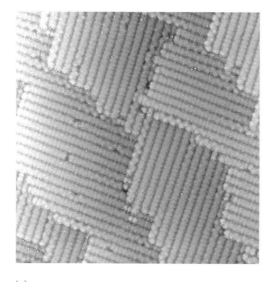

(a)

(b)

Figure 1.7.   (a) STM images of Si(001) (300 Å × 300 Å). Monatomic steps zig-zag diagonally across the image and the dimer rows of the (2 × 1) reconstruction are clearly visible as stripes. The direction of these rows rotates by 90° across a monatomic step, as discussed in Section 1.5.2.2. (Courtesy B. Voigtländer) (b) STM image of GaAs(001) (100 Å × 100 Å). The rows of As-dimers of the (2 × 4) reconstruction and the individual dimers on the top layer are resolved in this image. (Courtesy G. R. Bell) In both images, the shading darkens with increasing depth from the surface.

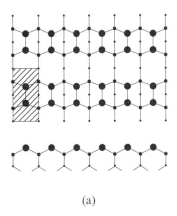

(a)

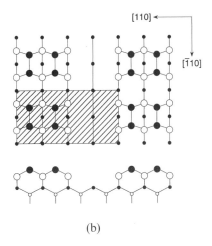

(b)

Figure 1.8. Plan and side views of ball-and-stick models of (a) Si(001) with the (2 × 1) reconstruction and (b) GaAs(001) with the (2 × 4) reconstruction. In (b) the filled circles represent As and the unfilled circles represent Ga. In each panel, the size of a circle indicates the proximity of that atom to the surface and the surface unit cells are indicated by the shaded regions.

maintained between the tip and the surface with a feedback loop similar to that used in the operation of an STM.

The AFM complements the STM in several ways. Because the STM relies on a tunnelling current for its operation, it is sensitive mainly to the density of electronic states near the Fermi level of the sample. Thus, this density of states must be non-zero, i.e. the sample must be *conducting*. However, since the AFM tip responds to interatomic forces, which is a cumulative effect of *all* electrons, the sample need not be a conductor. Additionally, since the tunnelling current decreases exponentially with the tip–sample distance, an STM tip must be placed within a few ångstroms of the surface to maximize the resolution of the image.

The AFM most commonly operates in this mode (the 'contact' mode) as well, but it can also operate at much larger distances from the surface (50–150 Å) for samples susceptible to damage or alteration by being in close proximity to the tip (the 'non-contact' mode). But, even in the contact mode, attaining atomic resolution is much more demanding technically than with the STM. Thus, many applications of the AFM involve scanning large areas (up to microns) to image the gross morphology of the sample. This has the advantage of not requiring a UHV environment and AFMs often operate in ambient atmosphere or in a liquid (Quate, 1994).

1.5   Atomistic Processes during Homoepitaxy

The fabrication of heterostructures requires growing crystalline materials on the surfaces of *different* materials, a process which is known as *heteroepitaxy*. But a useful starting point for appreciating the complexity of heteroepitaxial phenomena (Fig. 1.4) is a conceptual and computational framework for *homoepitaxy* – the growth of a material on a substrate of the *same* material. Many atomistic processes that occur during heteroepitaxy have direct analogues in homoepitaxy. Thus, experiments can be carried out to identify the morphological signatures of particular atomistic processes in a homoepitaxial setting unencumbered by inherently heteroepitaxial effects such as segregation, alloying, and the relaxation of misfit strain. In this section, we describe the experiments that have established some of the fundamental tenets of homoepitaxial growth kinetics. Their theoretical interpretation is discussed in Section 1.6.

1.5.1   Growth Kinetics on Vicinal GaAs(001)

A crystal cleaved precisely along a low-index crystalline plane exposes a flat surface. Such a surface is called *singular*. If the cleavage plane is slightly misoriented from one of these crystallographic planes, the surface – now called *vicinal* – breaks up into a 'staircase' of terraces of the low-index surface and steps that accommodate the misorientation (Fig. 1.9). Experiments which first highlighted the role of steps during MBE were performed at Philips Research Laboratories in Redhill in the mid-1980s (Neave *et al.*, 1985). Growth was carried out on GaAs(001) misoriented by approximately 2.25° from [001] toward [011] and RHEED measurements were taken with the electron beam directed perpendicular to the staircase of steps. These measurements showed an intriguing effect (Fig. 1.10): the RHEED oscillations seen at the lowest temperatures diminish and eventually disappear as the temperature is raised (with fixed As_2 and Ga fluxes).

To explain this observation we must understand how temperature affects the mobility of an *adatom*, i.e. a single atom on a surface. Adatoms are able to move along a surface as a result of 'kicks' from the thermal vibrations of the surface.

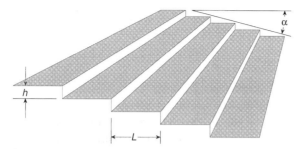

Figure 1.9.   Schematic diagram of an ideal stepped surface showing a regular array of terraces (indicated by shading) and straight steps. The vicinal angle α is given in terms of the step height h and terrace length L by $\alpha = \tan^{-1}(h/L)$. The smaller the misorientation angle, i.e. the closer the cleavage plane is to the low-index plane of the terraces, the greater the distance between the steps, and the closer the surface approaches the singular limit.

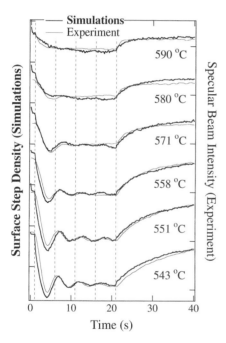

Figure 1.10.   RHEED specular intensity for a GaAs(001) vicinal surface misoriented by $2°$ toward [011] direction at the growth rate of 0.20 ML/s. Also shown is the step density obtained from a simulation of the growth kinetics (Šmilauer and Vvedensky, 1993).

Adatoms on a cold surface are not very mobile and tend to remain on the terraces for a long time, where they collide and bind with other atoms to form islands, a process known as *nucleation*. These islands then trap other migrating atoms to form three-atom islands, four-atom islands, and so on. Surface growth thus proceeds by the nucleation, aggregation and eventual coalescence of these islands into a complete layer. The morphology of the surface, as probed by the electron beam, reflects these changes. As the islands form, their edges interrupt the periodicity of the initially flat terraces, so the surface appears to roughen, which causes the reflected intensity of the specular beam to *decrease*. Then, as the islands coalesce, their edges disappear, the surface becomes smoother, resulting in an *increase* of the specular intensity. The repetition of this process during the growth of successive layers is manifested by oscillations of the RHEED specular intensity seen in Fig. 1.6.

The scenario at higher temperatures is substantially different. Adatoms deposited onto the surface now have sufficient mobility to migrate directly to, and form bonds with, the step. There are now fewer adatoms on the terraces than at lower temperatures, so the likelihood of adatoms colliding to form an island diminishes accordingly. This results in a type of growth known as 'step flow' because most of the deposited atoms are incorporated into the surface at the terrace edges. The surface thereby grows by the advancement of the staircase of steps, with there being little discernible change in the morphology of the surface. Hence, the RHEED specular intensity remains approximately constant.

The foregoing explanation was put forward by Neave *et al.* (1985) based solely on their RHEED data. Subsequent atomistic simulations (Shitara *et al.*, 1992*a*; Šmilauer and Vvedensky, 1993) supported this explanation and even suggested that the relation between the number of step edges and the RHEED specular intensity could, under appropriate circumstances, be quantitative (Fig. 1.10). STM studies in which the lengths of step edges were enumerated explicitly at fixed increments of deposited material and compared with the corresponding RHEED specular intensity supported these ideas (Sudijono *et al.*, 1993; Holmes *et al.*, 1997).

The RHEED traces shown in Fig. 1.10 were taken during the growth of four Ga-As bilayers, after which the Ga source is closed (but the As source remains open), whereupon the RHEED specular intensity increases toward its pre-growth value. This behaviour, called *recovery*, provides direct evidence that, during growth, the surface is maintained away from equilibrium by the molecular beams. For appropriately chosen diffraction conditions (Zhang *et al.*, 1987) and with all other factors being equal, we can associate a greater RHEED specular intensity with a smoother, more ordered surface. Thus, allowing the surface to equilibrate after a period of growth results in a smoothing of the as-grown surface (Ide *et al.*, 1992). However, there is a competing effect which limits the effectiveness of recovery. Even in a UHV environment, there is a background concentration of impurities. Leaving a surface exposed for extended periods of time can lead to an unacceptably large concentration of these impurities accumulating on the surface. This not only renders the material of limited use in any potential device applications, but can also affect the growth of subsequent layers of material. In practice, recovery periods rarely exceed a few minutes.

1.5.2   Anisotropic Growth and Surface Reconstructions

Surface reconstructions play an important role in many aspects of epitaxial growth. They produce preferred directions for the diffusion of atoms on the terraces, affect the binding of atoms at steps and island edges, and the size of the surface unit cell influences the energetics and kinetics of kinks at steps. In multicomponent systems, surface reconstructions can provide an indication of local stoichiometry. In this section, we will examine how anisotropy affects growth on GaAs(001) and Si(001).

1.5.2.1   Vicinal GaAs(001)

We will concentrate here on surfaces that exhibit the (2×4) reconstruction, which covers a large region of the surface phase diagram of GaAs(001) (Däweritz and Hey, 1990). The combined effects of the zinc blende structure of GaAs and the (2×4) reconstruction result in two quite different step structures (Fig. 1.11). For misorientations along the [110] direction, the steps are perpendicular to the As-dimers, i.e. they are Ga-terminated. These are called 'A' steps. For misorientations along the [$\bar{1}$10] direction, the steps are parallel to the As-dimers, i.e. they are terminated by As-dimers. These are called 'B' steps. A third type of misorientation, along the [010] direction, i.e. at 45° to the [110] and [$\bar{1}$10] directions, results in

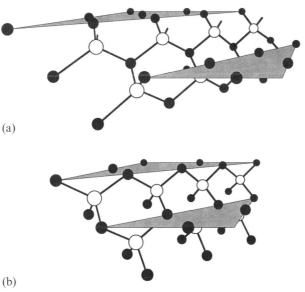

(a)

(b)

Figure 1.11.   Atomic structure (without surface reconstructions) near (a) type-A (Ga-terminated) and (b) type-B (As-terminated) steps of GaAs(001). The filled circles represent As atoms and the unfilled circles represent Ga atoms. The As-terminated terraces above and below each step are indicated by shading.

what are sometimes called 'C' steps, though for many purposes they may be regarded as mixtures of A and B components.

These step structures lead to an appreciable dependence on the misorientation direction of the temperature T_c at which, for the same growth conditions, the RHEED oscillations on vicinal GaAs(001) disappear: $T_c(A) > T_c(C) > T_c(B)$ (Shitara *et al.*, 1992b). The fact that the B surface exhibits the lowest T_c is due to the stronger bonding of Ga atoms to B steps compared with that to A steps. Thus, Ga atoms that attach to a B step are less likely to subsequently detach than atoms bonded to an A step. The concentration of Ga adatoms is therefore lower on terraces of B surfaces, so the likelihood of island nucleation is correspondingly lower which, in turn, promotes growth by step flow at a lower temperature than that on A surfaces. The STM images in Fig. 1.12 support this interpretation. Shown are $1000\,\text{Å} \times 1000\,\text{Å}$ sections of A and B surfaces of GaAs(001) misoriented by $1°$ after the deposition of $0.2\,\text{ML}$ at $560\,°\text{C}$. The rows of the (2×4) reconstruction are clearly discernible in both images (cf. Fig. 1.8(b)), but most

Figure 1.12. STM images of (a) A-type and (b) B-type vicinal GaAs(001) ($1000\,\text{Å} \times 1000\,\text{Å}$) misoriented by $1°$ (Bell *et al.*, 1999). Note the corrugation of the (2×4) reconstructing on the terraces and the smooth and rough edges on the A and B steps, respectively. In both (a) and (b), $0.2\,\text{ML}$ of material has been deposited at $560\,°\text{C}$ which, on the A surface, results in the appearance of several islands. (Courtesy G. R. Bell)

apparent is the absence of islands on the *B* surface, while there are a number of islands of different sizes on the *A* surface.

Another characteristic of *A* and *B* surfaces on GaAs(001) revealed by these images is the difference in the roughness of the steps. This can also be explained in terms of attachment and detachment processes. Atoms attached to an *A* step can explore more configurations and so can more easily find and attach to an energetically favourable site, such as a kink, than atoms at a *B* step. This leads to the *A* step being smoother (Fig. 1.12) and having a lower energy of formation than the *B* step (Heller *et al.*, 1993; Zhang and Zunger, 1995, 1996).

1.5.2.2 Vicinal Si(001)

Most of the basic characteristics of homoepitaxy on Si(001) can be traced to the formation of bonds between adjacent surface atoms to reduce the surface free energy of the bulk-terminated structure. STM studies of Si(001)-(2 × 1) surfaces (Hamers *et al.*, 1986) have shown that the dimers align in rows with a very low misalignment density, giving the surface a corrugated appearance, but have also revealed a relatively large number of missing dimer defects. Similar conclusions have been reached by *in situ* STM studies of Si(001) grown by MBE (Hoeven *et al.*, 1989; Mo *et al.*, 1989).

As is the case for GaAs(001), the surface reconstruction has a strong influence on the step structure of vicinal Si(001) and produces similar effects to those discussed in the preceding section. The orientation of surface dimers rotates by 90° on successive layers, so the direction of dimerization changes across a monatomic step. This produces two types of step on the *same* surface if there are only monatomic steps: one where the dimers on the terrace above the step are perpendicular to the edge, and one where the dimers are parallel to the edge (Fig. 1.7(a)). Following standard notation (Chadi, 1987), these *single-height* steps are referred to as S_A and S_B steps, respectively, with an analogous labelling for the terrace above each step. The S_B step has a higher energy than the S_A step (Chadi, 1987), which makes the S_B step more likely to roughen. This has a marked effect on the growth of this surface, since adatom attachment occurs preferentially at S_B steps (Hoeven *et al.*, 1989; Mo *et al.*, 1989), which produces a surface dominated by type-*A* terraces (Fig. 1.13).

For large misorientation angles ($\approx 2°$ or larger), i.e. for sufficiently short terraces, a vicinal Si(001) surface with monatomic steps can be forced to form *biatomic* steps by annealing the surface at sufficiently high temperatures ($\approx 1000°$C) for extended periods of time (≈ 1 hour). These biatomic, or *double-height* steps are labelled in the same way as the monatomic steps, but with a '*D*' replacing the '*S*'. Calculations (Chadi, 1987) have shown that the D_B steps have an energy of formation that is lower than the sum of the energies of formation of the S_A and S_B steps. This means that, by prolonged high-temperature annealing, pairs of monatomic steps can be made to form stable biatomic steps, producing a surface comprised only of type-*B* terraces. However, the inability to form

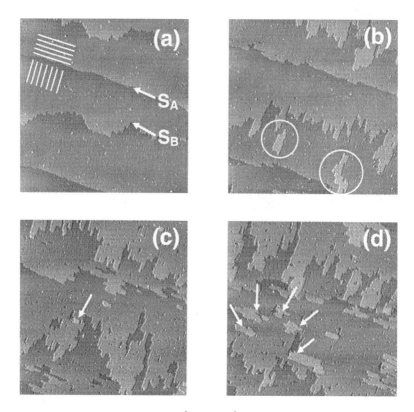

Figure 1.13.   STM images (2900 Å × 2900 Å) during growth on vicinal Si(001) at 725 K at 8 ML/h (Voigtländer *et al.*, 1997). The starting surface is shown in (a) and the same region of the surface after the deposition of 0.22, 0.53, and 0.94 ML is shown in images (b)–(d), respectively. The straight S_A step and rough S_B step are shown (a), with the white lines indicating the directions along which the dimer rows run along the two types of terrace. Islands formed during deposition are enclosed within the circles in (b) and islands of the next layer after the islands in (b) have coalesced with the advancing step are indicated by arrows in (c) and (d). (Courtesy B. Voigtländer).

double-height steps for misorientations below $\approx 2°$ is not a kinetic limitation; the anisotropy of the surface stress due to the dimers leads to the formation of (1×2) and (2×1) domains being energetically favoured over the formation of a single domain. A theory taking both this and step-edge energies into account (Alerhand *et al.*, 1990; Pehlke and Tersoff, 1991) predicts that for annealed vicinal surfaces there is a temperature-dependent phase transition between surfaces with single-height steps and those with double-height steps. At zero temperature, a transition from double- to single-height steps occurs for misorientation angles less than $\approx 0.05°$; at 500 K, the transition angle increases to $\approx 2.0°$ (Alerhand *et al.*, 1990).

The difference in energies of formation of steps parallel and perpendicular to the dimer bond-axis also affects island morphologies on Si(001) surfaces, though kinetic factors dominate the island shape during growth. STM studies have shown that, during growth, islands are highly anisotropic extending along the direction of the dimer row, with aspect ratios of up to 20 : 1 (Mo *et al.*, 1989). However, the *equilibrium* shapes of these islands, obtained after annealing, show a considerably reduced aspect ratio of approximately 3 : 1 (Mo *et al.*, 1989), which is a reflection of the energy difference of the S_A and S_B steps that form the island edges. The enhanced aspect ratio during growth is a kinetic effect caused by the large ratio (50 : 1) of adatoms sticking at the end of an island to sticking at the side of an island (Pearson *et al.*, 1996).

1.6   Models of Homoepitaxial Kinetics

The availability of a broad range of surface analytical techniques afforded by the UHV environment of MBE has had a profound impact on the development of models of epitaxial phenomena. The advent of the STM, in particular, has placed the emphasis in theoretical descriptions of MBE firmly on determining how particular atomistic processes affect the morphological evolution of epitaxial films (Villain, 1991; Gyure and Zinck, 1996).

In contrast, the relatively high pressures used for (MO)VPE has meant that the atomic-level characterization of this growth process has been much slower in development than that for MBE. Theoretical studies of (MO)VPE have been directed largely at the hydrodynamics of gas flow in reactors and near the surface. In the most sophisticated of such analyses (Jensen, 1993), the two are coupled to provide predictions for the growth *rate* as a function of the substrate temperature, the partial pressures of the reactants and the reactor design. Though an important ingredient in any theoretical model of (MO)VPE, this information provides little insight into materials issues that are dominated by atomistic processes. Understanding such processes is essential for many applications, including the fabrication of quantum-well lasers on patterned substrates (Bhat *et al.*, 1990), the growth of quantum wires on vicinal surfaces (Metiu *et al.*, 1992), and the control of compositional ordering (and, hence, the band gap) in compound semiconductors (Pearsall and Stringfellow, 1997).

In this section we develop two of the most common analytic descriptions of epitaxial kinetics: the Burton–Cabrera–Frank (BCF) theory and homogeneous rate equations. Although both approaches are usually employed to describe particular regimes of MBE, with suitable modifications, they can also be applied to more complex growth scenarios, including those in (MO)VPE.

1.6.1   The Theory of Burton, Cabrera and Frank

The BCF theory (Burton *et al.*, 1951) describes growth on a vicinal surface of a monatomic crystal in terms of the deposition and migration of single adatoms. The central quantity in this theory is, therefore, the adatom concentration $c(\mathbf{x}, t)$

at position $\mathbf{x}$ and time t. The processes which cause this quantity to change are the diffusion of adatoms, which have diffusion constant D, and the flux J of adatoms onto the surface from the molecular beam. We will assume that the desorption of the atoms from the surface can be neglected, but this can be readily included in the theory if required. In the simplest form of the BCF theory, the equation determining $c(\mathbf{x}, t)$ on a terrace is a one-dimensional diffusion equation with a source term:

$$\frac{\partial c}{\partial t} = D \frac{\partial^2 c}{\partial x^2} + J \tag{1.3}$$

This equation is supplemented by boundary conditions at the step edges which bound the terrace, e.g.,

$$c(0, t) = 0, \qquad c(L, t) = 0 \tag{1.4}$$

where L is the terrace length. These boundary conditions stipulate that adatoms are absorbed at a step and immediately incorporated into the growing crystal with no possibility of subsequent detachment. Less restrictive boundary conditions can also be chosen, as discussed in Exercise 5.

We will focus here on the steady-state (time-independent) solution of equation (1.3). Upon invoking the boundary conditions in (1.4), we obtain

$$c(x) = \frac{J}{2D} x(L - x) \tag{1.5}$$

which is a parabola with its maximum at the centre of the terrace and which vanishes at the terrace edges, as required by the boundary conditions. The scale of the adatom concentration is set by the growth conditions (substrate temperature and flux) through the ratio J/D (Exercise 6). This quantity is a measure of the competition between the deposition flux, which drives the surface away from equilibrium and *increases* the adatom density, and the relaxation of the surface toward equilibrium through adatom diffusion, which *decreases* the adatom density (Exercise 5). Since this theory neglects interactions between adatoms, the growth conditions must be chosen to ensure that the adatom concentration is maintained low enough to render their interactions unimportant. Thus, the BCF theory is valid only for relatively small values of J/D, i.e. high temperatures and/ or low fluxes, where growth is expected to occur by step flow (cf. Fig. 1.10).

1.6.2   Homogeneous Rate Equations

With increasing temperature or decreasing deposition rate, growth by the nucleation, aggregation and coalescence of islands on the terraces of a substrate becomes increasingly likely and the BCF picture is no longer appropriate. One way of providing a theoretical description of this regime within an analytic framework that complements the BCF theory is with equations of motion for the densities of adatoms and islands. These are called *rate equations* (Venables *et al.*, 1984).

In this section, we will consider the simplest rate equation description of growth, where adatoms are the only mobile surface species and the nucleation and growth of islands proceeds by the *irreversible* attachment of adatoms, i.e. once an adatom attaches to an island or another adatom, subsequent detachment of that adatom cannot occur (cf. equation (1.4)). We will signify the density of surface atoms by $n_1(t)$ and the density of s-atom islands by $n_s(t)$, where $s > 1$. The rate equation for n_1 is

$$\frac{dn_1}{dt} = J - 2D\sigma_1 n_1^2 - Dn_1 \sum_{s=2}^{\infty} \sigma_s n_s \qquad (1.6)$$

In common with most formulations of rate equations, the adatom and island densities are taken to be spatially homogeneous. In particular, there is no diffusion term, $D\nabla^2 n_1$, despite the fact that adatoms are mobile. This description is most suitable for singular surfaces, where there are no pre-existing steps to break the translational symmetry of the system and induce a spatial dependence in the adatom and island densities. But, even in this case, spatial effects cannot be neglected altogether. This will be discussed below.

The first term on the right-hand side of equation (1.6) is the deposition of atoms onto the substrate, which increases the adatom density, and so has a positive sign. The next term describes the nucleation of a two-atom island by the irreversible attachment of two migrating adatoms. This term decreases the number of adatoms (by two) and thus has a negative sign. The rate for this process is proportional to the *square* of the adatom density because two adatoms are required to form a two-atom island, and to D, the adatom diffusion constant, because these adatoms are mobile. The third term accounts for the depletion rate of adatoms due to their capture by islands. This term is proportional to the product of the adatom and total island densities and must also have a negative sign. The quantities σ_i, called 'capture numbers', account for the diffusional flow of atoms into the islands (Venables *et al.*, 1984; Bales and Chzan, 1994; Bartelt and Evans, 1996). We will discuss later how these quantities affect the solutions of rate equations.

The rate equation for the density of s-atom islands $n_s(t)$ is

$$\frac{dn_s}{dt} = Dn_1\sigma_{s-1}n_{s-1} - Dn_1\sigma_s n_s \qquad (1.7)$$

The first term on the right-hand side of this equation is the creation rate of s-atom islands due to the capture of adatoms by $(s-1)$-atom islands. Similarly, the second term is the depletion rate of s-atom islands caused by their capture of adatoms to become $(s+1)$-atom islands. There is an equation of this form for every island comprised of two or more atoms, so (1.6) and (1.7) represent an infinite set of coupled ordinary differential equations. However, since the density of large (compared with the average size) islands decreases with their size, in practice the hierarchy in (1.7) is truncated to obtain solutions for n_1 and the remaining n_s to any required accuracy. Notice that in writing (1.7) we have omitted any direct interactions between islands. This restricts us to a regime where there is no appreciable coalescence of these islands.

To illustrate the calculus of rate equations, we consider a limiting case where all of the capture numbers are set equal to unity. Then, by introducing the total island density, $N = \sum_{s>1} n_s$, using this definition in (1.6), and summing the equations in (1.7) over s, we obtain a closed set of two equations for n_1 and N:

$$\frac{dn_1}{d\theta} = 1 - 2Rn_1^2 - Rn_1N \tag{1.8a}$$

$$\frac{dN}{d\theta} = Rn_1^2 \tag{1.8b}$$

where $R = D/J$ and we have used the relation between the coverage and the flux in the absence of desorption, $\theta = Jt$, to replace the time by the coverage as the independent variable. This replacement is made because the coverage is the more 'natural' variable, since it can be measured directly (from an STM image) with greater accuracy than the product of the deposition time and the flux.

Equations (1.8) are straightforward to integrate numerically (Exercise 1.8) and their solutions are shown in Fig. 1.14. We will focus here on the initial and long-time behaviour of the adatom and island densities, where analytic solutions to (1.8) are easily obtained. At short times ($\theta \ll 1$),

$$n_1 \sim \theta, \qquad N \sim R\theta^3. \tag{1.9}$$

The density of adatoms initially shows a linear increase with coverage (or time), which is due entirely to the deposition flux. The islands are somewhat slower in their early development, showing a cubic time-dependence, because the adatom density is too low for there to be appreciable island nucleation. Equation (1.8b)

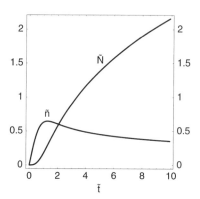

Figure 1.14.   The dimensionless adatom and island densities, denoted by $\tilde{n}$ and $\tilde{N}$, respectively, as a function of the dimensionless time, $\tilde{t}$, obtained by integrating the rate equations (1.8). See Exercise 1.8.

shows that N continues to increase for all later times, but equation (1.8a) indicates that, although n_1 increases initially, it eventually begins to decrease (Fig. 1.14). This continues until we reach a regime where $n_1 \ll N$ and $dn_1/d\theta \approx 0$. In this regime, we obtain *scaling laws* for the adatom and island densities (Exercise 8):

$$n_1 \sim \theta^{-1/3} R^{-2/3}$$

$$N \sim \theta^{1/3} R^{-1/3}$$

(1.10)

Notice that, just as in equation (1.5), the ratio D/J is the controlling parameter for quantities which characterize the surface morphology. In particular, the equation for N indicates that increasing the temperature (i.e. increasing D) and/or decreasing the flux J causes the island density to decrease, resulting in fewer, but larger, islands.

The expressions in equation (1.10) exhibit the correct scaling of N with D/J in the *aggregation regime* of island growth, i.e. where the island density has saturated and existing islands capture all deposited atoms. However, the scaling of n_1 and N with θ is not correct. Indeed, the island density in Fig. 1.14 does not exhibit saturation at all, which indicates that the approximation of constant capture numbers misses important aspects of island kinetics (Ratsch *et al.*, 1994; Bott *et al.*, 1996). The next level of approximation is to include the spatial extent of the islands in an average way by assuming that the local environment of each island is independent of its size and shape (Bales and Chzan, 1994). This produces the correct scaling of N with both D/J and θ, but still not the correct distribution of island sizes. The calculation of this quantity requires proceeding one step further by including explicit spatial information in the capture numbers to account for the correlations between neighbouring nucleation centres and the different local environments of individual islands (Bartelt and Evans, 1996; Bartelt *et al.*, 1998). In practice, only total island densities are analysed with rate equations; simulations are used for the quantitative analysis of measured island-size distributions (Ratsch *et al.* 1995).

1.6.3   Multilayer Growth on Singular Surfaces

In our discussion of the solution (1.5) to the BCF equation, we focussed on the roles of D and J in determining the conditions under which the growth of a vicinal surface proceeds by step flow. However, there is another quantity that is equally important in determining the growth mode on such a surface: the terrace length L. Suppose we fix the temperature (i.e. D) and the deposition flux J. Then, if L is small enough, the adatom density will be corresponding small, and growth proceeds by step flow. But for surfaces with larger terraces, the adatom concentration increases until, at some terrace width L^*, adatom interactions are no longer negligible, and the growth of islands becomes appreciable. This simple observation provides the basis of understanding the origin of multilayer growth on singular surfaces.

Consider deposition onto a singular surface. The probability of atoms eventually encountering one another is large, since there are no steps to absorb migrating adatoms. The growth of the first surface layer is initiated by the nucleation of islands, which grow laterally by capturing migrating adatoms at their edges. Thus, to an electron beam the surface appears to roughen, which causes the specular intensity of the beam to decrease. This roughness continues to increase until the islands begin to coalesce, at which point the surface appears to smooth, causing the specular intensity to increase. Once the new layer is formed this process is repeated, resulting in the oscillatory behaviour of the RHEED specular beam, as discussed in Section 1.5.1.

What is the origin of the decaying envelope seen in the RHEED oscillations shown in Figs. 1.6 and 1.10? The layer-by-layer growth just described is not perfect. Once the lateral size of an island becomes large enough, atoms deposited on top of this island can collide and initiate the growth of the next layer. This is easy to understand given our earlier observations. If we regard the top of a growing island as a terrace of length $L(t)$, then at early times, when the island is small, we have that $L^* \gg L(t)$. Thus, the growth of the island proceeds by 'step flow' in the sense that atoms which are deposited on top of the island migrate to the edge of the island, where they are incorporated into the lower layer. As the island grows laterally, however, the condition $L^* \ll L(t)$ is eventually reached. In this case, atoms deposited on top of the island are likely to collide and form a new island before migrating to the edge of the island. Thus, the next layer begins to form before the current layer is complete and, as this process continues in successive layers, the surface undergoes a gradual and progressive roughening – called *kinetic roughening* – whereby an increasing number of incomplete layers is exposed. The decay of the RHEED oscillations is indicative of this roughening. Comprehensive discussions of the theory of kinetic roughening and its experimental characterization may be found in the books by Yang *et al.* (1993) and Barabási and Stanley (1995).

The preceding argument can be extended to treat anisotropic islands, such as those discussed in Section 1.5.2.2 for Si(001). These islands will have terrace lengths of $L_\parallel$ and $L_\perp$ which are parallel and perpendicular, respectively, to the dimer rows within the island. The discussion in Section 1.5.2.2 indicated that $L_\perp \gg L_\parallel$, i.e. that islands on Si(001) comprise highly anisotropic 'string-like' rows of dimers aligned perpendicular to the dimer bond axis (Fig. 1.13). Atoms deposited on top of these islands will have a downward step nearby; this corresponds to the condition $L^* \gg L_\parallel$. Thus, since $L_\parallel$ increases much more slowly than $L_\perp$, the growth of the second layer is delayed to a much later stage of layer completion than in the case of isotropic growth. RHEED measurements support this conclusion. Vicinal Si(001) surfaces with only double-layer steps (which means the surface has only a single reconstructed domain) exhibit oscillations in the RHEED specular intensity with a bilayer period (for a suitably-chosen azimuthal orientation of the electron beam) and a very slow rate of decay (Sakamoto *et al.*, 1985).

1.7 Mechanisms of Heteroepitaxial Growth

The morphology that results during the growth of a material on the substrate of a different material is central to the fabrication of all quantum heterostructures. This morphology is determined by the surface and interface energies of the materials (Section 1.3), the manner in which strain is accommodated if the materials have different lattice constants, and any effects of alloying and segregation. Controlling the morphology during heteroepitaxy involves understanding the atomistic mechanisms by which these factors assert themselves.

1.7.1 Kinetics and Equilibrium with Misfit Strain

GaAs, AlAs and their alloys are the simplest semiconductor heteroepitaxial systems because of the very small lattice mismatch between AlAs and GaAs and the similar values of thermal expansion coefficients. But this situation is not typical. The fabrication of heterostructures from other combinations of materials with potentially attractive properties requires identifying (and possibly utilizing) the morphological and electronic consequences of any lattice mismatch.

There are abundant data available for several heteroepitaxial systems, but there is no theoretical approach with the generality of the BCF theory, rate equations, or simulations which captures the essence of morphological evolution if there is lattice misfit. There are two reasons for this. The kinetics of atomic processes on the surfaces of strained systems are not determined simply by the local environment of the atoms, as in the case of homoepitaxy, but may incorporate non-local information, such as the *height* of a terrace above the initial substrate or the *size* and *shape* of 2D and 3D islands. Then there is the issue of lattice relaxation and any resulting defect formation. The theoretical description of such effects at heterogeneous interfaces has relied largely on the minimization of energy functionals with various degrees of sophistication (Grabow and Gilmer, 1988; Dodson, 1990; Ratsch and Zangwill, 1993) to determine *equilibrium* atomic positions near the interface as a function of the lattice mismatch. However, there are few quantitative discussions about the *kinetic* manifestations of lattice mismatch, especially the appearance of misfit dislocations, except in a few special cases.

The Frenkel–Kontorova model (Frenkel and Kontorova, 1939) has been used to address several general aspects of the accommodation of misfit strain and the formation of dislocations in heteroepitaxial systems within a simply analytic framework. This model is described in the following section. A discussion of the broader theoretical issues of growth in the presence of lattice mismatch has been given by Markov (1995).

1.7.2 The Frenkel–Kontorova Model

In the Frenkel–Kontorova model, the equilibrium positions of atoms within the growing layer result from the competition between the preferred interatomic separation of these atoms, which interact through harmonic springs, and the periodicity imposed by the rigid potential of the substrate. This potential induces elastic strain in the epilayer and can result in the formation of misfit dislocations. This model has been applied to strained monolayer islands (Hamilton, 1997), multilayer islands (Ratsch and Zangwill, 1993), and films (Little and Zangwill, 1994).

Many of the characteristic features of strained islands can be captured by the simplest calculation of a one-dimensional monolayer island consisting of N adjacent atoms. The harmonic springs connecting these atoms have a natural length b, the lattice constant of the deposited material, and a force constant k. The interaction between the atoms within the island and the substrate is described by a rigid sinusoidal potential which has periodicity a, the lattice constant of the substrate:

$$V(x) = \tfrac{1}{2} W[1 - \cos(2\pi x/a)] \tag{1.11}$$

The ground state of this system is determined by calculating the energy as a function of the atomic positions within the island and then minimizing this expression with respect to these positions. We denote the distances from the origin to the nth and $(n+1)$th atoms by $X_n = na + x_n$ and $X_{n+1} = (n+1)a + x_{n+1}$, where x_n and x_{n+1} are the displacements of the atoms from the bottoms of their substrate potential troughs. The distance ΔX_n between the $(n+1)$th and nth atoms can then be written as

$$\Delta X_n = X_{n+1} - X_n = x_{n+1} - x_n + a \tag{1.12}$$

The strain $\varepsilon(n)$ of the bond between these atoms is

$$\varepsilon(n) = \Delta X_n - b = a(\xi_{n+1} - \xi_n - f) \tag{1.13}$$

where $\xi_n = x_n/a$ and $f = (b - a)/a$ is the misfit between the epilayer and the substrate. The energy of the N-atom island is now written as the sum of the potential energy in (1.11) and the strain energy due to the changes in the lengths of the springs in (1.13):

$$\frac{E}{W} = \ell_0^2 \sum_{n=1}^{N-1} (\xi_{n+1} - \xi_n - f)^2 + \tfrac{1}{2} \sum_{n=1}^{N} [1 - \cos(2\pi\xi_n)] \tag{1.14}$$

where

$$\ell_0 = \left(\frac{ka^2}{2W} \right)^{1/2} \tag{1.15}$$

is the ratio of the interaction energy between the atoms in the epilayer to that between the epilayer and the substrate.

Figure 1.15 shows the equilibrium configurations for islands of different size calculated with the parameters $f = 0.1$ and $\ell_0 = 10$, i.e. for atoms in the epilayer that are much more strongly bonded to each other than to the substrate and which have a 10% larger lattice constant than the substrate (Exercise 9). Several general issues can be discussed with reference to this figure.

- Atoms near the centre of the island adopt positions close to the minimum of the nearest potential energy trough. Atoms further from the centre of the island are correspondingly further away from their nearest minima. Thus, strain relaxation occurs predominantly at the edges of islands.
- As the number of atoms in an island increases, the strain energy within the island builds up and the energy difference between a coherent island and an island with a single dislocation diminishes until, at some critical size, a dislocation is formed.
- With the parameters we have chosen, an island with 12 or more atoms minimizes its energy by forming a dislocation, which is located in the centre of the island, as shown in Fig. 1.15(d). Since strain relaxation is largest at island edges, we would expect the dislocation to form at the island edge and to migrate

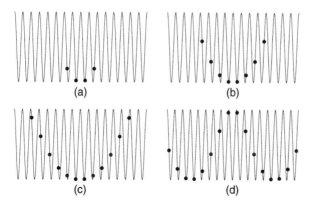

Figure 1.15. Equilibrium positions of atoms in a one-dimensional epilayer within the Frenkel–Kontorova model. Islands are shown with (a) 4 atoms, (b) 8 atoms, (c) 12 atoms, and (d) 16 atoms. The islands in (a), (b) and (c) are coherent, but the island in (d) has a dislocation located at its centre. The parameters used for this calculation are $f = 0.1$ and $\ell_0 = 10$.

toward the centre, with the interface between the epilayer and the substrate providing the one-dimensional slip 'plane'.

We conclude this section with a word of caution. Although the Frenkel–Kontorova model provides a useful starting point for understanding several aspects of heteroepitaxial systems, the balance between coherent island growth and dislocation formation can depend on competing mechanisms of strain relaxation which are not included in this model. One assumption which can be questioned is the treatment of the substrate as mechanically, chemically and kinetically inert. In fact, there is strain relaxation within the substrate, as well as the possibility of alloying between the epilayers and the substrate, either of which can alter the balance between different strain relaxation mechanisms.

1.8   Direct Growth of Quantum Heterostructures

The first observation of confinement in a quantum well (Dingle et al., 1974; Dingle, 1975) spawned an enormous world-wide effort to fabricate heterostructures that could confine carriers to one dimension (quantum 'wires') and to zero dimensions (quantum 'dots'). This was driven by the realization that the attractive optical and transport properties of low-dimensional heterostructures stem from the fundamental changes in the densities of states and the Coulomb interaction between carriers as their effective dimensionality is reduced (Chapter 2). For example, the characteristics of quantum wire and quantum dot lasers are predicted to show a significant improvement over those of quantum-well lasers in terms of threshold current, modulation dynamics, and spectral properties, provided there is no strong coupling between neighbouring wires or dots (Arakawa and Yariv, 1986). This can be traced to the enhanced peaks in the density of states with decreasing carrier dimension. The expectation (Sakaki, 1987) of high carrier mobility in quantum wires because of the strong suppression of elastic scattering is also due to the reduced dimensionality of the carriers. This effect led to the suggestion (Sakaki, 1987) that these heterostructures could be used as components of high-speed electronic devices. Among the potential technological applications of quantum dots (Eberl, 1997), which are discussed in Chapter 9, are devices which allow single electron (Wharam et al., 1988) or single photon (Imamoglu and Yamamoto, 1994) counting and as the gain medium of vertical-cavity surface-emitting lasers (Saito et al., 1996), which are discussed in Chapter 8.

The confinement of carriers to one or zero dimensions is relatively easy to achieve when the lateral dimensions of the confining region within a heterostructure is of the order of 500–1000 Å. This is the domain of *mesoscopic* structures. The techniques that permit the fabrication of such structures are based on (electron or ion) lithographic processing. Unfortunately, the lithographic process often produces interfaces with a high defect density and even damage to the bulk material itself. Moreover, carrier confinement within such lateral dimensions leads to a limited separation of subband energies (typically a few meV), so that in most cases the thermal broadening of the levels has been greater than their separation. Thus, most of the interesting physical behaviour in such heterostructures is observed only at very low temperatures ($T \ll 4\,\mathrm{K}$). This property distinguishes mesoscopic structures from true *quantum* heterostructures, in which the inter-subband separation is larger than the typical Coulomb interaction energy between carriers. For these reasons, fabricating quantum wires and quantum dots directly with *in situ* growth techniques provides an attractive alternative to *ex situ* processing with lithography.

Quantum heterostructures must aim to satisfy four important criteria. (i) Their lateral size(s) must be of the order of $100\,\text{Å}$ or less to achieve genuine quantum confinement in the sense discussed above. (ii) Their size distribution must be relatively narrow. This is especially important for optical applications, where the inhomogeneous broadening of electronic transitions across the ensemble of heterostructures must be minimized. (iii) Their interfaces must have a high degree of definition and produce an abrupt carrier-confining potential. (iv) They must have structural and chemical stability, i.e. the interfaces must retain their chemical integrity and be stable against the formation of structural defects such as dislocations. Attempts at producing heterostructures that satisfy these criteria have revolved largely around two basic strategies: direct growth on singular or slightly misoriented substrates, and growth on patterned substrates. These will be discussed in turn in this section and in Section 1.9.

1.8.1   Quantum Wells and Quantum-well Superlattices

The vast majority of quantum-well structures have utilized interfaces between GaAs and $Al_xGa_{1-x}As$, where x varies typically between 0.1 and 0.4. These semiconductors are closely matched in lattice spacing (0.08% misfit), so growth proceeds in the Frank–van der Merwe mode, allowing high-quality interfaces to be achieved. The electronic properties of these materials are also well-suited to the quantum-well architecture. For the stated range of x, the conduction band minimum of $Al_xGa_{1-x}As$ is approximately $x\,\mathrm{eV}$ *above* that of GaAs, while the valence band maximum is approximately $0.3x\,\mathrm{eV}$ *below* that of GaAs. This produces a discontinuity of the potential for the carriers in structures with abrupt interfaces between $Al_xGa_{1-x}As$ and GaAs, so the GaAs regions form quantum wells for both types of carriers and the $Al_xGa_{1-x}As$ forms the barriers between the wells. The ability to control the Al concentration in the growth direction allows the form of the confining potential to be tailored to a particular application.

As epitaxial growth techniques have matured, the range of materials available for quantum wells has expanded accordingly (Smith and Mailhiot, 1990). Some of these systems exhibit physical effects which are qualitatively different from those of GaAs/$Al_xGa_{1-x}As$. The InAs/GaSb system is almost lattice-matched (a 0.6%

misfit), but the relative energies near the band edges are quite different from those of $GaAs/Al_xGa_{1-x}As$. This results in the electrons being confined in the InAs region and the holes confined in the GaSb. In the $In_{1-x}Ga_xAs/GaSb_{1-y}As_y$ system, the alloy composition can be used to adjust (subject to the lattice matching of the alloys) the conduction-band edge of $In_{1-x}Ga_xAs$ to be higher or lower in energy than the valence-band edge of $GaSb_{1-y}As_y$.

Relaxing the constraint of lattice matching considerably enlarges the number of materials pairs that can be used in heterostructures. But this also introduces new factors, since the growth need no longer proceed in the Frank–van der Merwe mode. Nevertheless, planar structures can be grown for a number of lattice-mismatched systems, with the strain energy stored either in the epilayers or relaxed through a network of dislocations. Moreover, the strain-induced changes in the band gaps, as well as other characteristics of the electronic structure, lead to new physical effects resulting from the band line-ups. Taken together, the variety of materials, growth orientations, strain-induced effects, and atomic-scale control over the deposition of material provide a rich palette for varying the electronic properties of quantum heterostructures. Their applications are discussed extensively in later chapters of this book.

1.8.2   Quantum Wire Superlattices

A natural extension of the growth of quantum wells on singular surfaces is the growth of quantum wires on vicinal surfaces (Petroff *et al.*, 1984). The basic procedure is illustrated in Fig. 1.16. Two materials are deposited alternately onto the substrate (Fig. 1.16(a)) with the growth conditions chosen to ensure that all of the deposited material accumulates at the step edges, i.e. both materials grow by step flow (Figs. 1.16(b, c)). As this cycle of deposition is repeated (Fig. 1.16(d)), the steps induce a modulation of the lateral composition across

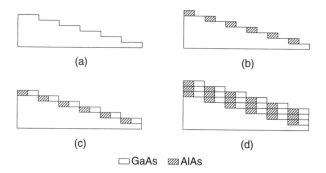

(a)   (b)

(c)   (d)

□ GaAs   ▨ AlAs

Figure 1.16.   Schematic cross-sectional view of the growth of a quantum wire on a vicinal surface. (a) The initial surface. (b) The deposition of a half-monolayer of AlAs, followed by (c) the deposition of a half-monolayer of GaAs. (c) Repetition of this deposition cycle yields a quantum wire superlattice, with the original vicinal surface having acted as a template.

the substrate. If this modulation is of sufficient integrity (see below), then the vicinal surface effectively acts as a template for the formation of a lateral array of quantum wires – a quantum-wire superlattice. This procedure has been applied to AlGaAs/GaAs quantum wires with both MBE (Petroff *et al.*, 1984; Tsuchiya *et al.*, 1989) and MOVPE (Fukui and Saito, 1987).

If a total of one monolayer of material is deposited during each cycle, then the quantum wires are formed normal to the terraces (Fig. 1.17(a)). But if *more* than one monolayer of material is deposited during a cycle, then the quantum wires tilt *away* from the step (Fig. 1.17(b)), while if *less* than a monolayer is deposited, the quantum wires tilt *toward* the step (Fig. 1.17(c)). An example of a tilted AlGaAs/GaAs quantum wire superlattice is shown in Fig. 1.18. One problem encountered in growing such superlattices is that the tilt angle and, thus, the coupling between adjacent wires depends sensitively on the local deposition rate. Variations in this rate of only a small percentage, which can occur even if the substrate is rotated, can lead to quite large (and uncontrolled) changes in the local tilt angle. Thus, while the tilt angle can be maintained effectively constant over relatively small regions of a substrate (Fig. 1.18), the variation in deposition rates across a typical substrate (Exercise 4) can degrade the integrity of the quantum wire ensemble.

However, the variation of the tilt angle with deposition rate forms the basis for a novel quantum wire superlattice (Miller *et al.*, 1992) that overcomes these difficulties (Fig. 1.17(d)). Let y_n denote the lateral position of the heterogeneous interface after the $(n-1)$th cycle of growth and let x_n denote its vertical position. The material deposited (in units of monolayers) during this cycle is θ_n. The coordinates of the interface in each layer then satisfy the following recursion relations:

$$x_{n+1} = x_n + h$$
$$y_{n+1} = y_n + (\theta_n - 1)L \tag{1.16}$$

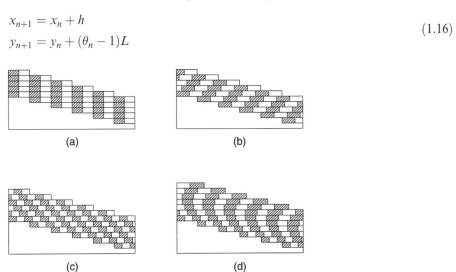

(a)   (b)

(c)   (d)

Figure 1.17.   Idealized cross-sectional view of quantum wire superlattices on a vicinal surface: (a) a vertical superlattice, as in Fig. 1.16(d); (b) and (c) tilted superlattices; (d) a serpentine superlattice.

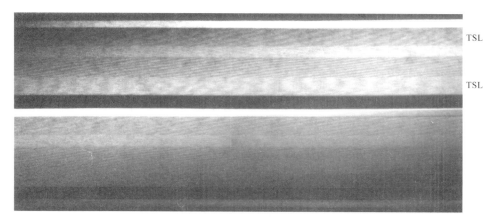

TSL

TSL

Figure 1.18. Cross-sectional transmission electron micrograph of a GaAs/AlAs tilted superlattice. The imaging conditions are such that the GaAs regions appear as the dark areas. The substrate misorientation is $2°$ and the period of the superlattice is $80\,\text{Å}$. (Courtesy P.M. Petroff)

where h is the step height and L is the terrace length. Suppose that θ_n is varied *linearly* between a minimum $\theta_{min} < 1$ and a maximum value $\theta_{max} > 1$:

$$\theta_n = \theta_{min} + (\theta_{max} - \theta_{min})\frac{n}{N} \tag{1.17}$$

where $n = 0, 1, \ldots, N$. The interface profile of the resulting heterostructure obtained by solving the recursion relations (Exercise 10) is a *parabola*,

$$y_n = \frac{L\Delta\theta}{2Nh^2}x_n^2 \tag{1.18}$$

where $\Delta\theta = \theta_{max} - \theta_{min}$ and the origins of x_n and y_n have been chosen in the layer for which $\theta_n = 1$. The curvature is seen to depend only on quantities that can be controlled. In particular, a systematic change in the deposition rate leads only to a translation along the growth direction; the curvature is unaffected.

An example of such a heterostructure, called a *serpentine superlattice*, is shown in Fig. 1.19(a). Carrier confinement in the y-direction of a serpentine superlattice is provided in the usual manner by potential barriers. The confinement in the x-direction, however, comes about because away from the apex of each parabola there is a narrowing of the lateral potential well to which carriers are confined. This causes the probability densities of the carriers to be greatest near the apices of the parabolas (Figs. 1.19(c, d)).

Evidence of carrier confinement in quantum wires has come from photo-luminescence excitation studies, which show a strong anisotropy in the intensity ratio of the electron-light hole and electron-heavy hole exciton peaks, depending on the orientation of the polarization with respect to the direction of the quantum wires (Miller *et al.*, 1992). However, the laser characteristics of these structures are susceptible to coupling between the wires and the roughness of the wire

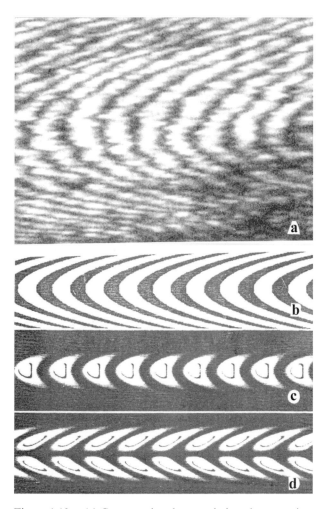

Figure 1.19.   (a) Cross-sectional transmission electron micrograph of a single crescent of a serpentine superlattice. The dark regions are GaAs-rich while the light areas are AlAs-rich. (b) Simulated serpentine superlattice and the electron density distributions in the (c) ground state and (d) first excited state within the superlattice (Yi *et al.* 1991). (Courtesy P. M. Petroff)

boundaries. Thus, while there has been substantial progress made in growing quantum wires on vicinal surfaces (Laruelle *et al.*, 1996), most attempts at producing these heterostructures now revolve around other strategies (Bhat *et al.*, 1990; Pfeiffer *et al.*, 1993; Koshiba *et al.*, 1994; Nötzel *et al.*, 1996).

1.8.3   Self-organized Quantum Dots

Although there are several ways of fabricating quantum dot structures *in situ*, most of these fail to fulfil the requirements of size uniformity of the dot ensemble

and high interface definition of the individual dots. In this section, we will describe
a method for producing quantum dot arrays that has attracted considerable
attention in recent years. The basic principle behind this approach is the formation of *coherent* 3D islands during the Stranski–Krastanov growth (Section 1.3) of
a highly strained system. The prototypical cases are InAs on GaAs(001)
(Madhukar and Rajkumar, 1990) and Ge on Si(001) (Eaglesham and Cerullo,
1990), both produced by MBE, though several other materials combinations,
some of which are grown with MOVPE, also exhibit this morphology (Petroff
and DenBaars, 1994; Seifert *et al.*, 1996). When the 3D islands are embedded
within epitaxial layers of a material that has a wider band gap, the carriers within
the islands are confined by the potential barriers that surround each island, forming an array of quantum dots. Because these quantum dots are obtained directly
by growth, with no additional processing, they are referred to as *self-organized* or
self-assembled structures.

1.8.3.1 Stranski–Krastanov Growth of InAs on GaAs(001)

Quantum dots formed from 3D islands during the Stranski–Krastanov growth of
InAs on GaAs(001) (7% misfit) have a number of fascinating and potentially
useful characteristics (Leonard *et al.*, 1993; Moison *et al.*, 1994). (i) They are
small enough to exhibit quantum effects in the confined carriers. The average
diameter of the base of the dots is typically near 300 Å and the average height
of the dots is near 50 Å. (ii) The dispersion about these averages is, typically,
±10% for the base diameter and ±20% for the height. (iii) The dot shapes are
elongated truncated pyramids with well-oriented sidewalls – called *facets* – along
one direction (e.g. Fig. 1.5). This suggests that the structures of the individual dots
are strongly influenced by thermodynamics, though the appearance of different
facets suggests that kinetic factors cannot be neglected. (iv) Since one of the
intended uses of these structures is for optical applications, one of their most
important properties is that they are *coherent* (Fig. 1.5).

Interest in quantum dots produced in this way was stimulated originally by the
observation of intense photoluminescence (PL) from small InAs islands on
GaAs(001). Optical studies of these structures have found that *individual* quantum
dots exhibit high optical quality in terms of a narrow ($\ll kT$) PL linewidth
(Marzin *et al.*, 1994). These sharp linewidths, together with their temperature-
dependence (Raymond *et al.*, 1995), are consistent with a density of states consisting of a series of delta functions and an enhanced oscillator strength, both of
which are expected of structures with 3D carrier confinement. However, *arrays* of
10^6–10^7 dots show a large inhomogeneous broadening ($\gg kT$) of the PL emission
due to the distribution of dot sizes and composition. Thus, identifying the kinetics
of 3D island nucleation and the subsequent development of their size distribution
is a central focus of current research.

The growth of InAs on GaAs(001) proceeds first by the nucleation of 2D
islands which coalesce into coherently strained layers. These are the 'wetting'

layers in the conventional Stranski–Krastanov description (Section 1.3). Coherent 3D islands are formed after the deposition of close to two InAs bi-layers, though the precise point at which this occurs depends on the growth conditions. The transition to growth by 3D islands is quite abrupt, occurring over less than 0.1 monolayers. This transition can be followed by RHEED, which shows a change from a streaky pattern, characteristic of layer-by-layer growth, to a spotty pattern that corresponds to the transmission of the electrons through the 3D islands.

Figure 1.20 shows a sequence of STM images taken during the formation and evolution of 3D InAs islands on GaAs(001) at 450 °C. The 3D islands are first observed near 1.7 monolayers. Note the preferential location of these islands near steps. This is a general feature of 3D islands which was apparent already in early studies (Leonard *et al.*, 1994) and can be used to control the location of the islands (Seifert *et al.*, 1996). This will be discussed further below. As more material is deposited, the islands grow in number, but soon reach a saturation density.

A noteworthy feature of the growth of the 3D islands is revealed by estimating the volume of the dots directly from STM measurements as a function of deposition time at different temperatures (Joyce *et al.*, 1998). At low temperatures, the material within the 3D islands can be accounted for simply from the accumulation

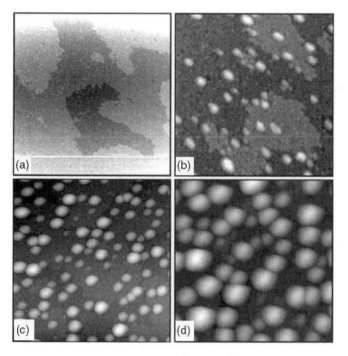

Figure 1.20.   STM images (1000 Å × 1000 Å of InAs deposited on GaAs(001)–(4 × 4) at 450 °C taken at coverages of (a) 1.4 ML, (b) 1.7 ML, (c) 2.0 ML, and (d) 2.7 ML. The change from layer-by-layer to 3D island growth, as determined by the changes in the RHEED pattern, occurs near 1.7 ML (Joyce *et al.*, 1998). (Courtesy G. R. Bell)

of deposited material. But, at higher temperatures, the volume of the islands far exceeds the volume of InAs actually deposited. The conclusion is that the additional material comes from the wetting layers, so the 3D islands are InGaAs alloys. Moreover, this demonstrates the active role of the wetting layers in the final stages of Stranski–Krastanov growth, which is in stark contrast to the inert wetting layers implied by the classical description of this growth mode (Section 1.3).

1.8.3.2 Controlled Positioning of Quantum Dots

The tendency of 3D islands to form near atomic steps suggests that their positions may be influenced by an appropriately modified substrate. The controlled positioning of individual quantum dots opens up new opportunities for utilizing their properties, for example, as the active components in single-electron and resonant tunnelling devices (Chapter 9). Two strategies that have been used to manipulate the positions of 3D islands are the strain-induced nucleation on homogeneous substrates (Xie et al., 1995) and either the local or large-scale modification of a substrate by etching a pattern on that substrate (Seifert et al., 1996).

Strain-induced nucleation provides a natural way of ordering 3D islands in the vertical (growth) direction. If two or more layers of 3D islands are grown sufficiently close to each other, then the islands on successive layers align, with the extent of the alignment decreasing with increasing interlayer separation (Xie et al., 1995). This is illustrated in Fig. 1.21. For the closest vertical separation, 36 layers, the 3D islands show a very high degree of vertical alignment. But, as the vertical separation is increased to 46 layers and then to 96 layers, this alignment diminishes and the positioning of the islands becomes more statistical.

One explanation (Xie et al., 1995; Tersoff et al., 1996) for this observation is that the strain relaxation of a 3D island is greatest in a region where the strain energy is a maximum, i.e. near the region of the substrate just above an underlying quantum dot (Section 1.7.2). Thus, just as the formation of 3D islands appears to be favoured near steps of the wetting layer, so the strain relaxation caused by underlying 3D islands creates preferred regions for island formation on subsequent layers.

1.8.3.3 Ge 'Hut' Clusters on Si(001)

The strain-induced 2D-to-3D transition during growth in the SiGe system (4% misfit) is manifested in a variety of surface morphologies whose characteristic feature size depends on the lattice misfit between the alloyed epilayer and the substrate. The growth of pure Ge on Si(001) follows the Stranski–Krastanov scenario, with 3D Ge islands called 'hut' clusters (Mo et al., 1990) appearing quickly after the deposition of the wetting layers. These Ge islands have several remarkable structural and kinetic characteristics. Structurally, they exhibit $\{105\}$ facets which never show partially complete layers, and they are elongated along the $\langle 100 \rangle$ directions, with rectangular bases having aspect ratios of up to 8 : 1, which

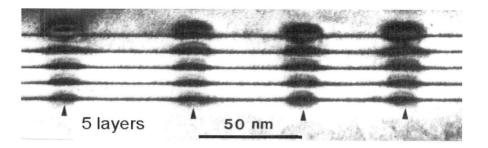

Figure 1.21. A $g = (200)$ dark field TEM image of InAs islands separated by 36 ML spacer layers of GaAs on a GaAs(001) substrate. The correlation in lateral positions of successive sets of islands diminishes with the thickness of the buffer layer (Xie *et al.*, 1995). (Courtesy A. Madhukar)

gives these islands a 'hut-like' morphology (Fig. 1.22). Their kinetic properties include metastability, since they dissolve upon annealing, and a self-limiting growth mechanism, which causes larger islands to grow more slowly than small islands. Thus, smaller islands can catch up in size, which results in a narrowing of the distribution of island sizes. This makes hut clusters natural candidates for producing arrays of quantum dots.

The presence of {105} facets appears to be a general feature of strain-induced roughening in the Ge/Si(001) system. Their stability can be understood by regarding each facet as composed of individual steps which interact through long-range elastic 'monopole' forces (Tersoff, 1995). This interaction is attractive for steps with the same orientation (up-up or down-down), but repulsive for steps with the opposite orientation (up-down). Thus, elasticity alone favours short terraces, i.e. steep facet inclinations. But balancing this tendency is the short-range repulsion due to the formation of unfavourable structural configurations, which are described by a 'force-dipole', and are present even in the absence of strain (Poon *et al.*, 1990). Thus, {105} facets result from the tendency to create steep inclines while minimising unfavourable structures near steps (Exercise 11).

The ideas just described can also be used to understand two of the other important characteristics of hut clusters: the absence of partially complete facets and the self-limiting growth of larger clusters. The addition of material at an arbitrary point on the facet will result in a local step spacing which differs from the most stable array and which, through the dipole repulsion, extracts a high energy cost. It is therefore most favourable to grow the next layer of the facet from the base of the facet upwards toward the apex. A calculation which incorporates the elastic monopole and dipole terms, as well the additional surface energy created (Jesson *et al.*, 1996, 1998), shows that, although the energy to form the next facet layer increases initially, for larger coverages, the release of elastic energy dominates and the facet layer will complete rapidly. Therefore,

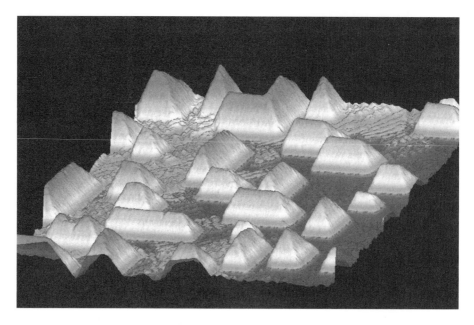

Figure 1.22.   STM image (1600 Å × 1600 Å) of Ge hut clusters grown in Si(001) at 575 K. (Courtesy B. Voigtländer)

this mechanism implies that there is a thermally activated barrier that must be overcome for an 'embryonic' facet layer to proceed to completion. Larger islands or higher lattice misfit cause this barrier to increase because the greater stress near the base of the islands increases the elastic energy associated with the nucleation of a new facet layer. Thus, larger islands grow more slowly than small islands, but even small islands will have their growth arrested for a sufficiently large misfit.

1.9   Growth on Patterned Substrates

The direct growth scenarios described in the preceding section produce macroscopically uniform arrays of heterostructures across a surface. There are, however, many situations where the growth of a heterostructure needs to be confined to a particular region of a substrate. Examples include the selective area growth required for integrating devices, such as surface emitting lasers or modulators on an electronics-bearing chip. In this section, we consider growth on specially prepared substrates that facilitate the formation of *individual* heterostructures within prescribed regions of that substrate.

1.9.1 Selective Area Growth

Selective area growth can be implemented in a number of ways, including ion- or laser-assisted growth on a planar substrate, and with masks, in which a regular array of windows that has been opened by lithography is placed on a substrate, so growth only occurs through the windowed region. The effect of selective area growth can be achieved by lithographically etching a substrate over which uniform growth has already occurred. However, as we have already discussed, this approach is not suitable for producing heterostructures which exhibit all of the benefits of true quantum confinement.

There is another way of achieving the growth of heterostructures in pre-determined areas of a substrate which has a number of attractive features for both strained and lattice-matched systems. This is through the etching of a substrate with a sequence of patterns or templates of particular shape, size and spacing. In lithographic etching, a pattern is first placed on the semiconductor surface by means of a mask and the uncovered regions are then cut, or etched, to a predetermined depth to transfer the pattern to the semiconductor. The lateral sizes of these regions are in the range 0.01–1 µm. The substrate is no longer planar because the etching process has exposed different facets in different regions of the substrate. In addition to providing control over the location of heterostructures, growth on patterned substrates offers additional strain-relaxation mechanisms compared with those available on singular or vicinal surfaces (Madhukar, 1993).

1.9.2 Quantum Wires on 'V-Grooved' Surfaces

Many of the characteristics of the growth morphologies on patterned substrates are due to different facets having different kinetic and chemical properties (Tsang and Cho, 1977). The migration of adatoms and other surface species, as well as the incorporation of material into the growth front, occur at different rates on different facets. These effects may be compounded by the slow transfer of material across the region where two facets meet (Exercise 12). All of these factors conspire to produce a growth rate which is not uniform across a patterned substrate (Ozdemir and Zangwill, 1992), so there can be adjacent regions with quantum wells of different thicknesses and composition, i.e. different quantum-well band gaps. This architecture produces two-dimensional lateral confinement by a mechanism similar to that in serpentine superlattices (Fig. 1.19(c, d)). If the lateral size of the confining region can be made comparable to the de Broglie wavelength of the carriers, then the carriers are effectively confined within a quantum wire.

The formation of quantum wires on surfaces which have been patterned with a series of 'V-grooves' provides one realization of the scenario just described (Bhat *et al.*, 1990). The starting surface for this procedure is typically GaAs(001) etched with a pattern of V-grooves along the [01$\bar{1}$] direction. AlGaAs cladding layers are

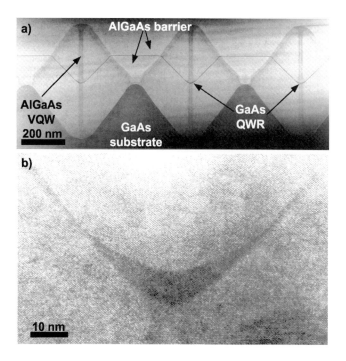

Figure 1.23.   (a) Dark field and (b) high resolution TEM cross-sections of V-groove GaAs quantum wires (QWRs) grown by low-pressure MOCVD on GaAs(001) patterned with 0.5 μm pitch [01$\bar{1}$] corrugations (Kapon et al., 1998). The crescent-shaped region in (b) corresponds to the GaAs core of the wire. Ga-rich regions forming vertical quantum wells (VQWs) are also indicated (Courtesy F. Lelarge and E. Kapon)

then deposited onto this structure to form a sharp apex bounded by two {111} planes. Subsequent deposition of GaAs by MOVPE produces a greater growth rate along the [100] direction near this apex, which results in the formation of a crescent-shaped quantum well (Fig. 1.23). The lateral tapering of the thickness of this quantum well away from the apex of the V-groove produces a lateral variation of the effective band gap, in this case because of the increase of the carrier confinement energy with decreasing quantum-well thickness (Kapon et al., 1989). This produces a potential well which confines carriers to the crescent-shaped region along the direction of the V-groove, i.e. this region forms a quantum wire. This procedure can be repeated to form a superlattice of quantum wires (Brasiol et al., 1998) or a tunnelling structure composed of two closely-spaced quantum wires (Weman et al., 1998).

1.9.3   Stranski–Krastanov Growth on Patterned Substrates

The abrupt transition to 3D growth with the amount of deposited 'wetting' layer material forms the basis of one method of controlling the lateral positioning of quantum dots on a patterned substrate. A corrugated substrate, with alternate concave and convex regions (i.e. 'hills' and 'valleys', respectively, in Fig. 1.24), will tend to flatten during epitaxial growth to minimize its surface free energy. This flattening occurs by faster growth in the convex regions because they offer more stable binding sites for migrating adatoms than the concave regions. Thus, when material forming the strained epilayer is deposited onto the corrugated substrate, the thickness in the convex regions will attain the point at which 3D islands appear *before* that in the concave regions. Hence, the 3D islands will appear preferentially in the convex regions of the substrate. An AFM image of 3D islands which have formed preferentially in the convex regions of a patterned GaAs(001) surface after the deposition of InAs is shown in Fig. 1.24 (Mui *et al.*, 1995).

There are other ways in which growth on patterned substrates can be used to control the positioning of quantum dots. The formation of 3D islands requires a 'wetting' layer of a certain thickness on group-V-terminated surfaces. But on other low-index surfaces, including group-III-terminated (001) surfaces, 3D islands do not form at all. Thus, substrates patterned to expose several low-index facets exhibit 3D islands only in regions of the group-V-terminated facet (Seifert *et al.*, 1996). The discussion in this section, combined with that in Section 1.8.3.2, indicates that, although the kinetics of 3D island formation are not completely understood, the phenomenology is sufficiently well-developed to envisage a

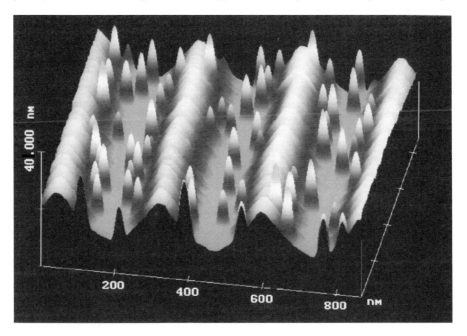

Figure 1.24.   AFM image of self-assembled 3D InAs islands which have been grown on a patterned GaAs(001) substrate by MBE (Mui *et al.*, 1995). The InAs islands, which are formed preferentially in the convex regions of the surface (the 'valleys'), have heights and diameters in the range of 10 nm and 20 nm, respectively. (Courtesy P. M. Petroff)

number of intriguing structures based on quantum dots, both for device applications and fundamental studies.

1.10   Future Directions

As discussed in the Introduction, devices whose operation is based on heterojunctions, such as quantum, well lasers and resonant tunnelling diodes, have already established themselves in the commercial marketplace. The expanding requirements of optoelectronic and mobile telecommunications, as well as emerging applications such as sensor technology, will ensure that these and related heterostructures will continue to maintain this position. More significantly, at least in commercial terms, is the advent of group III nitrides for red, blue and green light-emitting and laser diodes. This opens the way to mass-market commercial applications such as colour displays, lighting and optical data storage.

On the other hand, the impact of quantum heterostructures on new computer architectures is less certain. Since the 1960s, the number of transistors that semiconductor manufacturers can put on a chip has doubled every twelve to eighteen months. This is known as Moore's Law, named after Gordon Moore, one of the co-founders of Intel, who made this observation in a short article for *Electronics* magazine (Moore, 1965). This increase in transistor densities has been due largely to improvements in lithographic technology without corresponding increases in production costs. The result is more powerful, but less expensive, computers. However, the physical principles underlying chip production, especially lithography, suggest that there are obstacles to unabated miniaturization of electronic components. A possible way out of this impasse is the prospect of designing chips based on quantum heterostructures. This is where the ideas described in this and later chapters in this book would play a major role, though there are a number of formidable difficulties to be overcome before this scenario can be realized (Chapter 10).

One of the long-term goals of epitaxial technology is the development of total *in situ* processing, where sample preparation, growth and post-growth processing are all carried out under controlled conditions. While most studies of epitaxial processes have focussed on growth, the controlled removal of atoms from the surface also provides a way of manipulating surface properties and for etching at scales inaccessible by conventional (lithographic) means. One way of achieving atomic-scale etching is through chemical beams. Chemical etching of semiconductor surfaces can occur on a layer-by-layer basis (Tsang *et al.*, 1993) and there is evidence of atomic site-selectivity at lower temperatures (Kaneko *et al.*, 1995). This opens the way to removal processes being carried out with the same atomic-level precision as growth.

Finally, although our attention has been focussed here on semiconductor heterostructures, epitaxial techniques have also seen applications to metal–metal,

metal–semiconductor, and oxide–oxide materials pairs. For metals, in particular, the initiation of MBE in the late 1970s was motivated by the expectation that, in analogy with low-dimensional semiconductors, high-quality metallic thin-film structures would exhibit new magnetic phenomena which, in turn, might find application in high-speed, high-density magnetic data recording and storage (Farrow, 1998). This was, in fact, realized by several discoveries: antiferromagnetic interlayer exchange (Grunberg et al., 1986; Carbone and Alvaredo, 1987), enhanced magnetoresistance (Binasch et al., 1989), and giant magnetoresistance (Baibich et al., 1988). This again illustrates how low-dimensional materials produced by controlled epitaxial growth can have unexpected properties which not only expand our fundamental understanding of electron and phonon interactions in condensed-matter systems, but have commercial applications as well.

EXERCISES

1. Consider a gas in a container of volume V at a temperature T. Show that the number of particles dN striking a surface of area dA during a time dt is

$$dN = \tfrac{1}{4} n \bar{v} \, dA \, dt$$

where $n = N/V$ and $\bar{v}$ is the average velocity of the gas particles. If the velocity distribution is of the Maxwell form,

$$f(v) = 4\pi \left(\frac{m}{2\pi kT} \right)^{3/2} v^2 \, e^{-mv^2/2kT}$$

where m is the mass of the particles and k is Boltzmann's constant, show that the flux, $J = dN/(dA \, dt)$, is

$$J = \frac{P}{\sqrt{2\pi mkT}}$$

where the gas is regarded as ideal with pressure P. Calculate the flux for such a gas of Ga atoms at room temperature and at a pressure of 10^{-2} Torr. The atomic weight of Ga is 69.7.

2. Calculate the rate at which gas particles strike a sphere of radius R. Use this to obtain the following estimate of the mean distance between collisions between gas particles, called the *mean free path*:

$$\lambda \sim \frac{1}{\pi n d^2} = \frac{kT}{\pi d^2 P}$$

where d is the diameter of the gas particles. Estimate this quantity for the Ga gas in Exercise 1.

3. The molecular flow of material from a solid source is produced by a *Knudsen cell*, which is a sealed tube with a small orifice at one end. A vapour pressure is produced by heating the solid source which causes material to flow out of the cell through the orifice. To ensure that the flow is molecular, the diameter of the orifice must be less than the mean free path of the gas particles in the cell. By considering the number of particles per unit time emitted from the orifice which have a speed between v and $v + dv$ and make an angle between θ and $\theta + d\theta$ with respect to the normal to the plane of the orifice, show that the flux at a planar surface at a distance L from the orifice and parallel to the plane of the orifice is

$$J = \frac{AP}{4\pi L^2 \sqrt{2\pi mkT}}$$

where the orifice has area A and the Knudsen cell contains particles of mass m at a pressure P and temperature T.

4.   The growth rate of GaAs(001) under normal conditions is determined by the Ga flux. Assume that the plane of the orifice of the Knudsen cell and that of the substrate are parallel.

  (a) Calculate the growth rate of GaAs(001) in monolayers (ML) per second under the following conditions: $T = 1000°C$, $P = 10^{-2}$ Torr, and $L = 0.1$ m. For the diameter of the orifice, use half the value of λ calculated in Exercise 2.
  (b) Estimate the variation in flux across a circular substrate of radius 2.5 cm if the centre of the beam is at the centre of the substrate.

5.   The most general step-edge boundary conditions used for the BCF equation (1.3) are (Ghez and Iyer, 1988)

$$D \left.\frac{\partial c}{\partial x}\right|_0 = \beta(c - c_{eq})|_0, \qquad -D \left.\frac{\partial c}{\partial x}\right|_L = \beta(c - c_{eq})|_L$$

where β describes the detachment of adatoms at steps and c_{eq} is the equilibrium step concentration of adatoms. Show that the solution of (1.3) with these boundary conditions can be written as

$$c(x) = \frac{J}{2D} x(L - x) + c_{step}$$

where the adatom concentration at the step is

$$c_{step} = \frac{JL}{2\beta} + c_{eq}$$

What is the solution corresponding to $J = 0$?

6.   The adatom diffusion constant D is often taken to be of the following form:

$$D = \tfrac{1}{2}a^2 \nu \exp(-E_D/kT)$$

where a is the lattice constant, ν is an atomic vibrational frequency, and E_D is the energy barrier to diffusion and T is the temperature of the substrate. The factor of $\frac{1}{2}$ accounts for there being two equivalent sites to which the adatom can hop. The values of E_D and ν are typically $E_D \approx 1$ eV and $\nu \approx 10^{13}\,\mathrm{s}^{-1}$ and a is usually the distance between nearest-neighbour sites. For $T = 550°\mathrm{C}$ and $J = 1$ ML/s, compute

(a) the maximum density of atoms on a terrace of length $10a$ predicted by the BCF theory
(b) the density of atoms and islands on a singular surface at the time $t = 0.1$s.

7. To obtain the steady-state solution of the BCF equation when the step velocity v cannot be neglected, it is convenient to transform to a coordinate system that moves with the steps. This is accomplished with the coordinate transformation $x \to x - vt$. Show that $v = JLa^2$, where a^2 is the area of each site. By introducing the dimensionless concentration $n = a^2 c$ and distance $\xi = x/L$, show that the steady-state BCF equation can be written as

$$\frac{\mathrm{d}^2 n}{\mathrm{d}\xi^2} + \alpha \frac{\mathrm{d}n}{\mathrm{d}\xi} + \alpha = 0$$

where $\alpha = JL^2 a^2/D$. Provide a physical interpretation for α. Obtain the solution of this equation with the boundary conditions (1.4):

$$n(\xi) = \frac{1 - e^{-\alpha\xi}}{1 - e^{-\alpha}} - \xi$$

and show that the leading-order contribution to this solution as $\alpha \to 0$ is, in the original variables, given by (1.5).

8. Combine the diffusion constant D and the flux J to obtain a quantity which has the dimensions of area and one which has the dimensions of time.
(a) Use these quantities to write n_1, N, and t in dimensionless form and, hence, obtain the dimensionless form of the rate equations in (1.8):

$$\frac{\mathrm{d}\tilde{n}_1}{\mathrm{d}\tilde{t}} = 1 - 2\tilde{n}_1^2 - \tilde{n}_1 \tilde{N}$$

$$\frac{\mathrm{d}\tilde{N}}{\mathrm{d}\tilde{t}} = \tilde{n}_1^2$$

where $\tilde{n}_1$, $\tilde{N}$, and $\tilde{t}$ are the dimensionless forms of the adatom and island densities and the time, respectively.
(b) Use these equations to verify the short- and long-time limits in (1.9) and (1.10).
(c) Integrate the equations obtained in (a) to obtain the results in Fig. 1.14.

9. Beginning with the expression for the energy in equation (1.14), determine the equations that must be solved to obtain the equilibrium positions of the atoms in the Frenkel–Kontorova model. Use these equations with the parameters used to generate Fig. 1.15 to determine the energies of coherent islands and for islands with one and two dislocations as a function of the island size.

10. Suppose that in the recursion relations (1.16) θ_n takes on the same value θ for every layer. Show that a tilted superlattice (Fig. 1.17(b), (c)) is obtained with the tilt angle β given by

$$\tan \beta = \frac{\theta - 1}{\tan \alpha}$$

where $\alpha = \tan^{-1}(h/L)$ is the vicinal angle. Now solve the recursion relations in (1.16) and (1.17) to obtain the profile of the serpentine superlattice in (1.18).

11. The elastic energy of a hut cluster can be expressed in terms of step-step interaction energies. The energy E (per unit length parallel to the steps) for an array of steps at positions x_i can be written as (Tersoff, 1995)

$$E = \sum_i C_0 + \sum_{i \neq j} C_m s_i s_j \ln \left(\frac{x_i - x_j}{a} \right) + \sum_{i \neq j} C_d \left(\frac{x_i - x_j}{a} \right)^{-2}$$

where a is the lattice constant. The first term in this expression is the step forma-tion energy, C_0. The second term is the elastic step interaction energy, described by the interaction between 'force monopoles' created by the discontinuity of the 2D stress at a step on a strained layer. This interaction depends on the distance between the steps as well as on their orientation: $s_i = 1$ for up steps and $s_i = -1$ for down steps. The last term describes a repulsive short-range interaction between steps.

Consider the two-layer system shown in Fig. 1.25. By fixing the length L of the first layer, determine the energy E as a function of the length ℓ of the second layer. Discuss the effect of varying the relative magnitudes of C_m and C_d. You may regard C_0 as a constant.

12. As a simple model of how the growth rate can vary near the boundary of two facets, consider the following BCF-type equation for the steady-state adatom concentration c on one facet:

$$D \frac{d^2 c}{dx^2} + J - \frac{c}{\tau} = 0$$

where τ is the average time it takes an atom to be incorporated into the growing material. The growth rate R is therefore given by $R = c/\tau$. The condition at the boundary of the two facets, which is taken to be at the origin, is

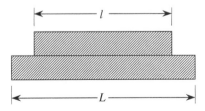

Figure 1.25.   Figure for Exercise 11.

$$-D \left. \frac{dc}{dx} \right|_{x=0} = I$$

where I is the flow of atoms from the facet in the region $x < 0$ to that in the region $x > 0$. Show that the growth rate for positive x is is given by

$$R = J + I\lambda^{-1} e^{-x/\lambda}$$

where $\lambda = \sqrt{D\tau}$. This exercise illustrates how the transfer rate between facets can affect the morphological evolution of a patterned substrate.

References

C. R. Abernathy, *Mat. Sci. Eng.* **R14**, 203 (1995).

O. L. Alerhand, A. N. Berker, J. D. Joannopoulous, D. Vanderbilt, R. J. Hamers and J. E. Demuth, *Phys. Rev. Lett.* **64**, 2406 (1990).

Y. Arakawa and A. Yariv, *IEEE J. Quantum Electron.* **QE–22**, 1887 (1986).

O. Auciello and A. R. Krauss, eds., *MRS Bulletin* **20**(5), 14 (1995).

M. N. Baibich, J. M. Broto, A. Fert, F. Nguyen Van Dau, F. Petroff, P. Etienne, G. Creuzet, A. Friederich and J. Chazelas, *Phys. Rev. Lett.* **61**, 2472 (1988).

G. S. Bales and D. C. Chzan, *Phys. Rev. B* **50**, 6057 (1994).

A.-L. Barabási and H. E. Stanley, *Fractal Concepts in Surface Growth* (Cambridge University Press, Cambridge, 1995).

M. C. Bartelt and J. W. Evans, *Phys. Rev. B* **54**, R17 359 (1996).

M. C. Bartelt, A. K. Schmid, J. W. Evans and R. Q. Hwang, *Phys. Rev. Lett.* **81**, 1901 (1998).

E. Bauer, *Z. Krist.* **110**, 372 (1958).

E. Bauer, *Rep. Prog. Phys.* **57**, 895 (1994).

E. Bauer, *Appl. Surf. Sci.* **92**, 20 (1996).

J. G. Belk, J. L. Sudijono, X. M. Zhang, J. H. Neave, T. S. Jones and B. A. Joyce, *Phys. Rev. Lett.* **78**, 475 (1997).

G. R. Bell, T. S. Jones and B. A. Joyce, *Surf. Sci.* **429**, L492 (1999).

R. Bhat, E. Kapon, J. Werner, D. M. Hwang, N. G. Stoffel and M. A. Koza, *Appl. Phys. Lett.* **56**, 863 (1990).

B. Binasch, P. Grunberg, F. Sauerenbach and W. Zinn, *Phys. Rev. B* **39**, 4828 (1989).

G. Binnig and H. Rohrer, *Rev. Mod. Phys.* **59**, 615 (1987).

G. Binnig, C. F. Quate and Ch. Gerber, *Phys. Rev. Lett.* **56**, 930 (1986).

G. P. Booker and B. A. Joyce, *Phil. Mag.* **14**, 301 (1966).

M. Bott, M. Hohage, M. Morgenstern, T. Michely and G. Comsa, *Phys. Rev. Lett.* **76**, 1304 (1996).

G. Brasiol, E. Kapon, Y. Ducommun and A. Gustafsson, *Phys. Rev. B* **57**, R9416 (1998).

W. K. Burton, N. Cabrera and F. C. Frank, *Phil. Trans. R. Soc. Lond*, A **243**, 299 (1951).

C. Carbone and S. F. Alvaredo, Phys. Rev. B **36**, 2433 (1987).

D. J. Chadi, *Phys. Rev. Lett.* **59**, 1691 (1987).

A. Cho, ed., *Molecular Beam Epitaxy* (American Institute of Physics, New York, 1994).

L. Däweritz and R. Hey, *Surf. Sci.* **236**, 15 (1990).

R. Dingle, W. Wiegmann and C. H. Henry, *Phys. Rev. Lett.* **33**, 827 (1974).

R. Dingle, *Current Reviews in Solid State Sciences* **5**, 585 (1975).

B. W. Dodson, *Critical Reviews in Solid State and Materials Sciences* **16**, 115 (1990).

D. J. Eaglesham and M. Cerullo, *Phys. Rev. Lett.* **64**, 1943 (1990).

K. Eberl, *Physics World* **10**(9), 47 (1997).

L. Esaki and R. Tsu, *IBM J. Res. Develop.* **14**, 61 (1970).

R. F. C. Farrow, *IBM J. Res. Develop.* **42**, 23 (1998).

J. S. Foord, G. J. Davies and W. T. Tsang, eds., *Chemical-Beam Epitaxy and Related Techniques* (Wiley, New York, 1997).

J. Frenkel and T. Kontorova, *J. Phys. USSR* **1**, 137 (1939).

T. Fukui and H. Saito, *Appl. Phys. Lett.* **50**, 824 (1987).

K. Georgsson, N. Carlsson, L. Samuelson, W. Seifert and L. R. Wallenberg, *Appl. Phys. Lett.* **67**, 2981 (1995).

R. Ghez and S. S. Iyer, *IBM J. Res. Develop.* **32**, 804 (1988).

M. H. Grabow and G. H. Gilmer, *Surf. Sci.* **194**, 333 (1988).

P. Grunberg, R. Schreiber, Y. Pang, M. B. Brodsky and H. Sowers, *Phys. Rev. Lett.* **57**, 2442 (1986).

M. F. Gyure and J. J. Zinck, eds., *Comp. Mat. Sci.* **6**, 113 (1996).

R. J. Hamers, R. M. Tromp and J. E. Demuth, *Phys. Rev. B* **34**, 5343 (1986).

J. C. Hamilton, *Phys. Rev. B* **55**, R7402 (1997).

E. J. Heller, Z. Y. Zhang, and M. G. Lagally, *Phys. Rev. Lett.* **71**, 743 (1993).

M. A. Herman and H. Sitter, *Molecular Beam Epitaxy: Fundamentals and Current Status*, 2nd edn (Springer, Berlin, 1996).

P. Hirsch, A. Howie, R. Nicholson, D. W. Pashley and M. J. Whelan, *Electron Microscopy of Thin Crystals* 2nd edn (Krieger, Malabar, Florida, 1977).

A. J. Hoeven, J. M. Lenssinck, D. Dijkkamp, E. J. van Loenen and J. Dieleman, *Phys. Rev. Lett.* **63**, 1830 (1989).

D. M. Holmes, J. L. Sudijono, C. F. McConville, T. S. Jones and B. A. Joyce, *Surf. Sci.* **370**, L173 (1997).

T. Ide, A. Yamashita, and T. Mizutani, *Phys. Rev. B* **46**, 1905 (1992).

A. Imamoglu and Y. Yamamoto, *Phys. Rev. Lett.* **72**, 210 (1994).

K. F. Jensen, in *Chemical Vapor Deposition Principles and Applications*, M. L. Hitchman and K. F. Jensen, eds. (Academic, London, 1993), pp. 31–90.

D. E. Jesson, K. M. Chen and S. J. Pennycook, *MRS Bulletin*, **21**(4), 31 (1996).

D. E. Jesson, G. Chen, K. M. Chen and S. J. Pennycook, *Phys. Rev. Lett.* **80**, 5156 (1998).

B. A. Joyce and R. R. Bradley, *Phil. Mag.* **14**, 289 (1966).

B. A. Joyce, R. R. Bradley and G. R. Booker, *Phil. Mag.* **15**, 1167 (1967).

P. B. Joyce, T. J. Krzyzewski, G. R. Bell, B. A. Joyce and T. S. Jones *Phys. Rev. B* **58**, R15981 (1998).

T. Kaneko, P. Šmilauer, B. A. Joyce, T. Kawamura and D. D. Vvedensky, *Phys. Rev. Lett.* **74**, 3289 (1995).

E. Kapon, D. M. Hwang and R. Bhat, *Phys. Rev. Lett.* **63**, 430 (1989).

E. Kapon, F. Reinhardt, G. Biasiol and A. Gustafsson, *Appl. Surf. Sci.* **123/124**, 674 (1998).

R. Kern, G. Le Lay and J. J. Metois, in *Current Topics in Materials Science*, E. Kaldis, ed. (North-Holland, Amsterdam, 1979), Vol. 3.

S. Koshiba, H. Noge, H. Akiyama, T. Inoshita, Y. Nakamura, A. Shimizu, Y. Nagamune, M. Tsuchiya, H. Kano, H. Sakaki and K. Wada, *Appl. Phys. Lett.* **64**, 363 (1994).

P. K. Larsen and P. J. Dobson, eds., *Reflection High-Energy Electron Diffraction and Reflection Electron Imaging of Surfaces* (Plenum, New York, 1988).

F. Laruelle, F. Lelarge, Z. Z. Wang, T. Melin, A. Cavanna and B. Etienne, *J. Cryst. Growth* **175/176**, 1087 (1996).

G. Le Lay and R. Kern, *J. Cryst. Growth* **44**, 197 (1978).

F. K. LeGoues, *MRS Bulletin* **21**(4), 38 (1996).

D. Leonard, M. Krishnamurthy, C. M. Reaves, S. P. Denbaars and P. M. Petroff, *Appl. Phys. Lett.* **63**, 3203 (1993).

D. Leonard, K. Pond and P. M. Petroff, *Appl. Phys. Lett.* **50**, 11 687 (1994).

S. Little and A. Zangwill, *Phys. Rev. B* **49**, 16 659 (1994).

A. Madhukar, *Surf. Sci.* **132**, 344 (1983).

A. Madhukar, *Thin Solid Films* **231**, 8 (1993).

A. Madhukar and K. C. Rajkumar, *Appl. Phys. Lett.* **57**, 2110 (1990).

I. V. Markov, *Crystal Growth for Beginners: Fundamentals of Nucleation, Crystal Growth and Epitaxy* (World Scientific, Singapore, 1995).

J. Y. Marzin, J. M. Garard, A. Izrael, D. Barrier and G. Bastard, *Phys. Rev. Lett.* **73**, 716 (1994).

H. Metiu, Y. T. Lu and Z. Y. Zhang, *Science* **255**, 1088 (1992).

M. S. Miller, H. Weman, C. E. Pryor, M. Krishnamurthy, P. M. Petroff, H. Kroemer and J. L. Merz, *Phys. Rev. Lett.* **68**, 3464 (1992).

Y.-W. Mo, B. S. Swartzentruber, R. Kariotis, M. B. Webb and M. G. Lagally, *Phys. Rev. Lett.* **63**, 2393 (1989).

Y.-W. Mo, D. E. Savage, B. S. Swartzentruber and M. G. Lagally, *Phys. Rev. Lett.* **65**, 1020 (1990).

J. M. Moison, F. Houzay, F. Barthe, L. Leprince, E. André and O. Vatel, *Appl. Phys. Lett.* **64**, 18 (1994).

G. E. Moore, *Electronics* **38**(8), 114 (1965).

D. S. Mui, D. Leonard, L. A. Coldren and P. M. Petroff, *Appl. Phys. Lett.* **66**, 1620 (1995).

S. Nakamura, *Science* **281**, 956 (1998).

S. Nakamura and G. Fasol, *The Blue Laser Diode – Nitride-Based Light Emitters and Lasers* (Springer, 1997).

J. H. Neave, B. A. Joyce, P. J. Dobson and N. Norton, *Appl. Phys. A* **31**, 1 (1983).

J. H. Neave, P. J. Dobson, B. A. Joyce and J. Zhang, *Appl. Phys. Lett.* **47**, 100 (1985).

R. Nötzel, M. Ramsteiner, J. Menninger, A. Trampert, H.-P. Schonherr, L. Däweritz and K. H. Ploog, *J. Appl. Phys.* **80**, 4108 (1996).

Y. Okuno, H. Asahi, T. Kaneko, T. W. Kang and S. Gonda, *J. Cryst. Growth* **105**, 185 (1990).

M. Ozdemir and A. Zangwill, *J. Vac. Sci. Technol.* **10**, 684 (1992).

D. W. Pashley, *Adv. Phys.* **5**, 173 (1956).

T. P. Pearsall and G. B. Stringfellow, eds., *MRS Bulletin* **22**(7), 16 (1997).

C. Pearson, M. Krueger and E. Ganz, *Phys. Rev. Lett.* **76**, 2306 (1996).

E. Pehlke and J. Tersoff, *Phys. Rev. Lett.* **67**, 465 (1991).

P. M. Petroff and S. P. DenBaars, *Superlat. and Microst.* **15**, 15 (1994).

P. M. Petroff, A. C. Gossard and W. Wiegmann, *Appl. Phys. Lett.* **45**, 620 (1984).

L. Pfeiffer, H. L. Stormer, K. W. Baldwin, K. W. West, A. R. Goni, A. Pinczuk, R. C. Ashoori, M. M. Dignam and W. Wegscheider, *J. Cryst. Growth* **127**, 849 (1993).

A. Pimpinelli and J. Villain, *Physics of Crystal Growth* (Cambridge University Press, Cambridge, 1998)

T. W. Poon, S. Yip, P. S. Ho and F. F. Abraham, *Phys. Rev. Lett.* **65**, 2161 (1990).

C. F. Quate, *Surf. Sci.* **299/300**, 980 (1994).

C. Ratsch and A. Zangwill, *Surf. Sci.* **293**, 123 (1993).

C. Ratsch, A. Zangwill, P. Šmilauer and D. D. Vvedensky, *Phys. Rev. Lett.* **72**, 3194 (1994).

C. Ratsch, P. Šmilauer, A. Zangwill and D. D. Vvedensky, *Surf. Sci.* **329**, L599 (1995).

S. Raymond, S. Fafard, S. Charbonneau, R. Leon, P. M. Petroff and J. L. Merz, *Phys. Rev. B* **52**, 17 238 (1995).

L. Royer, *Bull. Soc. Fr. Miner. Crystallog.* **51**, 7 (1928).

H. Saito, K. Nishi, I. Ogura, S. Sugou and Y. Sugimoto, *Appl. Phys. Lett.* **69**, 3140 (1996).

Y. Saito, *Statistical Properties of Crystal Growth* (World Scientific, Singapore, 1996).

H. Sakaki, *Jpn. J. Appl. Phys.* **19**, L735 (1987).

T. Sakamoto, N. J. Kawai, T. Nagakawa, K. Ohta and T. Kojima, *Appl. Phys. Lett* **47**, 617 (1985).

W. Seifert, N. Carlsson, M. Miller, M.-E. Pistol, L. Samuelson and L. R. Wallenberg, *Prog. Crystal Growth and Charact.* **33**, 423 (1996).

T. Shitara, Growth Mechanisms of GaAs(001) during Molecular Beam Epitaxy, PhD thesis (University of London, 1992).

T. Shitara, D. D. Vvedensky, M. R. Wilby, J. Zhang, J. H. Neave and B. A. Joyce, *Phys. Rev. B* **46**, 6815 (1992*a*).

T. Shitara, D. D. Vvedensky, M. R. Wilby, J. Zhang, J. H. Neave and B. A. Joyce, *Phys. Rev. B* **46**, 6825 (1992*b*).

T. Shitara, T. Kaneko and D. D. Vvedensky, *Appl. Phys. Lett.* **63**, 3321 (1993).

P. Šmilauer and D. D. Vvedensky, *Phys. Rev. B* **48**, 17603 (1993).

D. L. Smith and C. Mailhiot, *Rev. Mod. Phys.* **62**, 173 (1990).

G. B. Stringfellow, *Organometallic Vapor-Phase Epitaxy* (Academic, Boston, 1989).

J. Sudijono, M. D. Johnson, C. W. Snyder, M. B. Elowitz and B. G. Orr, *Surf. Sci.* **280**, 247 (1993).

J. Tersoff, *Phys. Rev. Lett.* **74**, 4962 (1995).

J. Tersoff and D. R. Hamann, *Phys. Rev. Lett.* **50**, 1998 (1983).

J. Tersoff and F. K. LeGoues, *Phys. Rev. Lett.* **72**, 3570 (1994).

J. Tersoff, C. Teichert and M. G. Lagally, *Phys. Rev. Lett.* **76**, 1675 (1996).

H. C. Theurer, *J. Electrochem. Soc.* **108**, 649 (1961).

W. T. Tsang, T. H. Chiu and R. M. Kapre, *Appl. Phys. Lett.* **63**, 3500 (1993).

W. T. Tsang and A. Y. Cho, *Appl. Phys. Lett.* **30**, 293 (1977).

J. Y. Tsao, *Materials Fundamentals of Molecular Beam Epitaxy* (Academic, Boston, 1993).

M. Tsuchiya, J. M. Gaines, R. H. Yan, R. J. Simes, P. O. Holtz, L. A. Coldren and P. M. Petroff, *Phys. Rev. Lett.* **62**, 466 (1989).

J. P. van der Eerden, in *Handbook of Crystal Growth*, Vol. 1, D. T. J. Hurle, ed. (North-Holland, Amsterdam, 1993), p. 307.

J. M. Van Hove, C. S. Lent, P. R. Pukite and P. I. Cohen, *J. Vac. Sci. Technol. B* **1**, 741 (1983).

J. A. Venables, G. D. T. Spiller and M. Hanbucken, *Rep. Prog. Phys.* **47**, 399 (1984).

J. Villain, *J. de Physique* **1**, 19 (1991).

B. Voigtländer and A. Zinner, *Appl. Phys. Lett.* **63**, 3055 (1993).

B. Voigtländer, T. Weber, P. Šmilauer and D. E. Wolf, *Phys. Rev. Lett.* **78**, 2164 (1997).

H. Weman, D. Y. Oberli, M.–A. Dupertuis, F. Reinhart, A. Gustafsson and E. Kapon, *Phys. Rev. B* **58**, 1150 (1998).

D. A. Wharam, T. J. Thornton, R. Newbury, M. Pepper, H. Ahmed, J. E. Frost, D.G. Hasko, D. C. Peacock and C. T. Foxon, *J. Phys. C* **21**, L209 (1988).

Q. Xie, A. Madhukar, P. Chen, and N. Kobayashi, *Phys. Rev. Lett.* **75**, 2542 (1995).

H. Yamaguchi, J. G. Belk, X. M. Zhang, J. L. Sudijono, M. R. Fahy, T. S. Jones, D. W. Pashley and B. A. Joyce, *Phys. Rev. B* **55**, 1337 (1997).

H.-N. Yang, G.-C. Wang and T.-M. Lu, *Diffraction from Rough Surfaces and Dynamic Growth Fronts* (World Scientific, Singapore, 1993).

K. Yagi, *Surf. Sci. Rep.* **17**, 305 (1993).

J. C. Yi, N. Dagli and L. A. Coldren, *Appl. Phys. Lett.* **59**, 3015 (1991).

J. Zhang, J. H. Neave, P. J. Dobson and B. A. Joyce, *Appl. Phys. A* **42**, 317 (1987).

S. B. Zhang and A. Zunger, *Mat. Sci. Eng.* **30**, 127 (1995).

S. B. Zhang and A. Zunger, *Phys. Rev. B* **53**, 1343 (1996).

2 Electrons in Quantum Semiconductor Structures: An Introduction

E. A. Johnson

2.1 Introduction

Electronics has progressed from the large to the small. In the process, intuitive ideas gained from the ordinary world of classical physics have had to give way to those of the entirely different world of quantum mechanics. When electronic processes are taking place in structures with sizes of the order of centimetres, or even microns, one can describe what is happening in a continuous way. By changing conditions very slightly, one can expect the results to show only very slight changes. When the physical size of the system becomes smaller, however, quantum mechanical effects become important. Typically, a small enough system will be able to have only a few discrete energies. This is the obvious difference between the continuous possibilities that appear to be available to the electrons making up a current in a wire, and the picture one must use to describe what an individual electron must be doing when it is bound to an individual atom making up that wire. In the latter case, one must talk about the possible discrete energy levels which are permitted.

This used to be a clear-cut division. Nature makes the quantum world, while people manufacture the wires. What is new, however, is that people can now make *materials* and *structures* which may be large on the scale of atoms, but which are small enough that the graininess that comes with quantum effects is crucial to understanding their behaviour. In other words, it is now possible to do *atomic engineering*. Matter can be grown to order, to ever greater precision. It has thus become possible to create qualitatively new kinds of electronic materials and devices.

It is helpful to bear in mind some relevant orders of magnitude. For atoms in a solid, a typical size is that of the Bohr orbit of the hydrogen atom in vacuo, about half an ångstrom ($1 \text{ Å} = 10^{-1} \text{ nm} = 10^{-8} \text{ cm}$). For a semiconductor such as GaAs, the crystal spacing is about 5 Å, and a monolayer (the smallest layer of a single type of atom) is half of this (about 2.5 Å). In principle, semiconductor materials can be grown to order to an accuracy of one monolayer. The situation in practice can be more complicated, as discussed in Chapter 1. What is true, in any case, is that the epitaxial growth techniques have produced structures involving small-scale confinement of electrons within sizes of the order of a few atomic spacings.

In talking about small systems, one uses the term 'microscopic' to refer to individual atoms, or to engineering on a scale of no more than about 10 Å.

Mesoscopic systems are those involving large numbers of atoms, but in which quantum effects are still crucial (say, 500 Å to 1500 Å), while macroscopic systems, with sizes of microns or bigger, are expected to behave classically. The focus of the next two chapters is on the quantum electronic effects one can obtain in very small semiconductor systems. This chapter will develop an independent particle description of electrons in simple situations. The next chapter will consider interaction effects and treatments which are more appropriate to real systems.

2.2   Ideal Low-dimensional Systems

Quantum effects arise in systems which confine electrons to regions comparable to their de Broglie wavelength. When such confinement occurs in one dimension only (say, by a restriction on the motion of the electron in the z-direction), with free motion in the x- and y-directions, a 'two-dimensional electron gas' (2DEG) is created. Confinement in two directions (y and z, say), with free motion in the x-direction, gives a 'one-dimensional electron gas' (1DEG) and confinement of its x-, y-, and z-motions at once gives a 'zero-dimensional electron gas' (0DEG). In this section, we consider the description of *ideal* electron gases in these cases, i.e. electron gases in which there is no motion in the confining direction and where we neglect interactions between the electrons. We will then use these results in the following section to characterize the density of states in real low-dimensional structures, in which there is some degree of lateral mobility.

2.2.1   Free Electrons in Three Dimensions: A Review

An unconfined electron in free space is described by the Schrödinger equation

$$-\frac{\hbar^2}{2m}\nabla^2\varphi = -\frac{\hbar^2}{2m}\left(\frac{\partial^2\varphi}{\partial x^2}+\frac{\partial^2\varphi}{\partial y^2}+\frac{\partial^2\varphi}{\partial z^2}\right) = E\varphi \tag{2.1}$$

where m is the free-electron mass. The solutions of this equation,

$$\varphi_{\mathbf{k}}(\mathbf{r}) = \frac{1}{(2\pi)^3}\,e^{i\mathbf{k}\cdot\mathbf{r}} \tag{2.2}$$

are plane waves labelled by the wavevector

$$\mathbf{k} = (k_x, k_y, k_z) \tag{2.3}$$

and correspond to the energy

$$E = \frac{\hbar^2 k^2}{2m} = \frac{\hbar^2}{2m}\,(k_x^2 + k_y^2 + k_z^2) \tag{2.4}$$

The vector components of $\mathbf{k}$ are the quantum numbers for the free motion of the electron, one for each of the classical degrees of freedom.

The number of states in a volume $d\mathbf{k} = dk_x\,dk_y\,dk_z$ of $\mathbf{k}$-space is

$$g(\mathbf{k})\,d\mathbf{k} = \frac{2}{(2\pi)^3}\,d\mathbf{k} \tag{2.5}$$

with the factor of 2 accounting for the spin-degeneracy of the electrons. To express this density of states in terms of *energy* states, we use the fact that the energy dispersion (2.4) depends only on the magnitude of $\mathbf{k}$. Thus, by using spherical polar coordinates in $\mathbf{k}$-space,

$$d\mathbf{k} = k^2 \sin\theta\,dk\,d\theta\,d\phi \tag{2.6}$$

where the variables have their usual ranges $(0 \le k < \infty, 0 \le \phi < 2\pi$, and $0 \le \theta \le \pi)$ and, integrating over the polar and azimuthal angles, we are left with an expression that depends only on the magnitude k:

$$g(\mathbf{k})\,d\mathbf{k} = \frac{2}{(2\pi)^3}\,d\mathbf{k} = \frac{1}{\pi^2}\,k^2\,dk \tag{2.7}$$

By invoking (2.4), we can perform a change of variables to cast the right-hand side of this equation into a form involving the differential of the energy:

$$\frac{1}{\pi^2}\,k^2\,dk = \frac{1}{\pi^2}\left(\frac{2mE}{\hbar^2}\right)\frac{dk}{dE}\,dE = \frac{1}{2\pi^2}\left(\frac{2m}{\hbar^2}\right)^{3/2}\sqrt{E}\,dE \tag{2.8}$$

From this equation, we deduce the well-known density of states $g(E)$ of a free-electron gas in three dimensions:

$$g(E) = \frac{1}{2\pi^2}\left(\frac{2m}{\hbar^2}\right)^{3/2}\sqrt{E} \tag{2.9}$$

Notice the characteristic square-root dependence on the energy. This results from the fact that, in three dimensions, the surfaces of constant energy in $\mathbf{k}$-space are spheres of radius $\sqrt{2mE}/\hbar$.

2.2.2 Ideal Two-dimensional Electron Gas

An ideal 2DEG differs from free electrons in three dimensions in that the electrons have unrestricted movement in only two dimensions (x and y) with complete confinement in the z-direction, i.e. there is no freedom of movement at all in this direction. The energy of an electron in a 2DEG is therefore

$$E = \frac{\hbar^2 k^2}{2m} = \frac{\hbar^2}{2m}(k_x^2 + k_y^2) \tag{2.10}$$

as shown in the dispersion relation in Fig. 2.1. The number of states within an area in $\mathbf{k}$-space box of $d\mathbf{k} = dk_x\,dk_y$ is

$$g(\mathbf{k}) = \frac{2}{(2\pi)^2} \tag{2.11}$$

with the factor of 2 again inserted to account for the spin degeneracy of the electrons. We proceed as above, but now use *circular polar coordinates* (Fig. 2.2) to obtain

$$g(\mathbf{k})\, d\mathbf{k} = \frac{2}{(2\pi)^2}\, d\mathbf{k} = \frac{1}{\pi}\, k\, dk \tag{2.12}$$

where $k = k_\parallel = (k_x^2 + k_y^2)^{1/2}$ (Fig. 2.3). We again use the relationship between E and k in (2.10) to express the density of states in terms of the energy:

$$\frac{1}{\pi}\, k\, dk = \frac{1}{\pi}\, k\, \frac{dk}{dE}\, dE = \frac{m}{\pi\hbar^2}\, dE \tag{2.13}$$

The density of states $g(E)$ of a 2DEG is therefore given by

$$g(E) = \frac{m}{\pi\hbar^2} \tag{2.14}$$

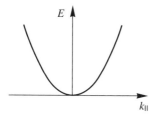

Figure 2.1.   Parabolic E–k relation in the (k_x, k_y) plane.

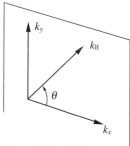

Figure 2.2.   Polar coordinates in the (k_x, k_y) plane.

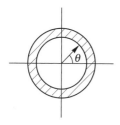

Figure 2.3.   The number of points, $g(k_\parallel)\, dk_\parallel$, within the shaded ring.

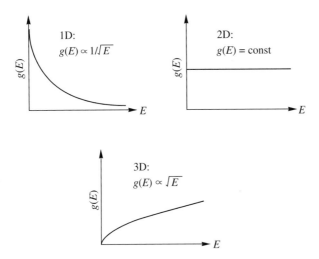

Figure 2.4.   Density of states for an ideal electron gas in 1, 2 and 3 dimensions.

Thus, for a 2DEG the density of states is a *constant*, i.e. *independent of the energy* (Fig. 2.4). This is one of the fundamental features of electrons in planar hetero-structures which make such structures useful for applications.

2.2.3   Ideal Zero- and One-dimensional Electron Gases

When an electron is allowed only one-dimensional motion (along, say, the x-direction), the energy is given by

$$E = \frac{\hbar^2 k^2}{2m} = \frac{\hbar^2 k_x^2}{2m} \tag{2.15}$$

A procedure analogous to that used in the preceding two sections (Exercise 1) then yields for the density of states the expression

$$g(E) = \frac{1}{\pi} \left(\frac{2m}{\hbar^2} \right)^{1/2} \frac{1}{\sqrt{E}} \tag{2.16}$$

This shows that the density of states of a one-dimensional electron gas (1DEG) has a square-root *singularity* at the origin.

An ideal zero-dimensional electron is one that exists in a single state of fixed energy E_0. The density of states is then given by (Exercise 1)

$$g(E) = \delta(E - E_0) \tag{2.17}$$

where $\delta(x)$ is the Dirac delta-function.

2.2.4   Quantum Wells, Wires, and Dots

The results in the preceding three sections show that the density of electronic states is a strong function of the spatial dimension. This has a strong influence on the transitions between different energy states, an effect which can be exploited in a number of ways, most notably in optical and transport properties in quantum heterostructures. But in real low-dimensional systems, confinement is never perfect. A restriction of the z-motion of an electron, for instance, will be a restriction to within a finite, but very small, range of z, comparable to the de Broglie wavelength of the electron. This imperfect confinement results in the quantization of electron energies where, in general, more than one such energy will be allowed. Only if the conditions are such that there are electrons in only a single such quantized energy level do we obtain one of the ideal cases mentioned above in the appropriate dimension. Thus, real low-dimensional systems exhibit departures from the 'ideal' behaviour of low-dimensional electron gases. This results in a hierarchy of quantum heterostructures in which electrons are confined in one or more dimensions, but are free in the other dimensions: quantum wells (2D electrons), quantum wires (1D electrons), and quantum dots (0D electrons).

2.3   Real Electron Gases: Single Particle Models

By a quantum well we mean any structure in which an electron (or hole) is strongly confined in one dimension. A practical example of great importance is obtained when a plane layer of GaAs lies within a sample of bulk $Al_xGa_{1-x}As$. These materials may be grown, e.g. by molecular-beam epitaxy (MBE) in a layer-by-layer fashion to form such a structure (Chapter 1). The materials are lattice-matched (the same lattice structure and very similar lattice spacing). Moreover, their band structures are qualitatively similar if the aluminum proportion x is less than approximately 0.4. However, the band gap of $Al_xGa_{1-x}As$ increases linearly with increasing x. What results (when $x \neq 0$) is a discontinuity in the conduction and valence band edges, E_c and E_v, as shown in Fig. 2.5. The precise proportion of the discontinuity taken up by the conduction band alone must be known beforehand, from experiment, as explained in Chapter 6, or else from theory. Quantum confinement of an electron within the thin layer of GaAs will happen if its energy is below that of the conduction-band edge in the AlGaAs. This is an example of a *compositional* quantum well.

Most simply, an electron confined by such a band-edge potential can be treated in the *effective-mass approximation*. The Bloch theorem tells us that the true wavefunction $\phi_{\mathbf{k}}(\mathbf{r})$ can be written as the product

$$\phi_{\mathbf{k}}(\mathbf{r}) \approx \phi(\mathbf{r})u_{\mathbf{k}}(\mathbf{r}) \tag{2.18}$$

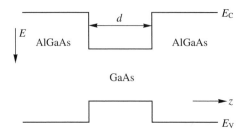

Figure 2.5.   Band-edge diagram for a typical AlGaAs/GaAs quantum well. The fraction, x, of Al is less than 0.35.

where $u_{\mathbf{k}}(\mathbf{r})$, the Bloch function, describes the rapidly varying crystal part of the wavefunction and $\phi(\mathbf{r})$, sometimes called the envelope function (as in Section 6.2), describes the part which is slowly varying on an atomic scale. The envelope function obeys a Schrödinger-like equation which, in the simplest materials, such as GaAs, and near $k = 0$, takes the form (cf. equation (6.5))

$$\left[-\frac{\hbar^2}{2m^*} \nabla^2 + V(\mathbf{r}) \right] \phi(\mathbf{r}) = E\phi(\mathbf{r}) \tag{2.19}$$

where E is measured from the conduction-band edge and V does not include the crystal potential. The entire effect of the crystal potential is to change the electron mass from m to m^*, the effective mass. (In GaAs, for instance, $m^* = 0.067\,m$.) The potential V in equation (2.19) contains the effect of all external potentials, and in particular, that due to changes in the conduction band edge.

2.3.1   Ideal Square Well

Most simply, the square-well potential produced in a compositional quantum well can be approximated by that of an infinite square well, i.e. that in which the potential is constant within the well and infinite outside the well:

$$V(z) = \begin{cases} 0 & \text{for } 0 \leq z \leq d \\ \infty & \text{otherwise} \end{cases} \tag{2.20}$$

This problem is studied at the beginning of most quantum mechanics textbooks (e.g. Liboff, 1980). Since motion is unrestricted in the y- and z-directions, the Schrödinger equation (2.19) is separable in rectangular coordinates, so the coordinate dependence of the wavefunction in the (x, y) plane can be separated from that in the z-direction. This results in plane-wave solutions for the motion of the electron in the x- and y-directions,

$$\phi(\mathbf{r}) = e^{ik_x x}\, e^{ik_y y}\, \phi(z) \tag{2.21}$$

where $\phi(z)$ obeys the one-dimensional Schrödinger equation for a particle in an infinite square well (Fig. 2.6):

$$\left[-\frac{p^2}{2m^*} + V(z) \right] \phi(z) = E\phi(z) \tag{2.22}$$

where $V(z)$ is given by (2.20).

As is well known, the general solution of (2.22) must be a linear combination of sines and cosines chosen to satisfy the boundary conditions imposed by the well (Fig. 2.7). Since ϕ must vanish at $z = 0$, the solutions must be of the form $\sin(kz)$ and, since it must also vanish at $z = d$, we must choose k to be $k = n\pi/d$, for any positive integer n (i.e. $n = 1, 2, \ldots$). This restriction results in the quantization of the energy. The allowed energies associated with the motion of the electron along the z-direction are

$$E_n = \frac{\hbar^2 \pi^2 n^2}{2m^* d^2}, \qquad n = 1, 2 \ldots \tag{2.23}$$

The total energy of the electron is the sum of this quantized energy and the kinetic energy due to its (x, y)-motion:

$$E = E_n + \frac{\hbar^2}{2m^*} (k_x^2 + k_y^2) \tag{2.24}$$

Figure 2.6.   Infinite square-well potential.

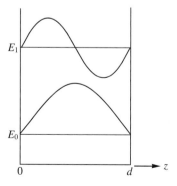

Figure 2.7.   Energies and wavefunctions for the first two quantized states in a square-well potential.

Already, differences from the ideal 2DEG case are evident: (i) there can be several different quantized energies E_n, i.e. several possible states of z-motion, and (ii) the electron wavefunctions have a finite spread in the z-direction.

The E–$\mathbf{k}$ dispersion relation for an infinite quantum well is thus a generalization of that shown in Fig. 2.1. One obtains instead the situation shown in Fig. 2.8, where from equation (2.23) with $n = 1$, E_0 is

$$E_0 = \frac{\hbar^2 \pi^2}{2m^* d^2} \tag{2.25}$$

For energies $E < E_0$ (A in Fig. 2.8), there are no states; for energies B ($E_0 < E < 4E_0$) the density of states (per unit area) is just that for a perfect two-dimensional electron gas, namely $g_0(E) = m/\pi\hbar^2$. For energies C the density of states (DOS) is $2 \times g_0$; energies D have $3 \times g_0$ for the DOS, and so forth.

To convert this $g(E)$, which is the density of states per unit *area* of real space, to a density of states per unit *volume*, one must divide by an appropriate length in the z-direction, in this case by the well width d. This three-dimensional DOS then rises in steps of $2m^*/\pi\hbar^2 d$, as shown in Fig. 2.9, where it is also compared with the

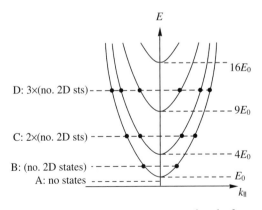

Figure 2.8.   Energy versus wavenumber $k_\parallel$ for an infinite square well.

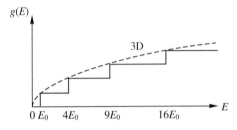

Figure 2.9.   Density of states for an infinite square well. The corresponding density of states for an unconfined 3D system is also shown for comparison (broken line). E_0 is defined in equation (2.25).

ordinary bulk DOS. If states in the well are filled up to some Fermi energy E_F, then states at the Fermi level (the ones of most interest for transport) will have different kinetic energies of motion in the x–y plane, and therefore different Fermi velocities, depending upon their quantum state in the well. Thus, for instance, if E_F lies above some level E_n, then the Fermi wavenumber for states in level n is given by

$$E_F - E_n = \frac{\hbar^2 k_F(n)^2}{2m^*} \tag{2.26}$$

with the corresponding Fermi velocity

$$v_F(n) = \frac{\hbar k_F(n)}{m^*} \tag{2.27}$$

related to k_F in the usual way.

2.3.2   Some Generalizations

2.3.2.1 Holes in Quantum Wells

Holes as well as electrons can be confined strongly in one or more dimensions. In a GaAs quantum well in the GaAs/AlGaAs system (Fig. 2.5), there is a quantum well for holes wherever there is a well for electrons. In other systems, for instance in delta wells (Section 3.3.2) or at heterojunctions (Section 3.2.4), one can create a well *just* for holes.

There is a three-fold degeneracy in the hole bands at the Γ-point, which is the highest point in the valence bands. In the bulk one deals with three sorts of holes. Spin-orbit splitting depresses one of these bands to create the spin-split-off band, which is then often ignored. The other two bands correspond to heavy and light holes (so called because of their greater or smaller effective mass); these states are degenerate in energy at the Γ-point. In the bulk one can treat the two (or three) sorts of holes separately. In quantum wells and other strongly confined systems, however, the confinement breaks the symmetry which caused the degeneracy in the first place. The hole states then mix and, in general, they will mix differently in different directions. These effects are complex and interesting, but they will not be discussed further here (see, for example, Bastard, 1988).

2.3.2.2 Non-parabolicity

For electrons with energies near the bottom of the conduction band, the $E(k)$ band structure is parabolic: $E = \hbar^2 k^2 / 2m^*$. For somewhat higher energies, this relation is no longer true (the importance of non-parabolicity, at a given energy, will depend on the material in question). However, one can still *define* an energy-dependent effective mass $m(E)$ by

$$E(k) = \frac{\hbar^2 k^2}{2m(E)} \tag{2.28}$$

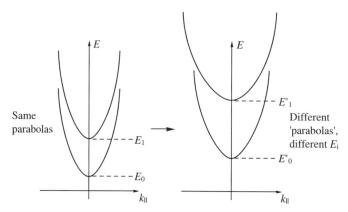

Figure 2.10.   Parabolic and non-parabolic E–k relations for an ideal square well (exaggerated).

(this neglects possible anisotropy of the bands). We can suppose that $m(E)$ is known in the bulk. Then, in an infinite square well, the quantized energies E_n will be given by

$$E_n m(E_n) = \tfrac{1}{2}\hbar^2 k^2 \qquad (2.29)$$

where $k = n\pi/d$ (cf. equation (2.23)). Again the full energy E for an electron in state n is given by

$$E = \frac{\hbar^2 k_{\parallel}^2}{2m(E)} + E_n \qquad (2.30)$$

The situation is illustrated schematically in Fig. 2.10.

Non-parabolicity will modify the quantized energies E_n' from their original value, though only negligibly for low energies near the conduction-band edge (i.e., for low n). Moreover, for each level the E–k relation will be parabolic near $k_{\parallel} = 0$, but the curvature of the parabolas will become broader as n becomes higher (non-parabolicity is known to increase $m(E)$ as E becomes higher). And finally, the E–k relation for each level will itself become non-parabolic as $k_{\parallel}$ becomes large. This is discussed further in Sections 6.4 and 6.9.

2.3.3   Finite Quantum Wells and Real Systems

The problem of a particle confined in a finite square well (Fig. 2.11) is another classical textbook problem (Liboff, 1980). One of the most crucial differences between this case and that of an infinite square well is that the electron wavefunction need not be zero in the barrier region. This fact, that electrons can penetrate into the barrier region, will be particularly important when it comes to a discussion of the physics of superlattices (Chapter 3).

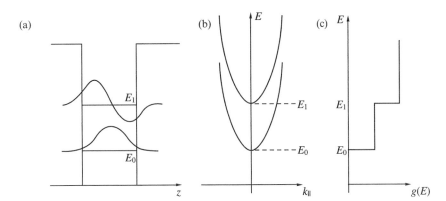

Figure 2.11. Schematic illustration of (a) wavefunctions, (b) energies and (c) density of states for a particle confined to a finite square well.

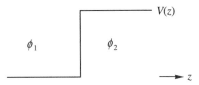

Figure 2.12. The wavefunction ψ at an ordinary potential step.

An electron in a finite square well is confined by two potential steps. (These could be the finite conduction-band discontinuities, for instance, in Fig. 2.5.) In the usual textbook case (Fig. 2.12), one solves for the wavefunction ϕ at a potential step using the following assumptions: (i) ϕ must behave suitably at infinity (usually, decaying to zero), (ii) ϕ must be continuous at the interface (say, at the potential step at $V = 0$), (iii) the first derivative of ϕ must also be continuous at the interface. These conditions can be stated as

$$\phi_1(0) = \phi_2(0), \qquad \left.\frac{d\phi_1}{dz}\right|_{z=0} = \left.\frac{d\phi_2}{dz}\right|_{z=0} \tag{2.31}$$

A semiconductor interface, however, is more subtle than an ideal potential step for various reasons. In spite of this fact, the first assumption one might make is that the envelope function ϕ for the electron responds to a conduction band offset in much the same way as the complete wavefunction to an ideal potential step, with the same matching conditions (2.31). This begs a number of questions, some of which we shall mention here. What one really wants to know is the correct matching conditions for the envelope function at a material interface (Fig. 2.13). (On an even more basic level, one wants to know whether it makes sense at all to talk in these terms, i.e. in an effective-mass approximation, for electrons near a material interface. Although there are many such questions still to

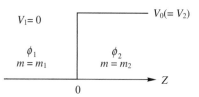

Figure 2.13.   Boundary conditions near a material interface. Electrons in the material on the left (right)-hand side have effective mass m_1 (m_2). The conduction band of material 2 is offset from that of material 1 by an amount V_0.

be answered, it is the case that a simple effective-mass approach is surprisingly good.)

Various prescriptions have been proposed for the matching conditions of a wavefunction at a semiconductor interface. Which is correct is still a matter of debate, and it may be the case that there is not a unique answer. We present here the matching conditions in common use, which are known as the Bastard conditions (Bastard, 1988); Bastard gives the credit, however, to Ben-Daniel and Duke (1955).

At an interface, the effective mass and the conduction band edge potential are effectively discontinuous. Using the notation of Fig. 2.13, we have approximately that

$$m^* = m(z) \sim m_1 + (m_2 - m_1)\Theta(z) \tag{2.32}$$

and

$$V = V(z) \sim V_0\Theta(z) \tag{2.33}$$

where the step function $\Theta(z)$ is defined by

$$\Theta(z) = \begin{cases} 0, & z < 0, \\ 1, & z > 0, \end{cases} \tag{2.34}$$

The usual effective-mass Hamiltonian

$$H = -\frac{\hbar^2}{2m^*}\nabla^2 + V \tag{2.35}$$

is not Hermitian for such a z-dependent effective mass, but must be made so for quantum-mechanical consistency. This can be done by using instead the Hamiltonian

$$H = -\tfrac{1}{2}\hbar^2\nabla \cdot \frac{1}{m(z)} \cdot \nabla + V \tag{2.36}$$

We assume that the envelope function $\phi(\mathbf{r})$ is continuous at $z = 0$. We next suppose that, in fact, $m(z)$ and $V(z)$ change very rapidly but continuously over a

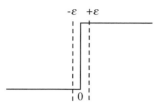

Figure 2.14. The quantities $m(z)$ and $V(z)$ show smooth but rapid variation over a small distance $\pm\epsilon$ on either side of the interface, at $z = 0$.

small distance $\Delta z = 2\epsilon$ (Fig. 2.14). We now take the wavefunction

$$\phi(\mathbf{r}) = e^{ik_x x} e^{ik_y y} \phi(z) \tag{2.37}$$

(since V is independent of x and y, ϕ is plane-wave-like in the x and y directions), and then integrate the effective mass equation $H\phi = E\phi$ from $z = -\epsilon$ to $z = +\epsilon$:

$$\int_{-\epsilon}^{\epsilon} \left\{ \frac{\mathrm{d}}{\mathrm{d}z} \left[\frac{1}{m(z)} \frac{\partial\phi}{\partial z} \right] \right\} \mathrm{d}z + V_0\phi(+\epsilon)\epsilon = E[\phi(+\epsilon) - \phi(-\epsilon)]\epsilon \tag{2.38}$$

We now let $\epsilon \to 0$. Since V, ϕ and E are all finite, the two terms multiplied by ϵ both approach zero as $\epsilon \to 0$. Thus, the first term on the left in (2.38) must also approach zero, giving

$$\frac{1}{m(\epsilon)} \frac{\partial\phi}{\partial z}\bigg|_{z=+\epsilon} = \frac{1}{m(-\epsilon)} \frac{\partial\phi}{\partial z}\bigg|_{z=-\epsilon} \tag{2.39}$$

This suggests that the proper boundary conditions, at the (plane) interface ($z = 0$) between materials 1 and 2, are (Bastard, 1988)

$$\phi_1(0) = \phi_2(0), \quad \frac{1}{m_1} \frac{\partial\phi_1}{\partial z}\bigg|_{z=0^-} = \frac{1}{m_2} \frac{\partial\phi_2}{\partial z}\bigg|_{z=0^+} \tag{2.40}$$

The boundary conditions in (2.39) and (2.40) can be shown to imply conservation of the probability current $\mathbf{J}$ for the full wavefunction ϕ,

$$\mathbf{J} = \mathrm{Re}\left[\phi^* \frac{\hbar}{im_0} \nabla\phi \right] \tag{2.41}$$

averaged over a small volume containing the interface. It also implies the existence of steady-state solutions, for the full wavefunction ϕ.

Since the conditions in (2.39) and (2.40) on the envelope function behave in a physically reasonable fashion, one may expect them to give good results *if* various other conditions hold (Bastard, 1988): (i) the amplitudes of the Bloch functions in each material should be comparable, and their k-dependence should be weak near $k = 0$, (ii) one should be near a band edge (k should be small), (iii) the same band edge should be under consideration in each material. (Thus, for example, it is less certain which matching conditions should be used at a GaAs–AlAs interface, since the GaAs conduction-band minimum is at the Γ-point, while that of AlAs is not.)

As to whether these are the correct ('true') boundary conditions to use, the situation has been clarified by Burt (1992) and by Foreman (1993, 1995), as discussed in a useful review by Meney *et al.* (1994). The conclusion is that as long as one does a good enough envelope calculation in general (e.g. an 8-band calculation) one can expect good agreement with the results obtained by a proper microscopic treatment of the interface. In practice, the Bastard boundary conditions, and the use of the envelope-function formalism in general, work surprisingly well even in situations (such as very abrupt interfaces, or in very small-period superlattices) in which one might have expected significant discrepancies from this simple approach. The presence of material interfaces has other consequences, too, several of which we include here for completeness.

2.3.4   Interface Effects

2.3.4.1   Effective Mass for Parallel Transport

The difference in effective mass between two semiconductor materials joined at a planar interface has several consequences, the best-known of which is the change in boundary conditions discussed in the preceding section. Two other consequences, less well-known, are also worthy of mention. The first of these has to do with electron transport parallel to the interfaces of a low-dimensional system.

Electrons in a quantum well (or wire) are generally assumed to have an effective mass m^* which is the same as the material making up the well (wire). When this material is embedded in some other material (a compositional low-dimensional structure), however, quantum mechanics predicts that the electron will tunnel into the barrier material, which has a different value for m^*. This is illustrated in Fig. 2.15, which shows the probability density for an electron in the lowest energy level of, say, a GaAs/AlGaAs quantum well. The total area under the curve is defined to be unit probability, and the unshaded area equals the total probability P that the electron is actually in the GaAs well. But it also has a probability $1 - P$ of being in the barrier, equal to the total shaded areas under the curve. What is the

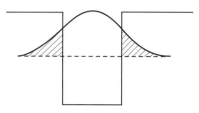

Figure 2.15.   Schematic probability density for an electron in the lowest confined level of a compositional quantum well. The shaded area under the curve gives the probability of finding the electron within the barrier region.

appropriate effective mass to use for electrons which spread into two different materials?

One finds (Priester *et al.*, 1984; Lässnig, 1985; Ekenberg, 1987; Bastard, 1988; Johnson and MacKinnon, 1988) that transport parallel to such a plane interface can still be described in effective-mass terms, but with an effective mass m related to the mass m_1 of the well-region and m_2 of the barrier according to the probability P,

$$P = \int_{QW} |\phi|^2 \, dz \tag{2.42}$$

of the electron being present in the quantum well (QW), or $1 - P$ of its being in the barrier, and is given by

$$\frac{1}{m} = \frac{P}{m_1} + \frac{1 - P}{m_2} \tag{2.43}$$

Thus, one consequence of higher effective masses in the barrier regions is that they act as a sort of drag, increasing the overall effective mass (for parallel transport) of electrons in the well to the degree that their wavefunction leaks into the barrier regions.

2.3.4.2 Effective-mass Correction to Conduction-band Discontinuities

Another result of the change in effective mass across an interface is a change in the apparent height of the potential barrier there which depends on the kinetic energy of the electron. It was first shown by Milanovic and Tjapkin (1982) that the existence of a difference of effective masses at a plane interface necessitates a k-dependent correction to the magnitude of the conduction-band offset,

$$V(z) = V_0 \Theta(z) \tag{2.44}$$

at that interface. To see this, we look at a state ϕ of energy E which satisfies the Schrödinger equation

$$\left[\frac{p^2}{2m^*} + V(z) \right] \phi = E\phi \tag{2.45}$$

We take ϕ to be the envelope function and write the energy as a sum of the quantum energy of confinement E_n (for energy level n) and the kinetic energy $E_\parallel$ of the electron's transverse motion. For the envelope function ϕ,

$$\phi(\mathbf{r}) = e^{ik_x x} e^{ik_y y} \phi(z) \tag{2.46}$$

E_n is obtained by solving the equation

$$\left[-\frac{\hbar^2}{2m_i} \nabla^2 + V(z) \right] \phi_n^{(i)} = E_n^{(i)} \phi_n^{(i)} \tag{2.47}$$

while $E_\parallel$ is given by

$$E_\parallel^{(i)} = \frac{\hbar^2}{2m_i}(k_x^2 + k_y^2) \tag{2.48}$$

(the energies in question have the label 'i' since the equations must be solved separately within each material). The total energy E and the transverse momenta (k_x, k_y) are good quantum numbers and therefore must be the same, whichever material one considers. In particular,

$$E = E_n^{(1)} + E_\parallel^{(1)} = E_n^{(2)} + E_\parallel^{(2)} \tag{2.49}$$

Other quantities, however, may be found to vary from one material to the other. If the effective masses across the interface are equal, the quantized energies E_n are the same on either side of the interface (the usual case). Therefore the transverse energy $E_\parallel$ must also remain constant across this interface. If, however, the effective masses across the interface are unequal, the kinetic energy $E_\parallel$ will change across the interface. From equation (2.48) this change is given by

$$E_\parallel^{(2)} = \frac{m_1}{m_2} E_\parallel^{(1)} \tag{2.50}$$

Equation (2.49) then implies that the quantum energies $E_n^{(i)}$ also must change as the interface is crossed from one material to the other. This effect, which is usually neglected, follows necessarily from the three-dimensional nature of low-dimensional systems. Fortunately, it can be taken into account by a change in the apparent height of the conduction-band discontinuity V_0 at the interface. Thus, instead of the potential $V(z)$ in equation (2.44) one has an effective potential $\widetilde{V}(z)$ which is given by

$$\widetilde{V}(z) = [V(z) - \Delta]\Theta(z) \tag{2.51}$$

where Δ is given by

$$\Delta = E_\parallel^{(2)} - E_\parallel^{(1)} = E_\parallel^{(1)}\left(1 - \frac{m_1}{m_2}\right) \tag{2.52}$$

Thus, at a potential step of height V_0, the *effective* barrier height is

$$\widetilde{V}_0 = V_0 - E_\parallel^{(1)}\left(1 - \frac{m_1}{m_2}\right) = V_0 - \Delta \tag{2.53}$$

The quantity Δ, the correction to the potential, is an energy-dependent potential which is attractive if $m_1 < m_2$ (the usual case – thus, at a GaAs/AlGaAs interface, for instance, the barrier height is lowered by this correction). The effect vanishes for $k = 0$, but becomes progressively more important as the transverse kinetic energy $E_\parallel$ increases. Its importance also increases with the magnitude of the difference in effective mass. Since the effective potential $\widetilde{V}$ depends on the solution, inclusion of this effect needs a self-consistent solution of Schrödinger's equation, which is perhaps one reason why it is usually neglected. Nevertheless, for the example of a 50 Å quantum well of GaAs in $Ga_{0.7}Al_{0.3}As$, the most energetic of the trapped electrons should see the barrier height lowered by as much as 40 meV by this effect.

2.3.5 Quantum Wires

Above we have considered confinement of electrons to two dimensions. To create systems of yet smaller dimension, it is usual to start with a well-confined two-dimensional electron gas, so tightly confined that electrons are present in only a single energy level. A new confinement is then imposed on this system, in a direction perpendicular to the original confinement (Fig. 2.16). A direct way to do this *lateral confinement* is by cutting the material containing the 2DEG, for instance by etching, to remove all but a thin strip of the 2DEG (a typical width L might be approximately 1000 Å. The electrons, now confined in two dimensions but free to move in the third, form a *quantum wire*.

The simplest theoretical picture of such a quantum wire is given by the confinement of an ideal (in the sense of Section 2.3) two-dimensional electron gas in an infinite square well. Figure 2.17 illustrates the wavefunctions of the first two quantum states of such a quantum wire. The original 2DEG lies in the plane $z = 0$, with the additional confinement in the x-direction. The energies of x-confinement are

$$E_n = \frac{\hbar^2\pi^2(n+1)^2}{2m^*L^2} \tag{2.54}$$

and the total energy (for x and y) is

$$E = \frac{\hbar^2}{2m^*}k_y^2 + E_n \tag{2.55}$$

Figure 2.18 shows the density of states $g(E)$ for such an ideal quantum wire. $g(E)$ shows the characteristic singularity in $E^{-1/2}$ which was derived for a 1DEG in equation (2.13). In a quantum wire, such a singularity will occur at each energy

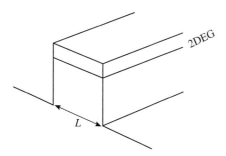

Figure 2.16.   A two-dimensional electron gas as the basic ingredient for forming a quantum wire.

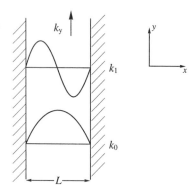

Figure 2.17.   Wavefunctions for an electron in an ideal quantum wire. An ideal 2DEG in the plane $z = 0$ has undergone additional confinement by infinite potential steps in the x-direction.

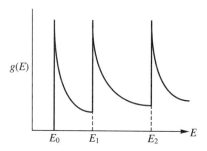

Figure 2.18.   Density of states for an ideal one-dimensional quantum wire.

E_n of quantization in the x-direction. For real quantum wires, the spacing of the quantized energies E_n, and the corresponding wavefunctions, will depend on the precise shape of the potential $V(y, z)$, just as they depended, for a 2DEG, on the shape of the potential $V(z)$.

2.3.5.1 Quantum Point Contacts

An important example of a quantum wire is a *quantum point contact*. This is a quantum wire which is produced electrostatically rather than mechanically, by a *split-gate* device (Fig. 2.19). On the plane surface of the material containing the 2DEG, a layer of material is deposited which will act as a gate. Electrons will be attracted to or repelled from the 2DEG according to the potential put on the gate. The gate material itself is selectively etched away (say, within a thin strip). When a repulsive potential is applied to the gate, electrons will be repelled *preferentially* from under the gate. This can be thought of instead as an *attractive* potential acting underneath the region where the gate material is missing, as shown in Fig. 2.20. Electrons will be present in the region in which the Fermi energy is

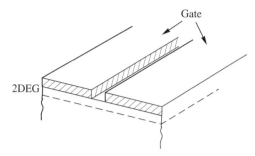

Figure 2.19.   Section of a split-gate device.

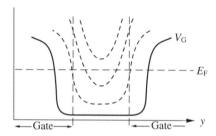

Figure 2.20.   Change in the effective potential felt by electrons in a split-gate device as the potential on the gate is changed.

greater than the band-edge potential set up by the gate. This device is therefore one in which the width of the wire is variable, by simply changing the gate potential.

Unlike the case of an etched wire, where a great deal of disorder can be introduced into the x-confinement by the etching process, a quantum wire produced in a split-gate device is one in which the confining potential is particularly smooth (much smoother than could be created in an etched wire). It is thus particularly good for observing effects involving the precise quantum states of the electrons.

2.3.6   Quantum Dots

Electrons can be confined in all three dimensions in a 'dot' or 'quantum box'. The situation is analogous to that of a hydrogen atom: only discrete energy levels are possible for electrons trapped by such a zero-dimensional potential. The *spacing* of these levels depends, as always, on the precise shape of the potential. Generalizations of the split-gate method described in the previous section are an important way of producing quantum dots (Sikorski and Merkt, 1989; Kotthaus, 1991) and other one-dimensional structures.

The development and application of quantum dot systems is an increasingly important research topic at the time of writing for a number of reasons, both technological and theoretical. Various aspects of quantum dots and wires will be discussed in more detail in Chapter 3 and later chapters. More extensive treatments of quantum dots can also be found in the books by Bányai and Koch (1993) and by Haug and Koch (1993).

EXERCISES

1. Show that the density of states $g(E)$ of an ideal one-dimensional electron gas is given by

$$g(E) = \frac{1}{\pi}\left(\frac{2m}{\hbar^2}\right)^{1/2}\frac{1}{\sqrt{E}}$$

and that the density of states of an ideal zero-dimensional electron gas occupying the energy level E_0 is

$$g(E) = \delta(E - E_0)$$

2. Consider two GaAs quantum wells that have been grown far apart in $Al_xGa_{1-x}As$ with the same Al composition x ($x \leq 0.3$). The GaAs layers have plane parallel interfaces with the AlGaAs barrier material. In well A the GaAs thickness is L, while in well B it is $2L$. Now approximate the conduction bands in wells A and B by ideal quantum wells between infinitely high potential barriers. Find the ratio E_0^A/E_0^B, where E_0^A and E_0^B are the lowest quantized energy levels in well A and well B, respectively.

3. Suppose that the quantum wells of Exercise 2 contain electrons and that both wells have the same Fermi energy $E_F = 3.0E_0^A$ as measured from the bottom of each well, where E_0^A is defined in Exercise 2. Assuming the approximation of ideal square wells between infinite potential barriers,

  (a) how many subbands in each well will contain electrons at zero temperature?
  (b) what is the two-dimensional charge density N_A, N_B, in each well? (Give the answer in terms of known physical quantities such as GaAs material parameters, $\hbar$, L etc.)
  (c) what is the (two-dimensional) Fermi velocity of the electrons in each filled subband in well A? In well B?

4. Consider a quantum wire formed from an ideal two-dimensional electron gas (2DEG). In the 2DEG the electrons are free to move in the x- and y-directions, and are all in the lowest conduction subband of the 2DEG. An extra confinement is then imposed, restricting the electrons to a straight narrow channel, free to move in the x-direction but tightly confined in the y-direction. Assume that the wire can be modelled by a long, narrow strip of width W with straight sides parallel to the x-axis, which give additional confinement in the y-direction (a hard-wall potential). Write down an expression for the allowed energies E_i associated with the additional confinement of these electrons in the y-direction. Express E_i in terms of the lowest such allowed energy E_0. Sketch the density of states for electrons in this 'one-dimensional quantum wire', including at least three subbands in this diagram. Label all points of interest and indicate differences from an ideal one-dimensional wire.

5. A quantum wire can be used to carry current if a bias is applied along its length. For each subband (given a small bias ΔV between the ends of the wire), the net current is carried by electrons in the small energy range ΔV at the Fermi energy. The magnitude of this current, in each subband, is

$$j = \frac{2e}{h} \Delta V$$

where h is Planck's constant and ΔV is the energy range over which electrons can flow into the wire from one reservoir but not from the other. The 'one-dimensional quantum wire' of Exercise 4 is maintained with one end at a potential of $(9.05\, E_0/e)$ and the other at a potential of $(8.90\, E_0/e)$, where E_0 is the lowest allowed energy in the wire.

(a) Determine the net current which flows in the wire.
(b) Describe one method for creating a quantum wire in practice, and list three ways in which the electron states or behaviour will differ in that system from the idealized system of Exercise 4.

References

T. Ando, A. B. Fowler and F. Stern, *Rev. Mod. Phys.* **54**, 437 (1982).

L. Bányai and S. W. Koch, *Semiconductor Quantum Dots* (World Scientific, London, 1993).

G. Bastard, *Wave Mechanics Applied to Semiconductor Heterostructures* (Halsted Press, New York, 1988).

G. Bauer, F. Kuchar and H. Heinrich, eds., *Two-Dimensional Systems, Heterostructures and Superlattices*, Springer Ser. Sol. St. Sc. **53** (1984).

D. J. Ben-Daniel and C. B. Duke, *Phys. Rev.* **98**, 368 (1955).

M. G. Burt, *J. Phys. Cond. Mat.* **4**, 6651 (1992).

U. Ekenberg, *Phys Rev B* **36**, 6152 (1987).

B.A. Foreman, *Phys. Rev. B* **48**, 4964 (1993).

B. A. Foreman, *Phys. Rev. B* **52**, 12241 (1995).

J. J. Harris, J. A. Pals and R. Woltjer, *Rep. Prog. Phys.* **52**, 1217 (1989).

H. Haug and S. W. Koch, *Quantum Theory of the Optical and Electronic Properties of Semiconductors* (World Scientific, London, 1993).

E. A. Johnson and A. MacKinnon, *J. Phys. C.* **21**, 3091 (1988).

J. P. Kotthaus, in *Granular Nanoelectronics* (Plenum, New York, 1991) pp. 85–102.

R. Lässnig, *Phys Rev B* **31**, 8076 (1985).

C. R. Leavens and R. Taylor, eds., *Interfaces, Quantum Wells, and Superlattices* (Plenum, New York, 1988).

R. L. Liboff, *Introductory Quantum Mechanics* (Addison-Wesley, Reading, MA, 1980).

A. T. Meney, B. Gonul and E. P. O'Reilly, *Phys Rev B* **50**, 10893 (1994).

V. Milanovic and D. Tjapkin, *Phys. Status. Solidi. B* **110**, 687 (1982).

R. E. Prange and S. M. Girvin, eds., *The Quantum Hall Effect* (Springer, Berlin, 1987).

C. Priester, G. Bastard, G. Allan and M. Lannoo, *Phys Rev B* **30**, 6029 (1984).

Ch. Sikorski and U. Merkt, *Phys. Rev. Lett.* **62**, 2164 (1989).

3   Electrons in Quantum Semiconductor Structures: More Advanced Systems and Methods

E. A. Johnson

3.1   Introduction

In Chapter 2, low-dimensional systems were discussed in terms of a single-electron picture, and the behaviour of an electron was examined in the case that it is acted on by various potentials in semiconductors. Those potentials have been supposed to be externally imposed, for instance by a discontinuity in the band gap at an interface between two materials. But an electron will also feel the effect of other electrons in the system in which it finds itself.

There are circumstances in which these many-electron effects can be ignored, for example, in an undoped semiconductor with very few free charges. But in many cases, effects due to the presence of other electrons can be extremely important. Some of the most interesting low-dimensional systems involve many charges: there can be many free electrons, and there will often be in addition some distribution of fixed charges (*space charge*). To study such systems properly, we must discuss how to take into account the presence of such charges. The problem is one of *self-consistency* because we are trying to predict the behaviour of electrons (or holes), while that behaviour will itself depend upon those charges whose behaviour we are trying to predict: in other words, the problem itself depends upon the solution to the problem.

3.2   Many-body Effects

3.2.1   The Hartree Approximation

Consider the reaction of conduction electrons to the presence of a potential well $V_0(z)$ (we suppose this to be an externally determined well, e.g. a finite square well, which restricts electrons into a two-dimensional region). Available electrons will be attracted to the well. Any one electron will react both to V_0 and to the presence of all the other free electrons in the system. (If the system responds in such a way as to leave net fixed charges in some regions of space, the electrons will also interact with those charge distributions; we ignore this effect in the first instance, but it can easily be included later.)

The simplest approximation which takes into account the presence of many electrons is to assume that the electrons as a whole produce an average electrostatic potential energy function V_{sc} and that a given electron feels the resulting total potential, which is the sum of the original potential and this electrostatic potential

$$V = V_0 + V_{sc} \tag{3.1}$$

This is the *Hartree* approximation. Since the external potential acts in the z-direction only, the electron gas will be confined in z but will be uniformly distributed in the x- and y-directions, so that $V = V(z)$. The electron wavefunction is then obtained from the Schrödinger-like envelope function equation (2.9)

$$\left[\frac{p^2}{2m^*} + V(z) \right] \phi(z) = E\phi(z) \tag{3.2}$$

with V given by equation (3.1).

The mobile electrons in the system all obey equation (3.2). They therefore form a static charge distribution $\varrho(z)$ which is constructed from their wavefunctions. It is this distribution of charge which is responsible for V_{sc}, the self-consistent part of the potential. Classically, the relation between a charge distribution ϱ and the electrostatic potential energy function V_{sc} arising from that charge is given by Poisson's equation,

$$\nabla^2 V_{sc}(\mathbf{r}) = \frac{e}{\epsilon_0} \varrho(\mathbf{r}) \tag{3.3}$$

where ϵ_0 is the dielectric constant of free space and e is the magnitude of the electronic charge. Note that, in semiconductor physics, by convention, an increase in the magnitude of the electron energy is taken as positive. In the present case, the $\mathbf{r}$-dependence specializes to a z-dependence, while the fact that the electrons are in a semiconductor rather than in free space is taken into account by inclusion of the static dielectric constant $\epsilon = \epsilon_r \epsilon_0$ of the medium (ϵ_r is the *relative* dielectric constant of the medium in question). The relevant equation is thus

$$\frac{d^2 V_{sc}}{dz^2} = \frac{e}{\epsilon_r \epsilon_0} \varrho \tag{3.4}$$

The self-consistent potential depends on the charge distribution ϱ, but that charge distribution depends on ϕ:

$$\varrho = e \sum_i |\phi_i|^2 \tag{3.5}$$

where the sum is over all occupied states. This means that one must sum over each occupied level n of the quantized system, and then integrate over k_x and k_y up to the Fermi energy E_F, for the level in question (i.e., over the energy range $(E_F - E_n)$ for the level n). If there is also some distribution of fixed charges, ϱ_d, in the system then the potential depends on the distribution ϱ of all these charges,

$$\varrho = e \sum_i |\phi_i|^2 + \varrho_d \tag{3.6}$$

The space (or depletion) charge density ϱ_d will usually be the charge density of ionized donors or acceptors in the system.

Thus, ϱ will determine V_{sc} through equation (3.4), V_{sc} will determine ϕ through equation (3.2), ϕ will determine ϱ through (3.5) or (3.6), and so on. To solve this self-consistent problem, one should start with some reasonable guess. It is common to start with ϕ_0, the solution to the problem of a single electron moving in the external potential V_0. The wavefunctions ϕ_0 give a first approximation to the charge density using, say, (3.5). An approximate self-consistent potential function V_{sc} is then obtained from Poisson's equation (3.3), new wavefunctions are calculated from (3.2) using the improved potential, and the process is repeated. The use of computers makes it a straightforward matter to continue this process until convergence is obtained, i.e. until the electrostatic potential generated from the wavefunctions is the same potential, to within a certain tolerance, as that appearing in the Schrödinger equation for which these wavefunctions are solutions.

3.2.2   Beyond the Hartree Approximation

The Hartree approximation treats the many-electron problem at the simplest level and is adequate for many purposes. The true solution to the problem, however, is given by the potential

$$V = V_0 + V_{sc} + V_{xc} \tag{3.7}$$

Here, V_{xc} is the correction to the potential due to exchange and correlation effects. That such effects exist is evident simply from the Pauli exclusion principle—no two electrons can exist in the same quantum state. The presence of an electron in a given state *automatically* excludes the possibility of another electron being in the same state, and it can be thought of as exerting a sort of repulsion on any other electron. These exchange forces have not been included in the discussion above.

In practice, V_{xc} stands for all those many-electron effects not included in the Hartree approximation. There is no exact theory from which V_{xc} can be derived. Exchange effects alone (those due to the Pauli principle) can be treated by the Hartree–Fock approximation, to be found in standard textbooks, but these corrections are rather cumbersome to calculate. A more productive approach seems to be that of the Thomas–Fermi approximation, or its modern extension, density functional theory. Since these corrections are often small in practice, we do not consider them further here.

3.2.3   The 2DEG at a Heterojunction Interface

An interface between two different semiconductor materials can result in a naturally occurring quantum well and one that is extremely important in practice. Consider a *single* plane interface between GaAs and AlGaAs of the type previously discussed. We suppose that the AlGaAs is *n*-doped, while the GaAs is undoped (Fig. 3.1). This is known as *modulation doping*.

The situation shown in Fig. 3.1 is physically impossible in equilibrium, because in equilibrium the system as a whole must have a common chemical potential which, in this case, is the Fermi energy E_F. Thus, if such a system is created, it must be unstable. What will happen is that electrons will be thermally excited into the conduction band of AlGaAs. They will then migrate into the adjoining GaAs, since there they can achieve states of lower energy (they will lose energy, e.g. by collisions with phonons). These electrons will thus leave behind in the AlGaAs a *lack* of electrons, i.e. the ionized donors will no longer be screened by an equal number of electrons, and the AlGaAs will acquire a net positive charge, which will build up as more electrons move into the GaAs. The mobile electrons which are now in the GaAs will be attracted by the fixed positive charge in the AlGaAs, but will no longer have enough energy to recombine with their ionized donors. They will thus be trapped in the vicinity of the interface: the band-edge discontinuity will prevent them from moving to the left, while the Coulomb attraction of the net positive charge in the barrier material will keep them from moving very far to the right. The process continues until the system reaches equilibrium and the electrons have been trapped in a quantum well, forming a two-dimensional electron gas at the interface between the two materials.

The charge distribution at such a heterojunction would thus be that shown in Fig. 3.2. The forces resulting from this redistribution of charge are conventionally represented as a bending of conduction and valence bands (Fig. 3.3). The slope of the bands in the neighbourhood of the interface is proportional to the electric field

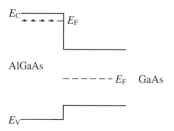

Figure 3.1.   Band-edge diagram for a GaAs–AlGaAs interface before redistribution of charge has taken place. Mobile electrons can easily be excited into the conduction band of the *n*-doped AlGaAs, while the GaAs is undoped. E_v and E_c refer to valence and conduction band edges, respectively.

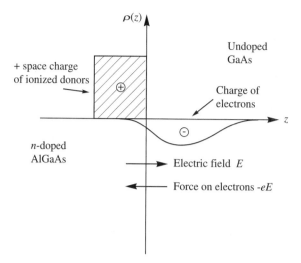

Figure 3.2.   Charge distribution in a simple heterojunction.

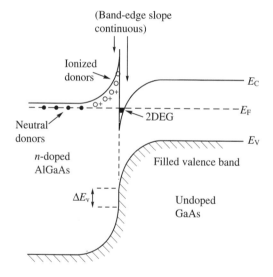

Figure 3.3.   A 2DEG in a heterojunction.

there. Donor states higher than the Fermi level must be unfilled (the region of positive space-charge in the AlGaAs). Conduction-band states below the Fermi energy must be filled (electrons in these states, in the GaAs, form the 2DEG). The electric field must be continuous across the interface, as shown by the equal band-edge slopes in the two materials there (we will ignore the small change in electric field caused by the slight difference between the dielectric constants of the two materials). However, the electric field must be zero far from the interface, since there the materials are required to have their original bulk properties. (We assume that no external potential has been imposed on this system.) We see that self-consistency is vital for a correct description of this system, since only if the charge due to the electrons themselves is included in the potential will the potential correctly go to zero at infinity.

This type of quantum well is among the most widely used and studied, occurring in slightly different forms with different names but with broadly similar characteristics. Two of these are the MOSFET (metal-oxide-semiconductor field effect transistor) and the MISFET (metal-insulator-semiconductor FET). These systems (Fig. 3.4) have the useful property that the number of electrons in the 2DEG can be controlled experimentally. The metal layer forms a gate, the potential of which can be varied so as to attract electrons to the surface of the p-doped Si. When the attraction is strong enough that the conduction band is bent well

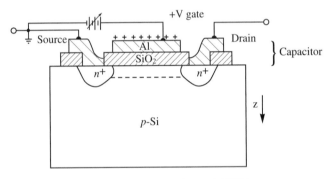

Figure 3.4.   Schematic diagram of a MOSFET.

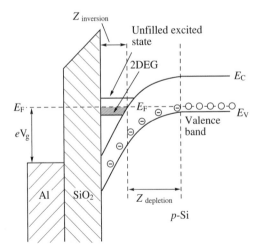

Figure 3.5.   Band-edge diagram for the MOSFET system of Fig. 3.4.

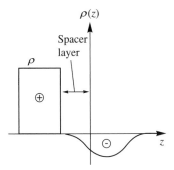

Figure 3.6.   Charge distribution in a MODFET (modulation doped field effect transistor).

below the Fermi energy, a quantum well is again formed and electrons will be trapped in a 2DEG, as shown in Fig. 3.5. Here, the depth of the well, and thus the number of electrons that can be trapped by it, is regulated externally by the potential on the gate. The electrons form an *inversion layer*, so called since the normal state of affairs in a *p*-doped semiconductor is for current to be carried by mobile holes, rather than electrons.

Other similar systems go by the names HEMT (high-electron-mobility transistor) and MODFET (modulation-doped field-effect transistor), and have been used to obtain 2DEGs of very high mobility. Such systems are particularly good for the investigation of phenomena such as the (fractional and integer) quantum Hall effect. One way in which the mobility of two-dimensional electrons in these systems has been improved is by inserting a spacer layer between the doped region and the 2DEG, as illustrated in Fig. 3.6. These devices are discussed later in Chapter 10.

3.2.4   The Ideal Heterojunction

Predicting the properties of a 2DEG at a heterojunction is usually a process that involves full-scale computation. Nevertheless, the appropriate starting point is in the simplest possible description of the system. The starting point usually chosen is that of an infinite triangular well (Fig. 3.7). Here the band-gap discontinuity at the interface, to the left, is approximated by an infinite potential step, while the conduction band edge is assumed to have a constant slope corresponding to the value of the confining electric field which is present at the interface. (The curvature of the band edge in the diagram is an artist's impression of the true band edge, once self-consistent effects have been taken into account.)

This approximation is simple enough to allow closed-form solutions, which are Airy functions. The resulting predictions are often referred to in the literature, and are discussed extensively in the classic review article of Ando *et al.* (1982). As one would expect, the electron energies are quantized, and the electrons themselves are confined to a narrow region to the right of the interface. The predicted electron

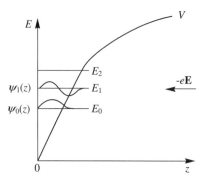

Figure 3.7.   The infinite triangular well. The rounding of the band edge to the right of the diagram is drawn to correspond to physical expectations, while electron energies and wavefunctions are calculated assuming a strictly linear potential, to this approximation; **E** is the electric field at the interface.

states spread out progressively in each higher-energy quantum level. All these properties are physically correct at a qualitative level, but they can also be taken to be a reasonable *physical* approximation if there is a relatively low electron density in the 2DEG.

Since even Airy functions involve the use of tables, or calculations, it is useful to have on hand an approximation technique which can often be used in conjunction with simple physical approximations for the system of interest, to obtain quick, approximate, and often simple numerical predictions. The following section discusses two such techniques.

3.3   Some Calculational Methods

In this section we consider two methods which can give surprisingly good approximate predictions for low-dimensional structures in which quantum mechanical effects are important: the Wentel–Kramers–Brillouin (WKB) approximation and the Thomas–Fermi approximation. They are simple enough to be used either separately or together, and can sometimes give analytic answers for quantum energies and wavefunctions in small structures.

The WKB approach is an approximate way of solving the Schrödinger equation and is appropriate for systems in which many-electron effects are either weak or absent. (It can also be useful, as a first approximation, even when many-electron effects are important.) The Thomas–Fermi approach (in the version presented here) is an approximate way of taking many-electron effects into account, giving what can be a very good approximation to the full self-consistent potential felt by an electron in the presence of an external potential together with the band-bending effects resulting from the presence of all the other electrons. An even better approximation can be obtained when these methods can be combined, using the WKB approximation to solve the Schrödinger equation for an electron in a self-consistent potential which has itself been obtained from the Thomas–Fermi approximation.

3.3.1 The WKB Approximation

We consider here a method for calculating approximate wavefunctions and energy levels which is 'semiclassical', but can nevertheless be quite powerful. This is the WKB approximation. A discussion of the method is to be found in most textbooks on quantum mechanics (e.g. Merzbacher, 1970). If many-body effects are neglected, the WKB approximation often leads to analytic results, or at least to closed-form expressions (in which the answer can be obtained numerically by doing a simple integral). Even in systems where many-electron effects are important, the WKB approximation can yield useful information about the way in which wavefunctions and energies depend upon the parameters of the system. Here, we do not justify the method in detail, but merely present the resulting approximation procedure in a way that can be applied to systems of interest.

The WKB approximation is semiclassical: in some sense it can be thought of as an expansion that is good when quantum effects are small. The procedure can be stated in terms of a classical picture of the motion of a particle in a potential (Fig. 3.8). Classically, if a particle of energy E moves in a potential V, the particle can be found only in regions where $E > V$. The *turning points* of its motion (a, b in the diagram) are the points at which $E = V$ (the particle must turn around, since it cannot proceed into a forbidden region). If V were constant in each region (e.g. a finite square well), the solution of the Schrödinger equation, ϕ, would be simply

$$\phi = \begin{cases} e^{\pm ikz} & \text{for } E > V \\ e^{\pm \kappa z} & \text{for } E < V \end{cases} \tag{3.8}$$

i.e. a travelling wave in the classically allowed region and a decaying exponential in the classically forbidden regions. If V is a slowly varying function of position, then one can try to approximate ϕ by a form similar to that of equation (3.8):

$$\phi = e^{iu(z)} \tag{3.9}$$

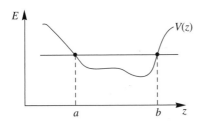

Figure 3.8.   Turning points a and b for a particle of energy E in a potential V.

where

$$u = \pm \int k(z)\,dz \qquad (3.10)$$

with

$$k = \begin{cases} \dfrac{1}{\hbar}[2m(E-V)]^{1/2} & \text{for } E>V \\[2mm] -i\kappa(z) & \text{for } E<V \end{cases} \qquad (3.11)$$

so k is real in the classically allowed regions and purely imaginary in the classically forbidden regions, as in equation (3.8). The WKB approximation consists in expressing ϕ as

$$\phi(z) \simeq \frac{1}{\sqrt{k(z)}} \exp\left[\pm i \int^z k(z)\,dz\right] \qquad (3.12)$$

According to this equation, we can write ϕ, for example, in terms of sines and cosines in the allowed region, and real exponentials in the forbidden regions. It is then necessary to connect these two sorts of solutions at the turning points a and b (for a derivation and detailed discussion, see, e.g. Merzbacher, 1970). The appropriate *connection formulae* at the **left-hand turning point** (a, in Fig. 3.8) are

Forbidden region   Allowed region

$$-\frac{1}{\sqrt{\kappa}}\exp\left(\int_z^a \kappa\,dz\right) \quad \Longleftarrow \quad \frac{1}{\sqrt{k}}\sin\left(\int_a^z k\,dz - \tfrac{1}{4}\pi\right)$$

$$\frac{1}{\sqrt{\kappa}}\exp\left(-\int_z^a \kappa\,dz\right) \quad \Longrightarrow \quad \frac{2}{\sqrt{k}}\cos\left(\int_a^z k\,dz - \tfrac{1}{4}\pi\right) \qquad (3.13)$$

and, for turning points like b (**right-hand turning points**),

Allowed region   Forbidden region

$$\frac{2}{\sqrt{k}}\cos\left(\int_z^b k\,dz - \tfrac{1}{4}\pi\right) \quad \Longleftarrow \quad \frac{1}{\sqrt{\kappa}}\exp\left(-\int_b^z \kappa\,dz\right)$$

$$\frac{1}{\sqrt{k}}\sin\left(+\int_z^b k\,dz - \tfrac{1}{4}\pi\right) \quad \Longrightarrow \quad -\frac{1}{\sqrt{\kappa}}\exp\left(\int_b^z \kappa\,dz\right) \qquad (3.14)$$

These connection formulae apply to turning points at which the potential is in some sense slowly varying. Other places at which one wants to apply the usual boundary conditions are at $z = \pm\infty$, where one specifies that ϕ cannot correspond to a state that is exponentially increasing, or at an infinite potential barrier (a 'solid wall'), where the wavefunction is required to vanish.

The arrows in equations (3.13) and (3.14) are not symmetric. This indicates that the connection between the indicated expressions is rigorous in one direction (double arrow) but not in the other. Thus, for example, an exponential with a negative coefficient to the left of the left-hand turning point (*a*) will *always* imply the existence of a cosine-type solution to the right of that turning point (the second of equations (3.13)), but the existence of a cosine-type component to the wavefunction to the right of *a* may not imply the existence of an exponential of negative coefficient to the left. Care is needed in the direction of the single arrows because the method is not sensitive to the existence of extremely small terms in forbidden regions (at any appreciable distance into the forbidden region, a growing exponential swamps a decaying one, so that, if a growing one is present, it is not possible to say whether a decaying one is there or not).

It is the matching conditions at the turning points and other boundaries that lead to quantization of the energy: only for certain discrete values of *E* will it be possible to find a solution ϕ which can be matched properly for *all* z. Since the matching conditions are approximate, however, the quantized values of *E* will also be approximate. Nevertheless, the approximation often proves to be a surprisingly good one. Note, too, that one has approximate wavefunctions (equations (3.13) and (3.14)) as well as approximate energies.

Example 1. A simple bound state (Fig. 3.8). In this case, ϕ must be a decaying exponential as $z \to \pm\infty$. This means that we must use the second of equations (3.13) and the first of equations (3.14). These formulae give two *different* cosine expressions for the wavefunction; the condition that these expressions must be the same imposes a condition on the arguments of the cosines, which produces the quantization condition,

$$\int_a^z k\,dz + \int_z^b k\,dz = \frac{1}{\hbar}\int_a^b [2m(E-V)]^{1/2}\,dz = (n+\tfrac{1}{2})\pi \tag{3.15}$$

where $n = 0, 1, 2, \ldots$ This condition is, in fact, the Bohr–Sommerfeld quantization condition, that a particle's orbit in phase space must equal a half-integer multiple of π. This is essentially a condition for a standing wave.

Example 2. A bound state in a triangular potential well as, for example, the simple picture of an inversion layer (Fig. 3.9), for which the potential energy function *V* is given by

$$V(z) = \begin{cases} e|\mathcal{E}|z, & z>0 \\ \infty, & z<0 \end{cases} \tag{3.16}$$

where $\mathcal{E}$ denotes the magnitude of an electric field. The wavefunction must decay exponentially for positive z in the forbidden region and in this case must vanish identically at $z = 0$. In the classically forbidden region where $z > 0$, we must use the first of equations (3.14). Then, for the cosine function to vanish at the origin,

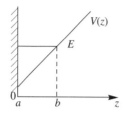

Figure 3.9.　Triangular potential well at an infinite potential barrier.

we must require that

$$\int_a^b k\,\mathrm{d}z = (n + \tfrac{3}{4})\pi \tag{3.17}$$

where $n = 0, 1, 2, \ldots$ from which it is simple to show that the quantized energies E_n are, in this approximation, given by

$$E_n = \left(\frac{\hbar^2 e^2 \mathcal{E}^2}{2m}\right)^{1/3} \left[\tfrac{3}{2}\pi(n + \tfrac{3}{4})\right]^{2/3} \tag{3.18}$$

The WKB approximation has given us not only approximate values for the energy levels, equation (18), but also approximate wavefunctions,

$$\phi(z) = \frac{1}{\sqrt{k(z)}} \sin\left[\int_a^z k(z')\,\mathrm{d}z' - \tfrac{1}{4}\pi\right] \tag{3.19}$$

The example of a triangular potential well is often used as the starting point for a treatment of two-dimensional electrons in an inversion or accumulation layer, discussed in Section 3.2.3. A different approximation for the ground-state wavefunction for this problem,

$$\phi_0(z) = 2(2b)^{-3/2} z\, \mathrm{e}^{-z/2b} \tag{3.20}$$

is worth noting. Equation (3.20) (Ando *et al.*, 1982) was obtained from a variational principle. A 'scale depth' b of about 30–50 Å seems to be appropriate for the problems considered in the review by Ando *et al.* (1982).

Finally, we note that the WKB method as discussed here does not include many-electron effects *per se*, since it assumes that the confining potential is known in advance. To generalize to many-electron systems, one can use the WKB method as part of an iterative solution together with the Poisson equation. Alternatively, one can use the WKB method to solve a Schrödinger equation with an approximate but self-consistent potential obtained from a Thomas–Fermi approximation, as will be described in Section 3.3.3.

3.3.2　The 2DEG in Doping Wells

So far we have discussed *compositional* quantum wells (Section 2.3.3) and quantum wells at heterojunctions (Section 3.2.3). Another important example is that of doping wells (Fig. 3.10). Consider an intrinsic semiconductor (e.g. GaAs), grown one plane atomic layer at a time, which has been uniformly n-doped in a slab of width d during growth. Suppose that all donors have become ionized. The resulting positive charge of these donors creates an attractive force which is then felt by the mobile (donated) electrons, which are trapped by that force. The attraction can easily be strong enough, over a short enough distance d, that quantum effects are important. This is yet another kind of two-dimensional electron gas.

The charge density of the donors is assumed to be

$$\varrho(z) = en_D(z) \tag{3.21}$$

where the donor doping density n_D is constant for $0 \leq |z| \leq \frac{1}{2}d$ and the potential energy function V these donors create is given by Poisson's equation:

$$\frac{d^2 V}{dz^2} = \frac{e}{\epsilon_r \epsilon_0} \varrho(z) \tag{3.22}$$

This equation is easily solved: if the second derivative of V is constant, it must be quadratic in z,

$$V = a + bz + cz^2 \tag{3.23}$$

We can choose a to be zero since the choice of a zero of energy is always arbitrary. Moreover, if we choose the doping slab to be centred at $z = 0$, as in Fig. 3.10, then $b = 0$ from symmetry. Thus,

$$V(z) = \frac{e^2}{2\epsilon_r \epsilon_0} n_D z^2, \qquad 0 \leq |z| \leq \frac{1}{2}d \tag{3.24}$$

which yields a parabolic potential well within the region of donor doping. The potential in the undoped regions is even more straightforward: in a region of zero charge, the right-hand side of equation (3.22) must be zero, and the potential must therefore be linear in z (and symmetric in z):

$$V = a' + b'|z| \tag{3.25}$$

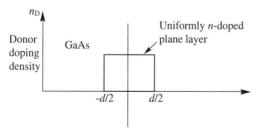

Figure 3.10.   Donor density profile: a uniformly n-doped slab creates a potential well within the host material.

Here a' and b' are determined from equation (3.24) at $z = \pm\frac{1}{2}d$ using the fact that the potential must be continuous. One easily finds that

$$a' = -\frac{e^2}{2\epsilon_r\epsilon_0}\,n_D(\tfrac{1}{2}d)^2, \qquad b' = \frac{e^2}{2\epsilon_r\epsilon_0}\,n_D d \tag{3.26}$$

The resulting quantum well is shown in Fig. 3.11.

If electrons are actually trapped in the well, then their presence will modify this potential (self-consistent effects). Ignoring these corrections, the energies and wavefunctions for electrons in the well are easily obtained, e.g. from the WKB approximation. In the region $|z| \leq \frac{1}{2}d$ the potential is that of a simple harmonic oscillator, for which the WKB approximation gives the exact answers (Merzbacher, 1970). Thus, for instance, the energies are given by

$$E_n = \hbar\omega(n + \tfrac{1}{2}), \qquad n = 0, 1, 2 \ldots \tag{3.27}$$

where the 'natural frequency' ω is

$$\omega = \sqrt{\frac{k}{m}} = \left(\frac{e^2 n_D}{2m\epsilon_r\epsilon_0}\right)^{1/2} \tag{3.28}$$

Self-consistent effects can modify this potential in important ways. First, $V(z)$ itself will be changed if electrons are present in the well. Secondly, the ionized donors themselves are more a random collection of point charges than a continuous uniform charge distribution. This randomness results in fluctuations in the potential $V(\mathbf{r})$. Electrons trapped in the well will tend to redistribute themselves in such a way as to reduce these potential fluctuations, thus screening out some of the effects of disorder.

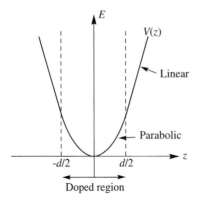

Figure 3.11. Quantum well created by the donor doping profile of Fig. 3.10. $V(z)$ is parabolic for $-\frac{1}{2}d \leq z \leq \frac{1}{2}d$, and linear elsewhere.

3.3.2.1 The Delta Well (Spike Doping)

An extreme example of a doping well is a *delta well*, formed when the layer of donors in a host material is as small as one atomic layer wide, approximately a delta-function distribution in z. (These are also called *delta-doped*, or *spike-doped*, systems.) One can therefore write

$$n_D(z) = D\delta(z) \tag{3.29}$$

where D is the number of ionized donors per unit *area* in the host material. Poisson's equation then gives the very simple potential energy function

$$V(z) = a + b|z| \tag{3.30}$$

with a chosen to be zero for convenience, and

$$b = \frac{e^2 D}{2\epsilon_r \epsilon_0} \tag{3.31}$$

(Fig. 3.12). In this case the WKB approximation gives the energy levels

$$E_n = \left(\frac{\hbar^2 b^2}{2m}\right)^{1/3} \left[\tfrac{3}{4}\pi(n + \tfrac{1}{2})\right]^{2/3} \tag{3.32}$$

with b given by equation (3.31). As always, this can be the starting point for a better approximation, which must include self-consistent effects if electrons are in fact trapped in the well. These effects can be quite important, particularly for high doping, as indicated by Fig. 3.13. Here, the number of electrons trapped by the well equals the number of donors, and the potential (heavy black line) is seen to differ considerably from the linear, 'bare' potential shown in Fig. 3.12.

We digress briefly here to note that delta-doping can be used, among other things, to create high electron mobility devices, starting from a basic accumulation or inversion 2DEG. Figures 3.1 and 3.2 show how electrons in such a 2DEG can come from nearby ionized donors. One is then often interested in the mobility of these electrons in the two-dimensional plane in which they are free to move.

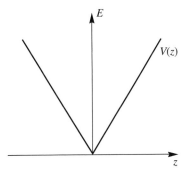

Figure 3.12.   Bare potential for a delta well.

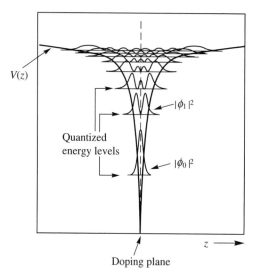

Figure 3.13.   Energies and electron probability-densities for a delta well in InSb (self-consistent calculation; probability densities are each normalized to unity).

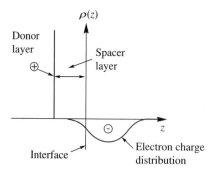

Figure 3.14.   Charge distribution in a high mobility electron gas created at an interface with delta donor doping.

This mobility can be severely limited by the scattering of electrons from ionized impurities. In the situation illustrated in Figs. 3.1 and 3.2, a large number of these impurities lie very close to the plane of the 2DEG and will strongly limit its mobility. The mobility of the 2DEG can be increased by placing a *spacer layer* between the donors and the plane of the 2DEG, thus making the scattering centres more distant. If the donor region is made into a delta-layer, this scattering can be reduced still further (Fig. 3.14). Here, any free charges to the left of the interface will be tightly bound to the donor layer, and will tend to screen out disorder due to the random positions of the donors. This will help to produce an approximately uniform charge distribution, further reducing ionized impurity scattering.

3.3.3   The Thomas–Fermi Approximation for Two-dimensional Systems

The Thomas–Fermi approximation is described in many textbooks on quantum mechanics (e.g. Ashcroft and Mermin, 1976). It allows one to take account of many-electron effects on the potential felt by a single electron (in other words, it can give an approximate way of calculating band-bending effects caused by the presence of many electrons or holes). This approximation can provide a very good description of the potential for the particular cases of heterojunctions, accumulation and inversion layers, and delta-doped systems in the presence of many-electron effects, and in some cases provides simple analytic formulae for the potential. Its use for other systems may be less straightforward.

When it can be used in this way, the Thomas–Fermi approach provides an approximate alternative to obtaining a self-consistent iterative solution of coupled Poisson and Schrödinger equations, as described in the preceding section. An early reference to this use of the method is given by Keyes (1976), whose approach we follow here.

We take the potential felt by an electron in a quantum well to be a sum of several parts:

$$V_{\text{tot}} = V_{\text{ext}} + V_{\text{fixed}} + V_{\text{e}} \tag{3.33}$$

where V_{ext} is the external potential (the band-edge potential in the absence of free charges, and of any applied electric field), V_{fixed} is the potential due to any fixed space-charge distribution in the system, and V_{e} the self-consistent contribution to the potential (arising from the presence of mobile electrons in the system).

In terms of the density $n_{\text{e}}(\mathbf{r})$ of mobile electrons, the self-consistent part of the potential obeys Poisson's equation

$$\frac{\mathrm{d}^2 V_{\text{tot}}}{\mathrm{d}z^2} = \frac{e^2}{\epsilon_r \epsilon_0} [n_{\text{e}}(\mathbf{r}) + n_A(\mathbf{r}) - n_D(\mathbf{r})] \tag{3.34}$$

(cf. (3.4)), where n_A and n_D are the densities of charged acceptors and donors, respectively. We next suppose that there are a number of mobile electrons in the system, and use the familiar relation between Fermi level (measured from the conduction band edge) and three-dimensional electron density, n_e

$$E_{\text{F}} = (3\pi^2)^{2/3} \frac{\hbar^2}{2m^*} n_{\text{e}}^{2/3} \tag{3.35}$$

which follows from the three-dimensional density of states in equation (2.2). Here, E_{F} is understood to mean the difference between the Fermi energy and the total band-edge potential V_{tot}. Since V_{tot} depends on position, we write equation (3.35) as

$$E_{\mathrm{F}} - V_{\mathrm{tot}}(z) = (3\pi^2)^{2/3} \frac{\hbar^2}{2m^*} [n_{\mathrm{e}}(z)]^{2/3} \tag{3.36}$$

If we choose our zero of energy to lie at E_{F}, we then obtain

$$-V_{\mathrm{tot}} = (3\pi^2)^{2/3} \frac{\hbar^2}{2m^*} n_{\mathrm{e}}^{2/3} \tag{3.37}$$

which can be rearranged to read

$$n_{\mathrm{e}} = \frac{1}{3\pi^2} \left(\frac{2m^*}{\hbar^2}\right)^{3/2} (-V_{\mathrm{tot}})^{3/2} \tag{3.38}$$

We can now obtain an approximation for V_{tot} if we restrict ourselves to external potentials which are either linear or constant in growth direction z. This includes inversion and accumulation layers and delta-doped systems when no space charge is present. (It can also be a reasonable first approximation even in the presence of space charge, since such charge is often distributed over length scales much larger than that of the quantum well in question, giving a relatively weak contribution to the band bending.)

We thus consider Poisson's equation (3.34) and note that, since V_{ext} is assumed to be a straight line and V_{fixed} is neglected here,

$$\frac{\mathrm{d}^2 V_{\mathrm{sc}}}{\mathrm{d}z^2} = \frac{\mathrm{d}^2 V_{\mathrm{tot}}}{\mathrm{d}z^2} \tag{3.39}$$

Thus we can combine equations (3.34) and (3.38) to eliminate n_{e} (which is not yet known), to obtain

$$\frac{\mathrm{d}^2 V_{\mathrm{tot}}}{\mathrm{d}z^2} = \frac{1}{3\pi^2} \left(\frac{e^2}{\epsilon_r \epsilon_0}\right) \left(\frac{2m^*}{\hbar^2}\right)^{3/2} (-V_{\mathrm{tot}})^{3/2} \tag{3.40}$$

This is a differential equation which can be solved for the self-consistent band-edge potential V_{tot}. Note that V_{tot} itself is negative: the approximation refers to systems which contain mobile electrons, and does not apply when there are none (i.e. when $V_{\mathrm{tot}} > 0$).

3.3.3.1 The Thomas–Fermi Approximation for Heterojunctions and Delta Wells

The Thomas–Fermi approximation of equation (3.40) is found to have a simple analytic form when applied to heterojunctions or delta wells with no background doping. It is easy to verify that (3.40) is satisfied for a potential of the form

$$V(z) = -b \frac{1}{(|z| + z_0)^4} \tag{3.41}$$

where b and z_0 are positive constants, and the interface (or the position of delta-doped layer) is at $z = 0$. A few simple manipulations show that in this case

$$b = \left(\frac{60\pi^2 \epsilon_r \epsilon_0}{e^2} \right)^2 \left(\frac{\hbar^2}{2m^*} \right)^3 \tag{3.42}$$

and z_0 is determined by a boundary condition (e.g. at $z = 0$) for the particular system of interest.

Since V_{tot} is given explicitly by (3.41) and (3.42), one also has an explicit form for the density n_e of mobile electrons from equation (3.38),

$$n_e = \frac{a}{(z + z_0)^6}, \qquad a = \left(\frac{2m^* b}{\hbar^2} \right)^{3/2} \tag{3.43}$$

This charge density can be a very good approximation to that obtained from a full self-consistent solution of Schrödinger's and Poisson's equations.

One can now obtain (approximate) energy levels and subband occupations by solving the Schrödinger equation, using the approximate potential V_{tot} obtained by this procedure. The exact self-consistent solution can also be obtained from this starting point, thus saving much computing time.

When the potential approaches zero as $z \to \infty$, one boundary condition is needed to fix the single unknown z_0. This may be, for instance, the electric field at the interface (i.e. the slope of the potential there), or the value of the band-edge potential at $z = 0$, or else the total two-dimensional density of mobile electrons in the system. In the more general case in which the system is not charge-compensated, for instance by removing mobile electrons from the system by some external mechanism, the situation is slightly more complicated, but an explicit analytic solution can still be found. A discussion of the application of this approach to delta-doped systems is given by Ioriatti (1990).

3.4   Quantum Wires and Quantum Dots

3.4.1   Quantum Point Contacts and Quantized Conductance Steps

In Section 2.3.5 it was noted that one way of confining electrons to low dimensions was the split-gate procedure. This produces a particularly smooth confining potential, which is very useful when one wants to look at precise quantum states of electrons. A split-gate constriction is often called a 'quantum point contact' and a typical system is shown in Fig. 3.15.

The uses of this device have been extended in a number of ways, and have led to the discovery of many new and interesting physical effects. These owe their discovery to the fact that low-dimensional structures created by split-gate techniques are 'perfect', i.e. free from disorder caused by impurities or randomness in modulation doping, or surface imperfections arising in etched structures. *Ballistic* electron effects, in which electrons preserve their quantum state over relatively large distances, are of especial interest. (For these, the mean free path for inelastic

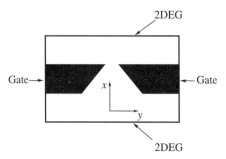

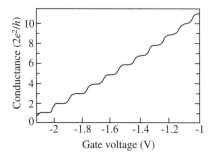

Figure 3.15.   A split-gate device or quantum point contact; the 2DEG lies in the plane of the diagram.

Figure 3.16.   Conductance through a quantum point contact, as a function of gate voltage (van Wees *et al.*, 1988). (Courtesy B. J. van Wees, L. P. Kouwenhoven)

collisions is long compared with the dimensions of the system by which the electrons are constrained.)

One of the earliest phenomena to be discovered using such a system is shown in Fig. 3.16. It was found by van Wees *et al.* (1988) and by Wharam *et al.* (1988) that when electrons were allowed through the channel between the gates, the conductance (i.e. conductivity per unit length) of the resulting current was quantized, rising in discrete steps of $2e^2/h$ in a regular way as the gate voltage was increased. To understand this result, note that, as electrons travel through the constriction, the constraint on their y-motion allows only electrons with discrete values of k_y to travel from one region of the 2DEG to the other. As the repulsion of the gate is decreased (by decreasing the gate voltage), more and more electrons are allowed to traverse the constriction, but only those with a k_y allowed by quantization (i.e. only if their k_y 'fits' within the constriction). The conductance is thus quantized as a function of gate voltage.

A formula for the conductance can be obtained simply, if approximately (cf. Harris *et al.*, 1989). We approximate the contact as a rectangular constriction and neglect any reflections or other end effects. A net current will flow through the contact if the Fermi energy of the two-dimensional electrons on one side of

(a)

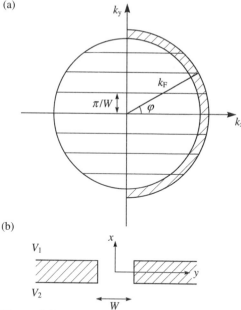

(b)

Figure 3.17. (a) Distribution of allowed k-vectors for an electron passing through a quantum point contact (after Harris *et al.*, 1989); the cross-hatched area gives the net excess number of electrons at the Fermi surface going in the $+x$ direction due to a small potential drop $\Delta V = V_2 - V_1$. (Courtesy J. J. Harris) (b) Current flow through a point contact. The current is carried by electrons leaving the lower reservoir with energies between V_2 and V_1.

the contact is slightly higher than that of electrons on the other side (Fig. 3.17). Of the range of k-values (all $k \leq k_F$) allowed in the unrestricted 2DEG, only electrons with values k_y represented by the horizontal lines in Fig. 3.17(a), $k_y = i\pi/W$, with i an integer and W the width of the constriction, will be allowed through. Suppose that the potential energy V_2 on one side of the constriction is slightly higher than that on the other (V_1). If ΔV is the difference in potential energy, then we will show below that each subband will carry current $2e(\Delta V)/h$.

We suppose ΔV to be a very small deviation from, say, the average Fermi energy of the 2DEG on either side of the constriction. Over most of the electron energy range there will be no net current flow, since electrons moving in opposite directions will balance one another. However, over the small range ΔV around the Fermi energy, electrons will be travelling in one direction only, towards the 2DEG held at the lower potential, giving a net current. These electrons carrying the current (in the $+x$-direction) will have a wavevector k_i of the correct magnitude to bring their total energy up to the Fermi energy. The density of states in each allowed state (open channel) i is

$$g(E) = g(k) \frac{dk}{dE} \qquad (3.44)$$

(cf. Section 2.2.2), where we note that $g(k)$ itself is a constant which is

$$g(k) = \frac{2}{2\pi} \qquad (3.45)$$

This is true for two spins, in one dimension, and for positive k (it is also true for negative k, but electrons with energies between V_1 and V_2 will only be travelling in the $+k$-direction). The current density in this open channel (i.e. corresponding to this allowed value of k_y) is the product of ev and the density of states $g(E)$, where v is the group velocity of the electrons:

$$v = \frac{1}{\hbar} \frac{dE}{dk} \qquad (3.46)$$

The current itself in this channel is obtained by integrating this density over the energy range ΔV in which electrons are travelling in one direction only. Thus, in each channel i,

$$j_i = \int_{V_1}^{V_2} ev\, g(E)\, dE$$

$$= e \int_{V_1}^{V_2} \frac{1}{\hbar} \frac{dE}{dk}\, g(k)\, \frac{dk}{dE}\, dE$$

$$= \frac{e}{\pi\hbar} \Delta V \qquad (3.47)$$

This equation may also be written

$$j_i = \frac{2e}{h} \Delta V \qquad (3.48)$$

The current in channel i is thus independent of the carrier velocity and independent of the density of states in that channel because these two dependencies exactly balance. Referring to Fig. 3.17, we see that only a finite number i_{max} of k-values will be allowed, where

$$i_{max} \equiv \text{the maximum integer} < k_F W / \pi \qquad (4.49)$$

k_F is the Fermi wavenumber. W increases as the gate voltage (repulsion under the gates) is decreased, allowing an increasing number of k_y values through the constriction. The conductance arising from each allowed k_y value is e^2/h, and the overall conductance $G = eI/(\Delta V)$ is

$$G = \frac{2e^2}{h} i_{max} \qquad (3.50)$$

G will thus increase in steps of equal size as W increases, as is indeed seen experimentally.

3.4.2   A Closer Look at Quantum Dots

We now wish to take a closer look at quantum dots (Bányai and Koch, 1993). The split-gate technique mentioned in the previous section and in Section 2.3.5 has been extended and refined in a number of ways. Some of these techniques have been very productive in producing quantum dots. Several different procedures, illustrated in Figs. 3.18 and 3.19, will be discussed in this section.

A two-dimensional array of dots of photoresistive material can be created on a planar semiconductor surface, as indicated in Fig. 3.18(a) (Sikorski and Merkt, 1989). This pattern was made by laying down a uniform plane layer of such material, shining onto it a laser interference pattern (parallel stripes of light), etching the material into a set of stripes, and then repeating the etching process with a perpendicular set of stripes. On top of this surface with its etched dots a layer of metal was then deposited (Fig 3.18(a)). A 2DEG near to this semiconductor

(a)

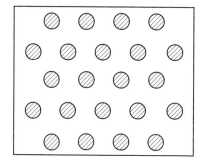

(b)

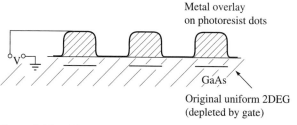

Metal overlay
on photoresist dots

GaAs

Original uniform 2DEG
(depleted by gate)

Figure 3.18.   (a) Schematic drawing of an array of dots of resistive material which can be produced by etching of a photodeveloped laser interference pattern (Sikorski and Merkt, 1989). (Courtesy U. Merkt) (b) Cross-section of a lattice of electron dots created by metal overlay on photoresist dots.

(a)

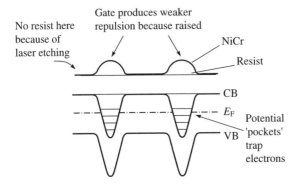

No resist here
because of
laser etching

Gate produces weaker
repulsion because raised

NiCr

Resist

CB

E_F

VB

Potential
'pockets'
trap
electrons

(b)

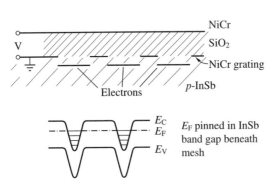

NiCr

SiO₂

NiCr grating

p-InSb

Electrons

E_C
E_F

E_V

E_F pinned in InSb
band gap beneath
mesh

(c)

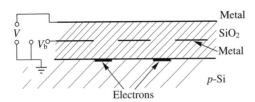

Metal

SiO₂

Metal

p-Si

V

V_b

Electrons

Versatile: dots (e or h);
anti-dots; nipi lattice

Figure 3.19.   (a) Cross-section of band structure within the 2DEG layer of Fig. 3.18(b). CB conduction band; VB valence band. (b) Quantum dot structure created by a NiCr gate separated by an insulator from a p-InSb plane surface overlaid by a NiCr mesh. (c) A two-gate system, the bottom gate being a plane metal grating embedded within the insulating layer which lies on the semiconductor surface.

surface can then be modulated by applying a voltage between the metal gate and the semiconductor (Fig. 3.18(b)). Electrons repelled by the gate potential will have reduced density (i.e. the gate depletes the 2DEG) beneath those areas where the gate is nearest the 2DEG, Hence, the electrons remain preferentially under the photoresist dots. For a strong enough repulsion, the effect on the conduction band edge is to produce a grid of one-dimensional quantum well 'pockets', one underneath each dot, with the 2DEG completely depleted elsewhere (Fig. 3.19(a)). One thus obtains an array of quantum dots.

Two other methods of producing quantum dots are depicted in Fig. 3.19(b, c). For the situation shown in Fig. 3.19(b), the Fermi energy is pinned within the band gap by the NiCr grating overlaying the semiconductor, but electrons can be attracted to form a layer of dots at the semiconductor interface by a strong enough attractive potential created by the overlying plane gate. The more versatile two-gate configuration illustrated in Fig. 3.19(c) can produce various kinds of charge confinement (Fig. 3.20). Depending on the biases of the gates (one of

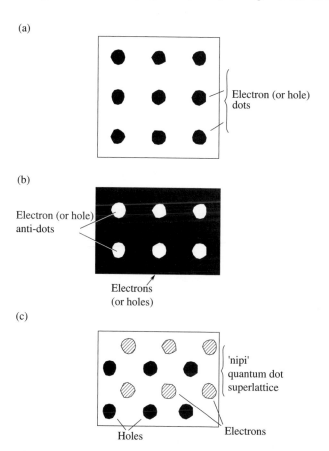

Figure 3.20.   Various possible zero-dimensional structures: (a) electron or hole dots, (b) electron or hole anti-dots, (c) *n-i-p-i* quantum dot superlattice.

which is embedded in the middle of the insulating layer), one can produce a lattice of electron dots or hole dots. A lattice of electron anti-dots (that is, of a 2DEG with a two-dimensional array of dots from which electrons are excluded) can also be produced (Fig. 3.20(b)). A two-dimensional lattice comprising alternating electron dots and hole dots (Fig. 3.20(c)) is yet another option in such a system. A more detailed discussion may be found in Kotthaus (1991).

3.4.3   The Coulomb Blockade and Single-electron Transistors

Quantum dots can be created by a different kind of split-gate method from that discussed in the previous section, and a simple way of doing this is shown in Fig. 3.21. One can pattern a metal gate not only to create a one-dimensional channel, but to create quantum dots connected by one-dimensional wires to 2DEGs (and more complicated structures). Moreover, by varying the gate voltage one has some control over the size of the dot (along with the width of the original one-dimensional channel).

Experiments with such quantum dots revealed another unexpected effect, which is illustrated in Fig. 3.22. The conductance through a quantum dot such as that of Fig. 3.21 is found to oscillate in an exceptionally regular fashion, over a very wide range of gate voltage. The reason for this behaviour is of particular interest, and follows from some simple considerations.

Consider what happens when one tries to send a current along a one-dimensional wire containing such a quantum dot. Each electron which goes through the dot must first tunnel into the dot (quantum-mechanically), and then out again; the conductance is a measure of how easily this process happens. But adding one electron to the charges already on the dot takes energy; how much is determined from elementary considerations, since the dot is essentially a capacitor. To add an

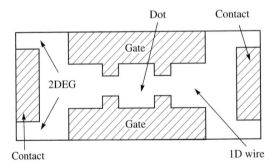

Figure 3.21.   Schematic split-gate device creating a quantum dot in a one-dimensional channel. The 2DEG lies in the plane of the diagram a small distance below the surface on which the gate is deposited.

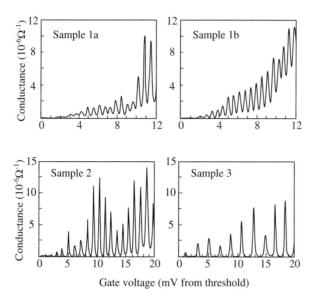

Figure 3.22. Conductance through a quantum dot as a function of gate voltage, for various samples (after Kastner, 1992). Samples 1*a* and 1*b* have the same geometry and the same period of oscillation; samples 2 and 3 have different lengths, and periods proportional to the inverse sample length. (Courtesy M. A. Kastner)

amount of charge Q to a capacitor with capacitance C takes energy

$$E = \frac{Q^2}{2C} \tag{3.51}$$

Thus to put one more electron onto the dot costs energy $e^2/2C$. Similarly for a hole to tunnel into the dot (i.e. for the electron to leave the dot) takes energy $-e^2/2C$. This means that electrons at the Fermi energy in the wire can only get into the dot if this energy is $e^2/2C$ higher than the lowest available electron state in the dot, and, once it is there, it can only get out again if it can lose at least $e^2/2C$ on the other side. If the voltage drop across the dot is less than a total of e^2/C, then current cannot flow ('Coulomb blockade'), but if it is precisely e^2/C, conditions are right for electrons to flow in, and back out again, with no loss of energy. This is shown schematically in Fig. 3.23.

The change in gate voltage changes the size of the dot, which changes the allowed energies of electrons relative to the Fermi level, and it is only when one of these is precisely $e^2/2C$ lower than the Fermi energy in the incoming channel that electrons can flow, through the dot. By comparing the period of the oscillations with the calculated capacitance of the dots, one finds that each oscillation is the result of adding *one* more electron to those residing on the dot. This Coulomb blockade had been seen some years earlier in metal particles, but is of special interest in low dimensional semiconductor systems because the fact that only

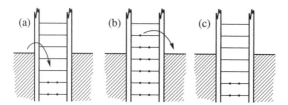

Figure 3.23.   Schematic picture of Coulomb blockade in a quantum dot (after Pals, 1995). (a) An electron tunnels into the dot via the left barrier to a level exactly $e^2/2C$ lower. (b) When the electron enters the dot, its electrostatic potential energy rises by an amount e^2/C. (c) The electron tunnels through the right barrier losing an energy $e^2/2C$, while the barrier returns to its former state. (Courtesy P. Pals)

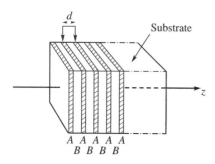

Figure 3.24.   A superlattice created by alternate growth of materials A and B in a periodic fashion. d is the period (repeat distance).

discrete energies are allowed in the dot makes the on/off nature of the conductance very precise. Single electron transport and Coulomb blockade are discussed in more detail in Chapter 9 and by Kastner (1992), and Beenakker and van Houten (1991).

3.5   Superlattices

A superlattice is a semiconductor structure created in such a way that periodicity is imposed on the system during growth (Fig. 1.1). This periodicity typically ranges from tens to thousands of ångstroms, so that it includes at least a few periods of the natural crystal structure, but is small enough so that quantum effects are important. These are thus *mesoscopic* structures. A simple example is a compositional superlattice, consisting of periodically alternating plane layers of, say, AlGaAs (*A*) and GaAs (*B*) (Fig. 3.24). The electronic periodicity is provided by the alternation of the conduction and valence band edges, as shown in Fig. 3.25. A superlattice creates a new kind of electronic raw material. The fact that it

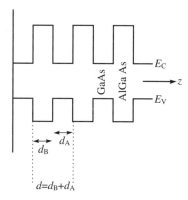

Figure 3.25.   The alternating band edges provide a periodic array of quantum wells for electrons (E_c) and holes (E_v).

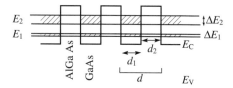

Figure 3.26.   Band edges and miniband energies for a typical superlattice, with alternating layers of GaAs (wells) and $Al_{0.11}Ga_{0.89}As$ (barriers). For this case $d_1 = d_2 = 90\,\text{Å}$. Miniband energies are $E_1 = 26.6$ meV and $E_2 = 87$ meV and respective bandwidths are $\Delta E_1 = 2.3$ meV, $\Delta E_2 = 20.2$ meV.

is a grown structure means that there is great freedom in creating a material with new sorts of electronic properties (Smith and Mailhiot, 1990).

3.5.1   Superlattices and Multi-quantum-wells

When a superlattice contains widely spaced quantum wells, so that electron tunnelling from well to well is essentially prohibited, one can treat the array as a set of isolated quantum wells. The energies and wavefunctions of electrons in each well will then be determined just by the properties of an individual well. Such arrays (multi-quantum-wells) are often used to enhance the signal obtainable from a single well. A true superlattice is a similar system, but with thinner barriers (more closely-spaced wells), so that there is electron tunnelling, and therefore good communication, from well to well. Some typical parameters are shown in Fig. 3.26.

There are two different ways of looking at a superlattice which are illuminating, and taken together give a good picture of the system. The first of these is the picture of a superlattice as a single bulk crystal with an additional modulation (periodicity) imposed on it. The second is the picture of a collection of equally-spaced quantum wells which are brought progressively closer together.

Crystal periodicity leads to the electron band structure observed in bulk crystals. Superlattice periodicity likewise gives a band structure, for the same reasons. However, since the superlattice spacing is *greater* than the crystal spacing, the superlattice **k**-space dimensions will be *smaller* than those of the crystal. This *new* band structure will be superimposed on the original bulk band structure, and will show up as a series of *minibands* and *minigaps* which will be superposed on the original band structure of the well material. These minibands and gaps result from *zone folding*: for a superlattice in which the wells contain M unit cells in the growth direction, there will be a new *super*lattice constant in this direction which is M times the atomic one. Associated with this new periodicity, there will be a new superlattice Brillouin zone with a size of $1/M$ times the crystal Brillouin zone. It is often the case that $1/M \ll 1$, so that many of the new *mini*-zones will fit into the original Brillouin zone. Thus, the band structure shows its new periodicity by breaking up into minibands and minigaps whose scale is determined by the size of the superlattice layers. Figure 3.27 illustrates this situation near the conduction band minimum of the well material, where the band structure can be taken to be parabolic to a good approximation.

One can also consider superlattices as collections of identical, isolated quantum wells which are brought closer together in such a way that they remain separated by equal distances. Each single well has its own set of discrete energy levels from electron confinement in the z-direction. If, for instance, one had started with only two such wells separated by a very large distance, there would be some common energy E_1, say, which an electron could have by being in one well or in the other (a two-fold degeneracy). As explained in elementary textbooks, as these quantum wells are brought closer together, interaction between the wells becomes possible, so that the levels are no longer degenerate, but have energies $(E_1 + \Delta)$ and $(E_1 - \Delta)$, where Δ increases from zero as the barrier width decreases. In place of a single degenerate level, one now has two levels, slightly split. The communication between wells causing this splitting comes about because of the fact that

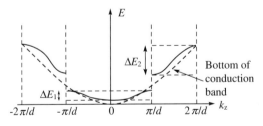

Figure 3.27.   Formation of minibands and minigaps near the conduction-band minimum of a direct-gap semiconductor. The minizone width $2\pi/d$ is typically much less than the Brillouin zone width $2\pi/a$, where a is a lattice constant.

an electron confined in one well can really be present in the barrier region as well, with a small probability. If the second well is near enough, the electron can also penetrate into the other well (tunnelling). A steady-state description is that of two possible eigenstates of a two-well problem. In each, the electron has equal probability of being in either well, and the wavefunction is symmetric or anti-symmetric, in the state with energy $E_1 - \Delta$ or $E_1 + \Delta$, respectively.

Since the energy splitting Δ is bigger if the amount of communication between wells is greater, one finds that the splitting of a higher-energy level $E_2 > E_1$ will be greater than that of a lower level. Higher-energy states have a higher probability of being present in the barrier regions: they have longer tails and thus can 'see' the presence of other wells more effectively. This in turn follows from the fact that, since the energy of such states is higher, the effective barrier $V - E$ through which they have to tunnel is lower.

Bringing many (N) identical wells together has a similar effect. In this case, a single-well level E_1 will be N-fold degenerate when the wells are far apart. As they are brought closer together in a uniform fashion, this degenerate level will split up into a *set* of closely spaced levels (N of them). This set of levels can be thought of as a continuum – a miniband. One will have other minibands, corresponding to each of the original levels of the single well. As in the two-well case, one expects higher-energy minibands to have a greater bandwidth than lower-energy ones. This can be a dramatic effect, as indicated in Fig. 3.26. One can say that level broadening increases as tunnelling becomes more effective.

Miniband broadening is also indicated in Fig. 3.28, which shows the effect on the density of states, for a set of square wells (as in Fig. 3.26) brought close together into a superlattice structure. Note the minibands (a–b), (c–d) and mini-gaps (b–c), and the increase of the bandwidth with miniband energy in Fig. 3.28.

3.5.2 Miniband Properties: The WKB Approximation

The WKB model can give good physical estimates for superlattice properties, if the system has relatively thick barriers. It also tends to give better than expected results for more general systems. We follow the discussion given by ter Haar (1964).

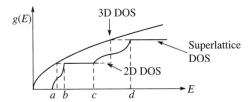

Figure 3.28. Superlattice density of states (DOS) in relation to that of a 3DEG and of a 2DEG in a square quantum well.

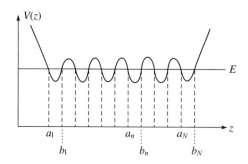

Figure 3.29.   Model of a superlattice potential, with N equally-spaced, equivalent quantum wells (edge effects should be negligible).

Consider the N-well system shown in Fig. 3.29. When one treats this problem in the WKB approximation, certain basic ingredients emerge:

$$\sigma = \int_{a_n}^{b_n} k(z)\,dz \tag{3.51}$$

(an integral over the allowed energy region in, say, the nth well), with

$$k = \frac{1}{\hbar}\left\{2m^*[E - V(z)]\right\}^{1/2} \tag{3.52}$$

and

$$\tau = \int_{b_n}^{a_{n+1}} \kappa(z)\,dz \tag{3.53}$$

(an integral over a single forbidden barrier region), with

$$\kappa = \frac{1}{\hbar}\left\{2m^*[V(z) - E]\right\}^{1/2} \tag{3.54}$$

Here, we want to describe the motion of an electron with energy E in a system with, say, the classical turning points a_n and b_n. We do this using the methods described in Section 3.3.1. As always, the confinement, and the requirement that the wavefunctions match at the classical turning points, results in quantization conditions, which may be written

$$\sigma = (n + \tfrac{1}{2})\pi + e^{-\tau}\cos\left(\frac{m\pi}{N+1}\right) \tag{3.55}$$

where $n = 0, 1, 2, \ldots$ and $m = 1, 2, \ldots, N$. (The derivation of (3.55) is rather tedious and will not be reproduced here; it is given in full by ter Haar (1964).) The first term on the right-hand side of (3.55) is the WKB approximation to the energy levels (labelled n) of an isolated well. The second term describes how these single levels split into N sublevels (labelled m), thus forming a miniband. This term is small because of the exponential in $-\tau$, so that the sublevels are closely spaced. Assuming such a small splitting of levels, one obtains from (3.55) the quantized energies

$$E_n = E_n^{(0)} + \frac{\hbar\omega}{\pi} e^{-\tau^{(0)}} \cos\left(\frac{m\pi}{N+1}\right) \tag{3.56}$$

Here, $E_n^{(0)}$ is the zero-order energy (coming from the first term on the right in (3.55)) and $\tau^{(0)}$ is calculated using the approximate energy $E_n^{(0)}$ in the definition (3.53). Finally, ω, the frequency of classical motion in a single quantum well, is defined by

$$\frac{\pi}{\omega} = \frac{m^*}{\hbar} \int_{a_n}^{b_n} \frac{dz}{k^{(0)}(z)} \tag{3.57}$$

where $k^{(0)}$ is calculated from (3.52) using the approximate energy $E_n^{(0)}$.

These results give, among other things, the WKB approximation for the miniband width, which may be read off from (3.56). Since $2\cos(\pi/N+1)$ is the maximum difference between the highest and lowest values of $\cos(m\pi/N+1)$, when $m = 1, \ldots, N$, we obtain the bandwidth

$$\Delta E_n = \frac{\hbar\omega}{\pi} e^{-\tau} \left[2\cos\left(\frac{\pi}{N+1}\right)\right] \tag{3.58}$$

From this equation one can see how the bandwidth depends upon the single-well energy level n from which that miniband arises. This dependence enters into the parameters ω and τ of equation (3.58). The latter is by far the most important dependence, because it appears as an argument of an exponential. Explicitly, one has that

$$\tau_n^{(0)} = \frac{1}{\hbar} \int_b^a \{2m[V - E_n^{(0)}]\}^{1/2} \, dz \tag{3.59}$$

It is evident that $\tau_n^{(0)}$ will be smaller (and the bandwidth greater) when E_n is greater (i.e. nearer to V). In this case the effective barrier $V - E$ is lower, and thus the integrand in (3.59) is smaller. As shown in Fig. 3.30, a second effect usually operates to increase the bandwidth with increasing n. The higher-energy states see a barrier which is also *thinner*, since at higher energies the limits b and a

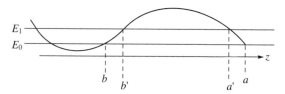

Figure 3.30.   A single well and barrier, for the superlattice of Fig. 3.29. E_0 and E_1 are unperturbed WKB energies for a single well with infinitely thick barriers.

are often closer together. Communication between wells is of course aided by thinner barriers.

We conclude this section by mentioning another approach that is sometimes used to derive typical superlattice properties. This is based on the Kronig–Penney model, which is a one-dimensional periodic array of square well potentials. We will not pursue this calculation here, but refer the reader to discussions in standard textbooks on quantum mechanics (Merzbacher, 1970).

3.5.3   Doping Superlattices

From what has already been said, it should be evident that a superlattice can be created by imposing a new periodicity of *any* sort on a semiconductor. One such example is the *doping superlattice*, first proposed by Döhler (1972) (see also Ruden and Döhler, 1983). The idea is to introduce a new periodicity into a semiconductor by doping it selectively, first by acceptors, then by donors, in repeated plane layers, as it is grown. The ionized impurities create repeated layers of negative (*n*) and positive (*p*) space charge in the conductor (Fig. 3.31), often separated by intrinsic (*i*), or undoped, layers. The periodically repeating electric fields which result create a superlattice in the semiconductor which, for obvious reasons, is called a *n-i-p-i* superlattice.

Figure 3.32 shows how such fields create a superlattice. The electric field of the ionized donors creates a parabolic quantum well for electrons within the *n*-doped region, as described in Section 3.3.2, while a parabolic well for holes (an inverted

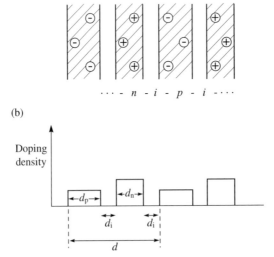

Figure 3.31.   (a) Ionized impurities introduced in alternate layers in a semiconductor. (b) Possible doping profile leading to the space charge shown in (a).

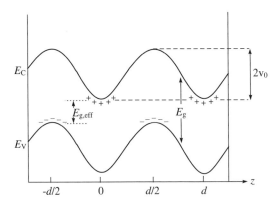

Figure 3.32. Conduction and valence band edges in a semiconductor with *n-i-p-i*-type doping. $E_{g,eff}$ is the effective band gap.

parabola) is created in the *p*-type regions. The charge density of the impurities may be written as

$$\varrho_0(z) = e[n_D(z) - n_A(z)] \tag{3.60}$$

where n_D is the number of donors per unit volume, n_A is the number of acceptors per unit volume and z is the growth direction. The band-edge potential $V_0(z)$ is easily calculated from Poisson's equation:

$$V_0(z) = \left(\frac{e^2}{2\epsilon_r\epsilon_0} n_D\right) z^2 \tag{3.61}$$

for $0 \le |z| \le \frac{1}{2}d_n$ (equation (24)), and

$$V_0(z) = 2v_0 - \left(\frac{e^2}{2\epsilon_r\epsilon_0} n_A\right)(\tfrac{1}{2}d - |z|)^2 \tag{3.62}$$

for $\frac{1}{2}d - |z| \le |z| \le \frac{1}{2}d_p$, where we have set $V = 0$ at the origin, which is taken to be the middle of an *n*-type layer, and spacings d_p etc. are defined in Fig. 3.31. In the intrinsic regions, where there are no ionized donors, the potential must be a linear function of z. We assume that this basic pattern is repeated indefinitely. The quantity $2v_0$ in equation (3.62) is the amplitude of the band-edge modulation shown in Fig. 3.32. Since the conduction band edge must be continuous (as shown), the parabolic regions in (3.61) and (3.62) must be joined by straight lines. This determines the depth of the modulation:

$$2v_0 = \frac{e^2}{2\epsilon_r\epsilon_0}[n_D(\tfrac{1}{2}d_n)^2 + n_A(\tfrac{1}{2}d_p)^2 + n_D d_n d_i] \tag{3.63}$$

Note that if the total number of ionized donors in one *n*-layer equals the number of ionized acceptors in a *p*-layer, the electrons that have been released by the donors will all reside on acceptors if the system is in its ground state. Thus, the potential that a free electron would feel is indeed given correctly by equations (3.61)–(3.63). There will be no free electrons in the wells to modify this potential with self-consistent effects.

The WKB approximation can be applied to the parabolic parts of *n-i-p-i* quantum wells to predict the superlattice energies. In particular, the miniband splittings and miniband widths can be calculated, as described in Section 3.5.3. The unperturbed energies (those for isolated wells), in particular, are given by

$$E_n^{(0)} = (n + \tfrac{1}{2})\hbar\omega \tag{3.64}$$

where, as before, ω given by (3.57), the miniband splittings by (3.56), and the miniband widths by (3.58). The calculation of these quantities is straightforward, but since the resulting expressions are rather unwieldy, they will not be reproduced here.

3.5.3.1 Delta-doped *n-i-p-i* s

A delta-doped *n-i-p-i* is essentially one in which the dopant layers are each only one atom wide. The resulting band-edge diagram is shown in Fig. 3.33, for *n* and *p* layers with equal numbers of ionized impurities. In this case, the conduction band edge is given by

$$V_0 = \frac{e^2 D}{2\epsilon_r\epsilon_0}|z|, \qquad |z| \leq \tfrac{1}{2}d, \dots \text{ and repeating} \tag{3.65}$$

(see (3.30) and (3.31)), where *D* is the number of ionized donors (or acceptors) per unit area (equation (3.29)), and the superlattice amplitude is given by

$$2v_0 = \tfrac{1}{2}d\left(\frac{e^2 D}{2\epsilon_r\epsilon_0}\right) \tag{3.66}$$

The WKB approximation can be used here, too, to obtain predictions for the properties of this superlattice, in the way already described; in particular the miniband parameters $\omega^{(0)}$ and $\tau^{(0)}$ are obtained in an analogous manner.

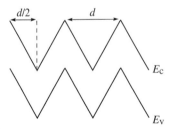

Figure 3.33.   Conduction and valence band edges for a *n-i-p-i* superlattice with equally-spaced *n* and *p* delta layers of equal strength.

3.5.3.2 Compositional and Doping Superlattices

For electronic devices, one of the most important semiconductor parameters is the value of the energy gap between the valence band maximum and the conduction band minimum. Many applications, particularly those that depend on optical properties, depend crucially on the value of the fundamental gap. One would like a wide range of effective band gaps, $E_{g,eff}$, for device use. The fundamental gap E_g is indeed modified in compositional quantum wells and superlattices, as is shown in Fig. 3.34. Once such a system is grown, its effective gap will be *greater* than E_g, and will moreover be *fixed*.

To obtain effective band gaps *smaller* than E_g, one must turn to something like a *n-i-p-i* system. Fig. 3.35 shows how different effective band gaps can be obtained

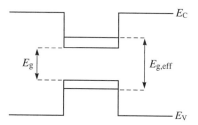

Figure 3.34.   Modification of the fundamental gap in an AlGaAs-GaAs quantum well.

(a)

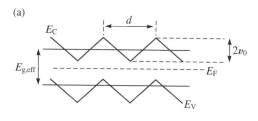

(b)

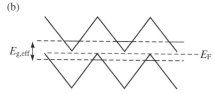

(c)

Figure 3.35.   Delta doping superlattices with (a) weak, (b) moderate, and (c) strong doping densities.

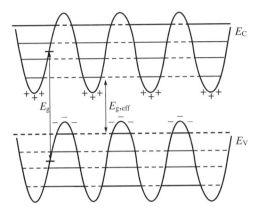

Figure 3.36.   A doping superlattice in its ground state.

using the same host material, with the same periodicity, but with different doping densities. From equation (3.66), the superlattice modulation is proportional to the product Dd. A small value of D results in a weak modulation of the band edge (Fig. 3.35(a)), and gives an effective gap not much smaller than that for the bulk semiconductor. The effective gap decreases, however, as doping strength D increases (b); and at high doping, as in (c), a semi-metal can result. The same principles hold for other doping superlattices.

Figure 3.35 shows the case of a doping superlattice in which there is complete compensation: the numbers of donors and acceptors are equal. Since the system is assumed to be in its ground state, all donated electrons reside on acceptors. Once such a system is grown, however, there is also freedom to *tune* it. Figure 3.36 shows such a many-subband doping superlattice in its ground state, with no free carriers. Electrons and holes are confined to separate regions of the superlattice (this is sometimes said to be a system with an 'indirect gap in real space'). One can now excite this system, for instance with light of the right frequency, to create electron–hole pairs. Free carriers have now been introduced into the system. Electrons will be attracted to the quantum wells in the n-layers, and holes to the p-layers. Since these regions are well separated in space, however, the electron and hole wavefunctions will be well separated. This small overlap means that electron–hole recombination will be very slow. Thus there will be an appreciable time during which these free carriers will reside in their respective wells, where they will screen the space charge already present. The net effect will be to reduce the superlattice modulation, and thus increase the effective band gap (Fig. 3.37). Devices based on these ideas are discussed in Chapter 7.

3.5.4   Other Types of Superlattices

Compositional superlattices of the GaAs/AlGaAs type, with low Al composi-tions, are called *Type I superlattices*. Materials such as the InAs-GaSb system

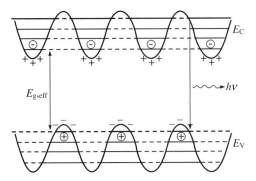

Figure 3.37.   A doping superlattice in an excited state.

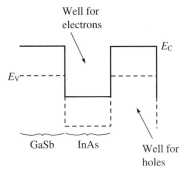

Figure 3.38.   Schematic band-edge diagram for a Type II superlattice.

offer a different possibility: the *Type II superlattice*. In Type II superlattices, electrons are confined to one material and holes to the other (Fig. 3.38).

Strained-layer superlattices are compositional superlattices in which the constituent materials are not perfectly lattice-matched to each other. If the materials are not too different, and if the layers of each material are not too thick, good growth of one material on another, in layers, is still possible. A given layer will then be compressed, or extended, in the plane perpendicular to the growth direction, by atomic forces arising from the layer onto which it is trying to grow. The effects of strain can also be useful. It breaks the degeneracy of hole states, and changes the band structure in other ways, all of which offer new possibilities for device development. Such devices are discussed in Section 8.4 and in Section 10.2.3.

Any way of imposing an artificial periodicity on a semiconductor system can in principle make a superlattice. The periodicity does not need to be imposed on the growth direction of a low-dimensional system. One superlattice of interest, for instance, is made by shining a laser interference pattern onto an already-created 2DEG. Extra electrons are generated (for instance in the surrounding medium) by the light, in a periodic way, thus creating a periodic electric field which modulates the 2DEG in its own plane. A similar way of modulating an existing 2DEG is with the help of acoustic waves. Here, the modulation is caused by the periodic electric field which is generated by the piezoelectric effect.

Still another kind of superlattice can be created using a regular array of closely spaced quantum dots. In this case, an artificial periodicity will be created in at least two different directions. These possibilities and many more exist to make life interesting for those who try to understand the physics as well as for those who invent devices based upon them.

EXERCISES

1. Consider the two $GaAs/Al_xGa_{1-x}As$ compositional superlattices, where $x \leq 0.3$, shown in Fig. 3.39.

(a) In each superlattice, the well width equals the barrier width. Neglecting miniband spreading, what differences would you expect to find between the energy levels of the first and second system (assume that self-consistent effects and effective-mass corrections can be neglected)?

(b) Suppose that the AlGaAs layers of a superlattice of this type are lightly and uniformly n-doped, while the GaAs layers remain undoped. Describe the distribution of the donated electrons in equilibrium. Sketch the potential felt by a single electron, including self-consistent effects, in both the well and barrier regions. Include all important labels in the diagram and note the key physical input and assumptions used.

2. For the undoped superlattices in Exercise 1, describe qualitatively, but in detail, the differences you would expect to find between (a) the wavefunctions and (b) the miniband spreading in system (a) compared with system (b). Justify your answers, e.g. by referring to relevant approximate results.

3. Consider a superlattice made from a large number of square quantum wells, each of depth $5E_0$ and width L. The barriers have width $B \gg L$, and E_0 is the lowest allowed energy of an electron measured from the conduction band edge in an isolated infinite square well of width L.

(a) Give a brief description (words and a sketch) of the miniband energies and band widths one might expect in this system, noting the reasoning used.

(b) *How* and *why* would this picture change if $B/L \leq 1$? Use qualitative arguments based on a WKB picture.

(a)   (b)

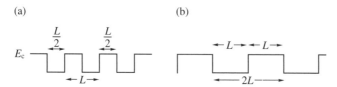

Figure 3.39.   Diagram for Exercise 1.

4. An electron of energy E is confined in the z-direction by a potential well $V(z)$ in a semiconductor with effective mass m^*. The potential $V(z) = \infty$ if $z \leq 0$ and the right-hand (classical) turning point of the motion of this electron is at $z = d$. Using the WKB approximation, write down the quantization condition which must be satisfied by the electron's allowed energies E_n in this quantum well. How is E_n related to the total energy of the electron?

5. A direct-gap semiconductor is bounded by an infinite potential barrier at $z = 0$. It is delta-doped, with a uniform plane monolayer of donors in the plane $z = a$ and a similar layer of acceptors in the plane $z = 2a$. The doping density of each layer is N atoms per unit area. Assume as a first approximation that the electric field is strictly zero for a small positive z, and that there are no surface states at $z = 0$. Assume also that the donors and acceptors are fully ionized, and that there are no free carriers or other charges in the system.

(a) Why should dV/dz go to zero as $z \to \infty$?
(b) Sketch and label the band-edge diagram as a function of z for this system.
(c) Find the potential energy difference, $\Delta = E_C(2a) - E_C(a)$, for this system, where $E_C(z)$ is the conduction band energy at position z.
(d) Give physical reasons why and under what conditions one should expect a two-dimensional electron gas in such a system.

6. A semiconductor sample has been doped so that it confines electrons within a parabolic potential well in the z-direction, $V(z) = Az^2$. In parts (a)–(d) below, ignore self-consistent (many-electron) effects.

(a) Express the quantized energies E_n for an electron confined in the z-direction by this potential well in terms of the well parameter A and three-dimensional doping density $n_D(z)$.
(b) Draw a graph of the two-dimensional density of states $g(E)$ as a function of E for this system. Include at least the first three energy levels E_0, E_1, E_2.
(c) Suppose the Fermi energy E_F lies exactly halfway between E_1 and E_2 (second and third energy levels). How many states per unit area in x, y are filled?
(d) Suppose such a system had been created by growing a doped slab of width W within the semiconductor (W is fairly wide). Find an expression for the doping density necessary to provide the number of electrons obtained in part (c), above.

7. A doping superlattice consists of alternating slabs of donor- and acceptor-doped GaAs. The width of each slab is ℓ. The three-dimensional doping density in each layer is constant, and of magnitude n_0 (Fig. 3.40).

(a) What is the period of the superlattice?
(b) Assuming all donors are fully ionized, sketch the conduction band-edge energy V in the growth direction z, assuming that $V = 0$ at $z = 0$.

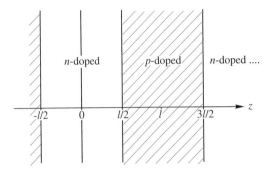

Figure 3.40.   Diagram for Exercise 7.

(c) Obtain formulae for V in terms of n_0, e, z and the dielectric constant of GaAs for z between $-\frac{1}{2}\ell$ and $\frac{1}{2}\ell$, and for z between $\frac{1}{2}\ell$ and $\frac{3}{2}\ell$, assuming that $V = 0$ at $z = 0$.

8. (a) For the superlattice described in Exercise 7, what is the amplitude of modulation V_0 in terms of n_0 and ℓ (that is, the difference between the maximum and minimum for the conduction band-edge energy)?

(b) Suppose n_0 and ℓ have been chosen so that the first miniband has energies less than $\frac{1}{2}V_0$. Use the WKB approximation to obtain an expression for the lowest allowed energy level E_0 for an electron in terms of n_0 and the electron effective mass m^*, considering the n-doped region as an isolated quantum well.

(c) Write down the WKB expression for the bandwidth ΔE of the lowest miniband for a superlattice with nine periods, and draw a diagram showing the relevant limits of integration.

9. Suppose that the donor and acceptor doping densities in the superlattice of Exercises 7 and 8 were increased from n_0 to $2n_0$.

(a) How would E_0 change?

(b) How would V_0, the amplitude of modulation, change?

(c) Give details of ways in which the bandwidth of the lowest miniband would be affected by this change. Do you expect the bandwidth to increase or decrease? Why?

10. A plane slab of GaAs of width L is grown in bulk $Al_{0.3}Ga_{0.7}As$. On either side of the GaAs slab, at distances d, are delta-doped planes of donors, each with a 2D doping strength N_δ. There is no other doping (Fig. 3.41).

(a) Assume that all donors ionize. Why will some donor electrons go into the GaAs?

(b) Suppose all donated electrons are trapped in the GaAs quantum well and that E_F is pinned to the conduction band edge at $z = \pm\infty$. Draw the con-

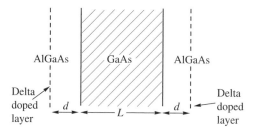

Figure 3.41.   Diagram for Exercise 10.

duction band-edge diagram for this structure. Include self-consistent effects. Show and label

(i) the band-edge behaviour far from the well

(ii) the Fermi energy E_F

(iii) the band-offset energy Δ at the GaAs/AlGaAs interfaces

(iv) the parts of the diagram which show self-consistent effects.

(c) Explain the reasons which lead in part (b) to

(i) the shape of the potential $V(z)$ in the well

(ii) the shape of $V(z)$ for $z \to \pm\infty$

(iii) the value of E_F in the well

(iv) the shape of $V(z)$ in the region between the delta layers and the well.

(d) Now approximate the GaAs quantum well by an ideal infinite square well of width $L = 100\,\text{Å}$. Given that the Fermi energy in the well is $E_F = 2E_0$, where E_0 is the energy of the lowest quantum level in the well, find the 2D electron density N_s and the Fermi energy (in meV) in the well.

(e) Given N_s, the 2D electron density in the well, could you estimate E_F by treating the well as an ideal square well? Explain. What might make such an estimate better or worse?

(f) Find an explicit formula for E_F in terms of the band offset Δ, doping strength N_δ, and relative dielectric constant ε_r for the system of part (b). The dielectric constant can be assumed uniform throughout the structure.

References

T. Ando, A. B. Fowler and F. Stern, *Rev. Mod. Phys.* **54**, 437 (1982).

N. W. Ashcroft and N.D. Mermin, *Solid State Physics* (Holt, Rinehart, and Winston, New York, 1976).

L. Bányai and S. W. Koch, *Semiconductor Quantum Dots* (World Scientific, London, 1993).

C. W. J. Beenakker and H. van Houten, in *Solid State Physics* **44**, H. Ehrenreich and D. Turnbull, eds. (Academic Press, London, 1991), pp. 1–228.

G. H. Döhler, *Physica Stat. Sol. (b)* **52**, 79 (1972).

L. Esaki and L. L. Chang, *Magnetism and Mag. Materials* **11**, 208 (1979).

J. J. Harris, J. A. Pals and R. Woltjer, *Rep. Prog. Phys.* **52**, 1217 (1989).

L. Ioriatti, *Phys. Rev. B* **41**, 8340 (1990).

M. A. Kastner, *Rev. Mod. Phys.* **64**, 849 (1992).

R. W. Keyes, *Comments Solid State Phys.* **7**, 53 (1976).

J. P. Kotthaus, in *Granular Nanoelectronics* (Plenum, New York, 1991) pp. 85–102.

E. Merzbacher, *Quantum Mechanics* (Wiley, New York, 1970).

P. Pals, PhD Thesis (University of London, 1995, unpublished).

P. Ruden and G. H. Döhler, *Phys. Rev. B* **27**, 3538 (1983).

Ch. Sikorski and U. Merkt, *Phys. Rev. Lett.* **62**, 2164 (1989).

D. L Smith and C. Mailhiot, *Rev. Mod. Phys*, **62**, 173 (1990).

D. ter Haar, ed., *Selected Problems in Quantum Mechanics* (Infosearch Ltd, London, 1964).

B. J. van Wees, H. van Houten, C. W. J. Beenakker, J. G. Williamson, L. P. Kouwenhoven, D. van der Marel and C. T. Foxon, *Phys. Rev. Lett.* **60**, 848 (1988).

D. A. Wharam, T. J. Thornton, R. Newbury, M. Pepper, H. Ahmed, J. E. F. Frost, D. G. Hasko, D. C. Peacock, D. A. Ritchie and G. A. C. Jones, *J. Phys. C* **21**, L209 (1988).

Phonons in Low-dimensional Semiconductor Structures

M. P. Blencowe

4.1 Introduction

One of the central themes of this book is how reducing the size of semiconductor structures down to mesoscopic and smaller scales brings the quantum wave nature of electrons into play, resulting in electronic and optical properties which are markedly different from those of bulk semiconductors. Chapters 6–10 describe how these properties can be exploited in device applications. But one of the key challenges facing physicists and engineers is how to make devices operate at room temperature. The main obstacle to achieving this goal is the unavoidable presence of phonons, the quantum vibrations of atoms making up a solid, and their ability to scatter electrons.

Phonons would thus appear to occupy an uncomfortable position in the study of low-dimensional structures, with an understanding of their properties required solely for the purpose of finding ways to reduce their interaction with electrons. As we shall see in this chapter, however, the physics of phonons in low-dimensional structures is sufficiently fundamental and non-trivial to be of interest in its own right. Just as for electrons, phonons can be confined within heterostructures and we would like to know how the dynamics of low-dimensional phonons differs from that of bulk phonons. We would also like to understand the effects of dimensionality on the electron–phonon interaction and hence such electron transport properties as the phonon-scattering-limited mobility.

In fact, finding ways to reduce the electron–phonon interaction is not the only reason for investigating phonons in low-dimensional structures. Being the vibrations of atoms making up a structure and given their interaction with electrons, phonons have proved to be an effective probe of the electronic and structural properties of low-dimensional semiconductors. Two such probe techniques are acoustic phonon pulse spectroscopy and Raman spectroscopy. Phonon pulse spectroscopy relies on the fact that acoustic phonons with wavelengths on the order of several hundred ångstroms can travel for relatively large distances in semiconductor crystals without scattering. By detecting the flux of acoustic phonons emitted from an electron gas heated above the lattice temperature (Rothenfusser *et al.*, 1986) or, alternatively, measuring the response (e.g. conductance) of an electron gas due to the interaction with an incident beam of acoustic phonons (Kent *et al.*, 1988), information about the electron gas can be obtained.

For example, phonon pulse spectroscopy can be used to study the electron current distribution of a two-dimensional electron gas in the quantum Hall regime (McKitterick *et al.*, 1994) and more complicated electron states, such as those occuring in the fractional quantum Hall regime (Digby *et al.*, 1996; see Chapter 5 for a basic description of the quantum Hall and fractional quantum Hall effects) and two-dimensional excitons (Akimov *et al.*, 1996). Acoustic pulse methods have also been used to study the materials properties of semiconductors, such as surface and interface roughness (Kozorezov *et al.*, 1996).

Raman spectroscopy has proven to be an extremely powerful probe of the dynamics of phonons in heterostructures (for a review, see Cardona and Güntherodt, 1989). This method takes advantage of the fact that energy conservation requires a photon which absorbs or emits a phonon to suffer a change in its frequency. This frequency change is called the *Raman shift*. By measuring the intensity versus Raman shift of laser light which is inelastically scattered from a solid, the energies and, hence, the frequencies of the allowed phonon modes of the solid can be measured. Since the phonon frequencies are determined by the interatomic forces and atom masses that comprise a material, Raman spectroscopy can be used to infer various materials properties of heterostructures.

However, before going on to study the various phonon spectroscopies and their interpretation in terms of heterostructure properties, we should first learn about the basic physics of phonons and electron–phonon interactions in heterostructures. This is the purpose of the present chapter.

4.2   Phonons in Heterostructures

In the classical approximation, the vibrations of the atoms making up a crystalline solid are most conveniently described by the function $\mathbf{u}_j(\mathbf{R}, t)$ which gives the displacement from the equilibrium position at time instant t of the jth basis atom in the unit cell located at $\mathbf{R} = n_1\mathbf{a}_1 + n_2\mathbf{a}_2 + n_3\mathbf{a}_3$. The displacement is specified by its magnitude and direction and thus is a vector. We have used boldface here to indicate vectorial nature, e.g. $\mathbf{u}_j \equiv (u_{jx}, u_{jy}, u_{jz})$. The displacement of a given atom as a function of time will be governed by its interaction with all the other atoms making up the solid. Because of the non-linear nature of the full equations of motion and the very large number of atoms involved, it is in practice impossible to find solutions for the atom displacements without first making some simplifying approximations. The most common approximation is to expand the interaction potential energy in the displacements and keep only terms to quadratic order. (Note that linear order terms vanish since the displacements are defined with respect to the equilibrium positions of the atoms.) The equations of motion are then linear in the displacements and we have the important result that any solution can be expressed as a sum of solutions with the property that all atoms vibrate at the same frequency:

$$\mathbf{u}_j(\mathbf{R}, t) = \mathrm{Re}\left[\sum_\alpha c_\alpha e^{-i\omega_\alpha t}\mathbf{u}_{j\alpha}(\mathbf{R})\right] \tag{4.1}$$

Because of the latter property, this is called the *harmonic approximation*.

Thus, to understand the classical dynamics of atomic vibrations in a crystalline solid, it is sufficient to find the single frequency solutions $e^{-i\omega_\alpha t}\mathbf{u}_{j\alpha}(\mathbf{R})$, the so-called *normal modes*. The quantum dynamics and interpretation in terms of vibration quanta – phonons – then follow directly. For example, the total energy of a crystal in a given quantum state is

$$E = \sum_\alpha \hbar\omega_\alpha(N_\alpha + \tfrac{1}{2}) \tag{4.2}$$

where N_α is the number of phonons present of mode-type α.

This picture of the dynamics, involving the concepts of classical normal modes and quantum phonons, is valid provided the higher order non-quadratic terms in the potential energy expansion – the anharmonic terms – are much smaller than the quadratic terms. Physically, this means that changes in the distances between neighbouring atoms due to their vibrations are small compared with their equilibrium separations. This is the case as long as we are well below melting temperatures, as we indeed are for the low-dimensional structure physics and applications discussed in this book. Note, however, that the presence of even small anharmonic terms will mean that these phonon modes have only a finite lifetime.

4.2.1   Superlattices

To gain some idea of the nature of phonon modes occurring in heterostructures, we shall focus our discussion on the modes of a special type of heterostructure – the superlattice. An account of the epitaxial growth and electronic properties of superlattices can be found in Chapters 1 and 3, respectively.

As an illustrative example, we consider a superlattice formed from alternating GaAs and AlAs slabs. Both GaAs and AlAs crystals have zincblende structure (see, e.g., Chapter 4 of Ashcroft and Mermin, 1976) and thus a superlattice grown on an (001) face of a GaAs crystal substrate will comprise alternating layers of atoms of a single species with a monolayer spacing equal to half the lattice constant. The lattice constant of GaAs is 5.65 Å, while that of AlAs is 5.62 Å, so that the respective monolayer spacings are 2.83 Å and 2.81 Å.

One class of modes which occurs in such a superlattice involves the vibration of entire (001) planes of atoms in the direction normal to the planes, i.e. the [001] direction. These are called *longitudinal* modes. Because the atoms in a given monolayer are all moving in unison, we can model the superlattice by using a one-dimensional system. Further simplifications follow if we make the harmonic approximation and include only nearest-neighbour interactions between atoms.

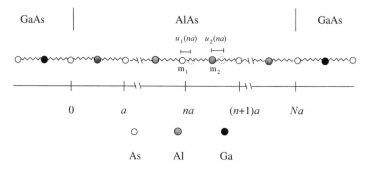

Figure 4.1.   Linear chain model for [001] longitudinal modes in a GaAs/AlAs superlattice. The atoms are shown in their equilibrium positions.

The resulting dynamics is identical to that of a mechanical system consisting of a chain of masses connected with springs (Fig. 4.1). As we shall now see, this linear chain model gives a reasonably good description of the [001] longitudinal modes. Our discussion follows that of Colvard *et al.*, (1985).

In the GaAs or AlAs slabs, the equations of motion take the form

$$m_1 \frac{d^2 u_1}{dt^2}(na, t) = -\kappa\{2u_1(na, t) - u_2(na, t) - u_2[(n-1)a, t]\} \tag{4.3}$$

$$m_2 \frac{d^2 u_2}{dt^2}(na, t) = -\kappa\{2u_2(na, t) - u_1[(n+1)a, t] - u_1(na, t)\} \tag{4.4}$$

where a is the monolayer spacing (i.e. the unit cell length), κ is the force constant and $u_j(na, t)$ denotes the displacement from equilibrium along the line of the chain at time t of one of the two atoms of mass m_1 and m_2 in the unit cell located at na. To obtain solutions for the entire chain, these equations must be supplemented by certain conditions on the displacements at the boundaries between GaAs and AlAs slabs. However, before we consider these full solutions, let us first address the simpler problem of a single slab with periodic boundary conditions. Substituting the trial mode solution $u_j(na, t) = A_j \exp[-i(\omega t - qna)]$, $q = 2\pi/\lambda$, into equations (4.3) and (4.4), we find that we can have a non-trivial solution (i.e. $A_j \neq 0$) only if the determinant of the coefficients of the two unknowns A_1 and A_2 vanishes. This results in the following relation between the angular frequency ω and wavevector q:

$$\omega^2 = \frac{\kappa}{m_1 m_2}\{m_1 + m_2 \pm [m_1^2 + m_2^2 + 2m_1 m_2 \cos(qa)]^{1/2}\} \tag{4.5}$$

The mode angular frequency dependence on wavevector is called the *dispersion relation* and the two roots in (4.5) are the *branches* of the dispersion relation. Before we can display the dispersion relations for GaAs and AlAs, we must determine their force constants. This can be done, for example, by measuring ω

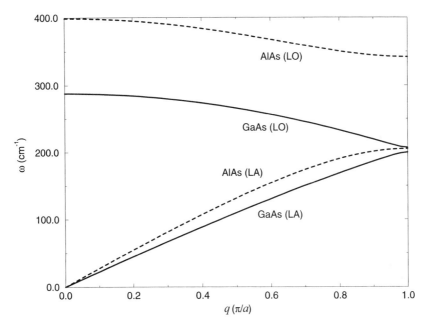

Figure 4.2.   Longitudinal [001] acoustic (LA) and optical (LO) phonon dispersion curves for GaAs (solid line) and AlAs (dashed line).

at $q = 0$ using Raman spectroscopy and then fitting the positive root at $q = 0$ in (4.5) to the measured value of ω. For GaAs, this gives the value $\kappa = 90.7\,\mathrm{N/m}$, while for AlAs we obtain $\kappa = 95.4\,\mathrm{N/m}$. In Fig. 4.2 we show the resulting dispersion curves for GaAs and AlAs. The dispersions are given as solid lines to guide the eye. In fact, they are a set of discrete points at $q = \pm 2\pi n/Na$, for $n = 0, 1, 2, \ldots$, where N is the number of monolayers. Altogether there are $2N$ distinct modes, N for each branch. Following the usual convention, the angular frequency has been divided by the factor $2\pi c$, where c is the speed of light in vacuum $(3 \times 10^{10}\mathrm{cm\,s^{-1}})$. This has the advantage that the Raman shift wavenumber corresponding to a given phonon mode can be inferred directly from the plot in Fig. 4.2. To convert to electron-Volts $(1\mathrm{eV} = 1.6 \times 10^{-19}\,\mathrm{J})$, the wavenumber value is multiplied by the factor $1.24 \times 10^{-4}\mathrm{eV\,cm}$. For a mode in a GaAs or AlAs positive root branch, we see that the Raman shift will be in the infra-red range. For this reason, the positive root branch is called the *optical branch*. The negative root branch extends down to zero frequency as q goes to zero and thus is called the *acoustic branch*.

If we expand the dispersion relation (4.5) with respect to q and keep only the lowest-order non-vanishing terms, we find for the optical branch:

$$\omega = \left[\frac{2\kappa(m_1 + m_2)}{m_1 m_2} \right]^{1/2} \tag{4.6}$$

and for the acoustic branch:

$$\omega = qa\left[\frac{\kappa}{2(m_1+m_2)}\right]^{1/2} \tag{4.7}$$

Thus, as q approaches zero, the optical *group velocity* ($v_g = d\omega/dq$) vanishes, while the acoustic group velocity becomes constant and non-zero, coinciding with the acoustic *phase velocity* ($s = \omega/q$). To determine how the atoms move for small q, we require the remaining part of the solution to equations of motion (4.3) and (4.4), namely the relation between the constants A_1 and A_2:

$$A_2/A_1 = -\frac{\kappa(1+e^{iqa})}{m_2\omega^2 - 2\kappa} \tag{4.8}$$

Substituting expression (4.6) for ω into (4.8), we find that for q tending to zero, the optical modes approach a standing wave with the Ga and As (or Al and As) atoms vibrating 180° out of phase: $A_2/A_1 = -m_1/m_2$. For the acoustic modes all the atoms vibrate in phase.

Substituting the force constant values given above into (4.7), we have a prediction for the small-q group velocity of longitudinal acoustic (LA) phonons propagating in the [001] direction. For example, for GaAs we obtain the value 4040 m s^{-1}, whereas the actual value is about 4770 m s^{-1} and therefore the linear chain model prediction is out by about 15%. This is quite reasonable given the simplicity of the model. A more accurate linear chain model will take into account interactions between next-nearest neighbour atoms as well, with the new unknown force constants determined by fitting to additional measured frequencies. The existence of a non-negligible acoustic group velocity gives rise in heterostructures to a range of acoustic phonon transport phenomena (Section 4.2.2) and, as mentioned in the Introduction, makes possible the field of acoustic phonon pulse spectroscopy.

Our discussion so far has been about the longitudinal [001] modes of a single GaAs or AlAs slab. Let us now consider the longitudinal [001] modes of the full GaAs/AlAs superlattice structure. An initial, basic understanding of the superlattice modes can be gained just by comparing the single-slab GaAs and AlAs dispersion relations (Fig. 4.2). The acoustic branches of AlAs and GaAs overlap in frequency and therefore the acoustic modes will extend throughout the entire superlattice. On the other hand, the AlAs and GaAs optical branches do not overlap in frequency – a consequence of the large difference between the Ga and Al atom masses – and hence we expect the superlattice optical modes will be confined either to the AlAs or GaAs slabs. An estimate of the extent to which an optical mode is confined to a given slab can be obtained by considering the imaginary wavevector solutions to the single-slab equations of motion (4.3) and (4.4). The details are worked out in Exercises 1 and 2 at the end of this chapter. We find that optical modes confined to the GaAs slabs extend between about 1 to 4.5 monolayers into the AlAs slabs, while AlAs confined modes extend between about 0.3 to 0.4 monolayers into the GaAs slabs, where the lower and upper limits are for zone centre ($q = 0$) and zone boundary ($q = \pm\pi/a$) modes, respectively. Thus the superlattice optical modes are well-confined.

The superlattice dispersion relation and mode solutions are determined by first constructing solutions in a single superlattice unit cell (which comprises an AlAs and GaAs slab) and then using Bloch's theorem to extend the solutions throughout the whole superlattice. The details can be found in Colvard *et al.* (1985). In Fig. 4.3, we show the longitudinal [001] dispersion relation for a superlattice with 4-monolayer GaAs slabs and 3-monolayer AlAs slabs – a (4, 3) superlattice. Also shown for comparison are the single-slab dispersion relations. A single slab with periodic boundary conditions has period a, the monolayer spacing, while an (M, N) superlattice has a larger period $d = (M + N)a$ (where we regard the AlAs and GaAs monolayer spacings to be the same). Thus, the superlattice Brillouin zone (boundaries at $q = \pm\pi/d$) is smaller than the slab Brillouin zone (boundaries at $q = \pm\pi/a$). Notice that the part of the dispersion curve below about $200\,\text{cm}^{-1}$ can be approximately obtained by 'folding' the bulk acoustic phonon branches into the smaller superlattice zone. For this reason, this part of the dispersion curve is called the *folded acoustic phonon branch*. A Raman spectrum of a superlattice is shown in Fig. 4.4, which gives clear evidence of folded LA phonons in the form of double peaks occurring at the predicted frequencies.

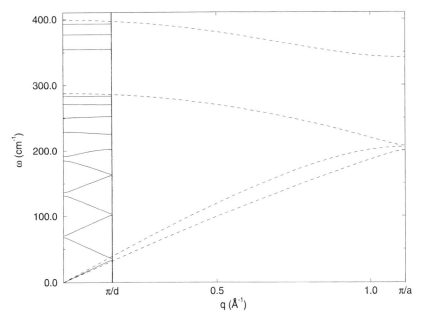

Figure 4.3.   Dispersion curve for [001] longitudinal phonons of a superlattice comprising alternating 4-monolayer slabs of GaAs and 3-monolayer slabs of AlAs. Also shown for comparison are the dispersion curves for bulk GaAs and AlAs (dashed lines).

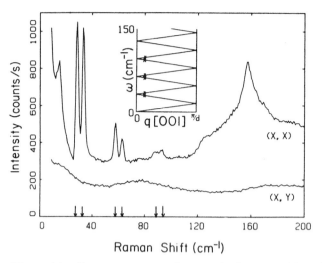

Figure 4.4.   Raman spectrum of a superlattice comprising alternating 42 Å GaAs slabs and 8 Å $Al_{0.3}Ga_{0.7}As$ slabs (Colvard *et al.*, 1985). The inset shows a model calculation of the folded acoustic branch as well as the location of the observed peaks. (Courtesy A. C. Gossard)

Several other types of mode can occur in the GaAs/AlAs superlattice. For example, the (001) planes of atoms can also freely vibrate in the [110] direction, *transverse* to the direction of propagation. These transverse modes can similarly be approximated using a linear chain model (see, e.g., Barker *et al.*, 1978). Another type of mode which can arise are the *interface* modes which are confined to the regions of the interfaces between the GaAs and AlAs slabs. These various mode-types will also occur in GaAs/AlAs superlattices grown on other GaAs substrate faces and also in superlattices made from other alloy materials (see, e.g., Cardona, 1991).

Although heterostructures such as quantum wells, wires and dots lack the layer periodicity of the superlattice, the basic features of the phonon modes are the same. For example, in a quantum well formed by sandwiching a GaAs slab between two AlAs slabs, the optical modes will be confined either to the GaAs or AlAs slabs, while the majority of the acoustic modes will extend throughout the whole structure. Interface modes will also occur.

The linear chain model used here is a *phenomenological* model since the force constants are determined by experiment. Other phenomenological models have also been developed to describe phonon modes in heterostructures. Discussion and references for superlattices can be found in, for example, Chamberlain *et al.* (1993) and for quantum wells, in Constantinou and Ridley (1994). Distinct from the phenomenological models are the *ab initio*, or *microscopic*, models which attempt to reproduce quantitatively the phonon dispersion relations and displacement functions starting from the fundamental quantum Hamiltonian of the interacting ions and electrons making up the solid. An introduction to the various *ab initio* approaches for bulk solids can be found in Chapter 3 of Srivastava (1990). *Ab initio* treatments of the phonon dynamics in heterostructures are given, for example, in Molinari *et al.* (1992) for superlattices, Rücker *et al.* (1992) for quantum wells and Rossi *et al.* (1994) for quantum wires.

4.2.2 Mesoscopic Phonon Phenomena

We have seen that acoustic phonons propagate, transporting vibrational energy. The usual probe of the transport properties of acoustic phonons is the thermal conductance. If acoustic phonons could be confined to well and wire-like electrically insulating structures having transverse dimensions of the order of the phonon wavelength, then at sufficiently low temperatures we would expect *phonon transport behaviour analogous to certain mesoscopic electron phenomena*. For example, just as for the electrical conductance (see Chapters 5 and 9), the combined properties of low dimensionality, the wave nature of phonons and phonon scattering due to unavoidable imperfections and impurities in the confining structures can result in weak or perhaps even strong phonon localization and a consequent suppression of the thermal conductance. While certain analogies can be drawn between various low-dimensional electron and acoustic phonon transport phenomena, there will also be significant differences in the physics of the respective phenomena, a consequence of the fundamentally different nature of phonons and electrons. For example, phonons obey Bose–Einstein statistics, while electrons obey Fermi–Dirac statistics. However, little is currently known about the physics of low-dimensional acoustic phonon transport. The main problem is the difficulty in confining acoustic phonons.

To obtain a better understanding of this problem, let us consider the example of a transverse acoustic wave incident on a GaAs/AlAs interface from the GaAs side, producing a GaAs reflected transverse wave and AlAs transmitted transverse wave, where the transverse direction is in the plane of the interface (Fig. 4.5). The

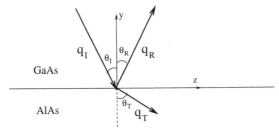

Figure 4.5. Incident, reflected and transmitted transverse acoustic waves at a GaAs/AlAs interface. The displacement vectors are normal to the *y*–*z* plane.

incident, reflected and transmitted displacement vectors are, respectively:

$$\mathbf{u}_I = A\hat{\mathbf{x}}\exp[-i(sq_I t - q_{Iy}y - q_{Iz}z)]$$

$$\mathbf{u}_R = B\hat{\mathbf{x}}\exp[-i(sq_R t - q_{Ry}y - q_{Rz}z)] \tag{4.9}$$

$$\mathbf{u}_T = B'\hat{\mathbf{x}}\exp[-i(s'q_T t - q_{Ty}y - q_{Tz}z)]$$

where we have assumed that the acoustic wavelengths are much larger than the interatomic spacing, allowing us to approximate the atomic lattice as a continuum, and where we have also neglected the anisotropy of the elastic properties of the continuum, so that the GaAs and AlAs transverse phase velocities s and s' are independent of direction. At the interface, the net GaAs displacement must match the AlAs displacement, i.e. $\mathbf{u}_I + \mathbf{u}_R = \mathbf{u}_T$ at $y = 0$, for otherwise the GaAs and AlAs slabs would become detached. Setting $y = 0$ in the displacements (4.9), we find that this matching condition is only possible provided $q_{Iz} = q_{Rz} = q_{Tz}$. Expressing these components in terms of the wavevector magnitudes and angles to the interface normal and using the condition that the frequencies must be the same, $sq_I = sq_R = s'q_T$, we find

$$\theta_I = \theta_R$$

$$\sin(\theta_I)/s = \sin(\theta_T)/s' \tag{4.10}$$

These relations are none other than *Snell's law* for acoustic waves. To have confinement, there must be an angle of incidence for which the resulting transmitted wave moves parallel to the interface. Below this critical angle, the incident wave will be partially transmitted (cf. the normal incidence superlattice modes discussed in the preceding subsection), while above the critical angle, the incident wave will be completely reflected, i.e. confined. From (4.10), the critical angle $\theta_c = \sin^{-1}(s/s')$, which is about $60°$ for the GaAs/AlAs example. While this result shows that it is indeed possible to have acoustic modes confined to GaAs wells and wires embedded in AlAs substrates (Nishiguchi, 1994), it is clear that there will be unconfined modes of similar frequency as well, so that such structures will not be genuinely low-dimensional for thermal conductance measurements. In fact, there are no perfect thermally insulating materials.

The only solution to this problem is to have a *free-standing* structure, i.e. a structure which is physically separated from the substrate for most of its length. For bulk acoustic phonons in thermal equilibrium, the phonon energy distribution has its maximum at $\omega \approx 3k_B T/\hbar$, corresponding to a wavelength $\lambda \approx hs/3k_B T$. This is called the dominant phonon wavelength. For $s = 5 \times 10^3$ m s^{-1}, the dominant wavelength is approximately 80 nm at $T = 1$ K and 8 μm at $T = 10$ mK. Thus, the transverse dimensions of the free-standing structures must be extremely small and temperatures very low in order that the dominant wavelength acoustic phonons confined to the structures be of lower dimension.

Several theoretical investigations have been carried out concerning the low-dimensional thermal transport properties of free-standing insulating wires and films, with particular emphasis on phonon localization due to the presence of impurities (Jäckle, 1981; Kelly, 1982; Seyler and Wybourne, 1990; Perrin, 1993; Blencowe, 1995). Pioneering experiments involving mesoscopic free-standing wires were carried out by Potts *et al.* (1990a, 1990b, 1992). A combination of electron beam lithography and etching techniques was used to fabricate free-standing GaAs wires of width $\approx 0.5\,\mu\text{m}$ and lengths on the order of microns. However, as is evident from the above estimates, these wires were not narrow enough to observe lower-dimensional phonons for the temperatures at which their experiments were performed ($\approx 1\,\text{K}$). Furthermore, the wires were electrically conducting, with most of the heat being transported by the electrons, making it difficult to resolve the phonon contribution. Considerable progress has since been made on further reducing the size of the transverse dimensions of free-standing wires (see, e.g., Yoh *et al.*, 1994). The most concentrated effort in this direction is being made by Roukes and coworkers, who have succeeded in constructing devices which can measure the low-dimensional thermal conductance of free-standing, electrically-insulating nanostructures (Tighe *et al.*, 1997) (Fig. 4.6).

In addition to probing the thermal conductance, the possibility has also been raised of detecting *single* phonons (Travis, 1994). One method of phonon detection might be to measure directly the vibrational amplitude of a free-standing structure. We now give some rough estimates of the amplitude for a single phonon. Recall that, for a mechanical oscillator with mass m, force constant κ and amplitude A, the frequency is $\omega = \sqrt{\kappa/m}$ and vibrational energy is $\epsilon = \kappa A^2/2$. Therefore, equating ϵ with $\hbar\omega$, the energy of a single phonon (see equation (4.2)),

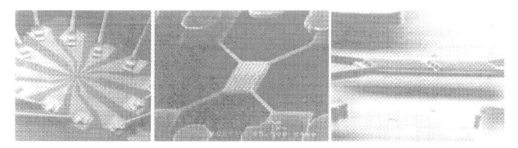

Figure 4.6. Scanning electron microscope micrograph of a suspended, monocrystalline device for mesoscopic thermal conductance measurements. The centre image, which is an enlargement of the central region of the image on the left, shows a semi-insulating GaAs reservoir with area $\approx 3\,\mu\text{m}^2$ which is suspended by four semi-insulating GaAs bridges with length $\approx 5.5\,\mu\text{m}$ and cross-section $\approx 200\,\text{nm} \times 450\,\text{nm}$. Deposited on top of the reservoir are two meandering Si-doped GaAs conductors, one of which serves as a heater and the other as a thermometer. The image on the right presents an edge view of the device which is approximately 300 nm thick and suspended about 1 μm above the substrate (Tighe, Worlock and Roukes, 1997). (Courtesy M. L. Roukes)

we obtain a rough estimate for the single phonon vibrational amplitude: $A \approx (\hbar^2/\kappa m)^{1/4}$. Thus, to increase the amplitude so that it may be detectable, we must decrease the mass and the force constant. One possible structure is a cantilever, where the vibrations of the free end are probed. The force constant for flexing displacements normal to the plane of a cantilever is much less than that for stretching displacements along the length of a cantilever, and therefore a flexural phonon mode will give a much larger free-end displacement than a longitudinal phonon mode. The frequency and vibrational energy of the lowest flexural mode of a cantilever are (see, e.g., Sections 18 and 25 of Landau and Lifshitz, 1986):

$$\omega \approx 3.5 \frac{t}{l^2} \sqrt{\frac{E}{12\rho}} \tag{4.11}$$

and

$$\epsilon \approx 0.1 E w t^3 A^2 / l^3 \tag{4.12}$$

where l, w and t are the length, width and thickness of the cantilever, respectively, A is the free-end displacement amplitude, ρ is the mass density and E is Young's modulus. Equating ϵ with the single phonon energy, we find

$$A \approx 3\hbar^{1/2} t^{-1} (l/w)^{1/2} (E\rho)^{-1/4} \tag{4.13}$$

To get an idea of the numbers, a GaAs cantilever ($E = 2.4 \times 10^{11}$ N m^{-2}, $\rho = 5.4 \times 10^3$ kg m^{-3}) having a width 50 Å and thickness 10 Å must be 2 microns long in order to give a free-end displacement of 1 Å. Furthermore, the temperature must be below 10 µK in order to have no more than a few flexural phonons present. Thus, the detection of a single phonon by direct measurement of the vibrational amplitude would be quite difficult to achieve. Note that, although we have assumed the same elasticity properties as in bulk, macroscopic cantilevers, the fourth root dependence on E in (4.13) will mean that possible modifications in E even as large as an order of magnitude will not change our estimate for A by very much. A relatively straightforward way to determine Young's modulus for these very small cantilever structures is to measure the mean-squared displacement amplitude of the free end due to thermal vibrations at high enough temperatures so that such vibrations can be resolved. Equating the vibrational energy ϵ in (4.12) with $k_B T$, we have a linear relationship between mean-squared amplitude and temperature, and Young's modulus can be obtained from the slope. This method has in fact been used to measure Young's modulus for individual carbon nanotubes up to several microns long and with diameters on the order of nanometers (Treacy et al., 1996).

The devices of Roukes and coworkers (Fig. 4.6) are in principle capable of detecting single phonons. Instead of measuring directly the vibrational amplitude of a free-standing structure, the change in resistance of an electrically conducting nanowire in thermal contact with the structure is measured. When the wire electron gas absorbs (emits) a phonon from the structure, its temperature increases (decreases) resulting in a change in the wire resistance.

Defect spectroscopy has been suggested as another possible use for mesoscopic, free-standing structures (Travis, 1994). The amplitude of a macroscopic vibrating structure (e.g. a plucked guitar string) will continuously decrease in time due to absorption of mechanical energy by the large number of defects present in the structure. Now, the typical absorption energy per defect is of the order of an eV. For a GaAs cantilever in the lowest flexural mode with dimensions $w = t = 100\,\text{Å}$ and $l = 1\,\mu\text{m}$, a vibrational energy of $1\,\text{eV}$ gives a free-end displacement $A \approx 200\,\text{Å}$. Thus, in contrast to the continuous damping of a macroscopic vibrating cantilever, we might expect that, for a cantilever having these much smaller dimensions, the comparable defect/vibrational energies will cause the amplitude to decrease in large, *discrete* amounts, allowing the possibility to probe the energetics of a *single* defect. Of some relevance is an experiment which measured the decay rates at mK temperatures of vibrating, free-standing Si wires having cross-sections on the order of a micron (Greywall *et al.*, 1996). Anomalies in the decay rates were observed and explained as being possibly due to thermally activated relaxational processes involving defects.

4.3   Electron–Phonon Interactions in Heterostructures

So far, we have discussed the dynamics of phonons. We would now like to understand the effect phonons have on the transport properties of low-dimensional electrons. The following discussion of the electron–phonon interaction will be largely qualitative. Let $\psi_a(\mathbf{r})$ denote the energy eigenstate of an electron confined within an ideal (i.e. random impurities and phonons neglected) heterostructure electrostatic potential. Now, suppose the vibrational degrees of freedom of the heterostructure are in some quantum state specified by N_α, the number of phonons present in each mode α. An electron initially in an eigenstate $\psi_a(\mathbf{r})$ will then experience a perturbing potential which is fluctuating in time and hence may scatter into a different eigenstate $\psi_b(\mathbf{r})$. If the potential experienced by the electron due to mode α is

$$V_\alpha(\mathbf{r}, t) = V_{+\alpha}(\mathbf{r})e^{i\omega_\alpha t} + V_{-\alpha}(\mathbf{r})e^{-i\omega_\alpha t} \tag{4.14}$$

where $V_{+\alpha}(\mathbf{r}) = V^*_{-\alpha}(\mathbf{r})$, then time-dependent perturbation theory (see, e.g., Section 42 of Landau and Lifshitz, 1977) gives, to first order, the scattering probability per unit time:

$$\frac{1}{\tau_a} = \frac{2\pi}{\hbar} \sum_b \sum_\alpha [|\langle\psi_b|\hat{V}_{+\alpha}|\psi_a\rangle|^2 (N_\alpha + 1)\delta(E_b - E_a + \hbar\omega_\alpha)$$

$$+ |\langle\psi_b|\hat{V}_{-\alpha}|\psi_a\rangle|^2 N_\alpha\delta(E_b - E_a - \hbar\omega_\alpha)] \tag{4.15}$$

The delta-functions enforce energy conservation for the scattering process, the first one describing (stimulated plus spontaneous) phonon emission and the second one phonon absorption. A necessary requirement for the validity of this approximation is that the perturbing potential matrix elements be much smaller than the electron energy:

$$E_a \gg |\langle \psi_b | \hat{V}_\alpha | \psi_a \rangle| / \sqrt{N_\alpha} \tag{4.16}$$

Another requirement (see page 58 of Paige, 1964) has to do with the fact that in the derivation of (4.15) we make the replacement

$$\frac{4\sin^2[(E_b - E_a \pm \hbar\omega_\alpha)t/2\hbar]}{(E_b - E_a \pm \hbar\omega_\alpha)^2 t} \rightarrow \frac{2\pi}{\hbar}\delta(E_b - E_a \pm \hbar\omega_\alpha) \tag{4.17}$$

While (4.17) is correct for $t \rightarrow \infty$, it is only meaningful to take t as large as τ_a, the time between scatterings. Therefore, the replacement (4.17) is valid provided the matrix elements in (4.15) change only slightly as E_b varies over the range $[E_a - \hbar/\tau_a, E_a + \hbar/\tau_a]$. This gives the requirement:

$$\tilde{\tau}_a \ll \tau_a \tag{4.18}$$

where $\tilde{\tau}_a$ is the duration of the electron–phonon scattering event:

$$\tilde{\tau}_a := \frac{\hbar}{|\langle \psi_a | \hat{V}_\alpha | \psi_a \rangle|^2} \frac{\partial}{\partial E_b} |\langle \psi_b | \hat{V}_\alpha | \psi_a \rangle|^2 \Big|_{b=a} \tag{4.19}$$

Thus, energy is accurately conserved for a single scattering event only if the duration of the event is much less than the time between events. We shall assume conditions (4.16) and (4.18) hold.

One of the distinguishing features of electron transport in low-dimensional structures is the enhanced mobility (see Chapters 9 and 10). As we shall soon learn, this is partly due to the severe constraints placed on scattering by energy-momentum conservation and lower phase space dimension. However, it is important to note that under such conditions even a small violation of a conservation law for electron–phonon scattering events, such as was just discussed above, can bring about a large change in the transport properties, whereas in a bulk system the properties would not be significantly affected. We shall give an example of this for quantum wires below.

To proceed with our discussion of the electron–phonon interaction, we must write down expressions for the electron energy eigenstates $\psi_a(\mathbf{r})$ and phonon mode potentials $V_\alpha(\mathbf{r},t)$. A description of the electron energy states of low-dimensional structures is given in Chapter 2. Since we only give a qualitative description of the electron–phonon interaction, we use the simple infinite barrier well model for the confined electrons. In the case of a quantum well with thickness d and infinite well-plane dimensions, the electron energy eigenstates and eigenvalues are

$$\psi_\mathbf{k}(\mathbf{r}) = Ce^{i(k_x x + k_y y)}\sin(\pi z/d) \tag{4.20}$$

and

$$E_k = \frac{\hbar^2}{2m^*}(k_x^2 + k_y^2)$$ (4.21)

where C is a normalization constant, m^* is the effective electron mass, and we consider only the lowest subband and define the energy with respect to the subband edge. For a quantum wire with width w, thickness d and infinite length along the x-direction, we have

$$\psi_k(\mathbf{r}) = Ce^{ikx}\sin(\pi y/w)\sin(\pi z/d)$$ (4.22)

and

$$E_k = \frac{\hbar^2 k^2}{2m^*}$$ (4.23)

where we again consider only the lowest subband. For a split-gate wire (see Chapter 9), we usually have $w > d$.

Theoretical investigations (Rücker et al., 1992; Tsuchiya and Ando, 1993) have found that in heterostructures the electron–phonon scattering rate with correct phonon mode solutions taken into account does not differ significantly from the rate assuming bulk phonon modes. For example, the electron–phonon scattering rate for an AlAs–GaAs–AlAs quantum well will be somewhere in between the rates for bulk GaAs and AlAs phonons, approaching the rate for bulk GaAs (AlAs) phonons as the well width increases (decreases). Only in free-standing structures, when the confined electrons and phonons are both of lower dimension, are the rates expected to be qualitatively different from those assuming bulk phonons (Bannov et al., 1994; Yu et al., 1994). In the following, we shall restrict ourselves to quantum wells and wires embedded in a substrate and make the simplifying approximation of bulk phonons. The qualitative differences in the electron–phonon rates between low-dimensional and bulk semiconductors are then a consequence of the reduced dimensionality of the electrons only.

At $T = 100$ K, the dominant phonon energy ($3k_B T$) is approximately 26 meV. Dividing this by the conversion factor 1.24×10^{-4} eV cm, we see from Fig. 4.2 that, at temperatures $T \ll 100$ K, mostly acoustic phonons will be present. Let us first consider this lower temperature regime where the effect of optical phonons on the electron–phonon rate can be neglected. Acoustic phonons give rise to a perturbing potential in two different ways (see, e.g., Chapter 3 of Ridley, 1993). In the first, small changes in the relative positions of the atoms perturb the electrostatic potential experienced by the electrons, resulting in a change in the electron energy. In the second way, changes in the relative positions of oppositely charged ions (such as Ga and As) produce an electric polarization and hence long-range electric field which again affects the electron energy. The perturbing potentials, known respectively as the *deformation* and *piezoelectric* potentials, have the following form:

$$V_{+\mathbf{q}}^{\mathrm{DP}}(\mathbf{r}) \approx C^{\mathrm{DP}} q^{1/2} \mathrm{e}^{-i\mathbf{q}\cdot\mathbf{r}} \tag{4.24}$$

$$V_{+\mathbf{q}}^{\mathrm{PE}}(\mathbf{r}) \approx C^{\mathrm{PE}} q^{-1/2} \mathrm{e}^{-i\mathbf{q}\cdot\mathbf{r}} \tag{4.25}$$

where C^{DP} and C^{PE} are the coupling strength constants and $\mathbf{q}$ is the phonon wavevector. We have neglected the screening due to the other electrons which tends to reduce the electron–phonon interaction strength. We have also neglected the dependence on longitudinal/transverse mode-type and have approximated the atomic lattice as an elastically isotropic continuum. For the considered temperature range, the dominant phonon wavelength is much larger than the interatomic spacing, justifying our use of the continuum approximation which replaces the discrete atomic lattice vector with continuous position vector $\mathbf{r}$.

For the perturbing potentials (4.24) and (4.25) and electron states (4.20) and (4.22), the quantum-well matrix elements are

$$|\langle \psi_{\mathbf{k}'} | \hat{V}_\pm(q) | \psi_{\mathbf{k}} \rangle|^2 \sim (|C^{\mathrm{DP}}|^2 q + |C^{\mathrm{PE}}|^2 q^{-1})$$

$$\times q_z^{-2}[(2\pi/d)^2 - q_z^2]^{-2} \sin^2(q_z d/2) \delta_{k'_x, k_x \mp q_x} \delta_{k'_y, k_y \mp q_y} \tag{4.26}$$

and the quantum wire matrix elements are

$$|\langle \psi_{\mathbf{k}'} | \hat{V}_\pm(q) | \psi_{\mathbf{k}} \rangle|^2 \sim (|C^{\mathrm{DP}}|^2 q + |C^{\mathrm{PE}}|^2 q^{-1}) q_y^{-2}[(2\pi/w)^2 - q_y^2]^{-2}$$

$$\times \sin^2(q_y w/2) q_z^{-2}[(2\pi/d)^2 - q_z^2]^{-2} \sin^2(q_z d/2) \delta_{k', k \mp q_x} \tag{4.27}$$

where $\mathbf{k}$ and $\mathbf{k}'$ are the electron wavevectors before and after scattering, respectively. In addition to energy being conserved during a scattering event, we see that the momentum components parallel to the quantum well plane and wire length are also conserved. On the other hand, the momentum components normal to the well plane and wire length are not conserved. Furthermore, the matrix elements are suppressed for normal phonon momentum components exceeding the inverse well or wire thickness.

Let us restrict ourselves to the case where the electrons and phonons are in thermal equilibrium at a common temperature T. The phonons are then described by the Bose–Einstein distribution:

$$N(q) = \frac{1}{\exp(\hbar s q / k_{\mathrm{B}} T) - 1} \tag{4.28}$$

and the electrons by the Fermi–Dirac distribution:

$$F(E) = \frac{1}{\exp[(E - E_{\mathrm{F}})/k_{\mathrm{B}} T] + 1} \tag{4.29}$$

When $k_{\mathrm{B}} T \ll E_{\mathrm{F}}$, the scattering of electrons by phonons is approximately elastic, as only electrons with energy near E_{F} can change their energy state because of the Pauli principle. From equation (4.21) and the relation between Fermi energy and areal electron density, $E_F = n\pi\hbar^2/m*$ [see equation (2.14)], we find that an AlAs/GaAs quantum well with $n = 5 \times 10^{11}$ cm^{-2} has Fermi energy in units of temperature, $E_F/k_B \approx 200$ K, where $m* = 0.067m_e$ for GaAs. A split-gate AlAs/GaAs quantum wire with a Fermi wavevector of the order of $k_F \approx \pi/w = 6 \times 10^7$ m^{-1} for width $w = 500$ Å has a Fermi energy $E_F/k_B \approx 30$ K [see equation (2.23)]. Such numbers are representative of those for actual structures and therefore we typically find for quantum wells that electron scattering is approximately elastic in the acoustic phonon-dominated temperature regime, while for quantum wires temperatures must be somewhat lower to have approximately elastic scattering. Energy–momentum conservation and the approximation of elastic scattering restrict the phonon momentum component in the plane of the well to be $q_\parallel^2 \approx 2k^2(1 - \cos\theta)$, where θ is the angle between $\mathbf{k}$ and $\mathbf{k}'$ (see Fig. 4.7(a)), while the phonon momentum component along the length of the wire is $q_\parallel \approx 2k$ or 0 (see Fig. 4.7(b)).

An informative probe of the effect of phonons on the motion of electrons is the phonon-limited *mobility*, defined as the average drift velocity of the electrons in unit electric field. Provided the scattering of electrons by phonons is approximately elastic, the mobility can be written as

$$\mu = \frac{e}{m*}\bar{\tau}_m \qquad (4.30)$$

where $\bar{\tau}_m$ is the thermally averaged electron-momentum relaxation time. Note that the momentum relaxation time is not the same as the scattering time given in equation (4.15). For the quantum well, $\tau_m(\mathbf{k})$ differs from $\tau(\mathbf{k})$ in the addition of the factor $1 - \cos\theta$ in the sum over $\mathbf{k}'$, so that large-angle scattering events have a larger weighting than small angle scattering events. For the quantum wire, $\tau_m(k)$ differs from $\tau(k)$ in that only the backscattered wavevector $k' \approx -k$

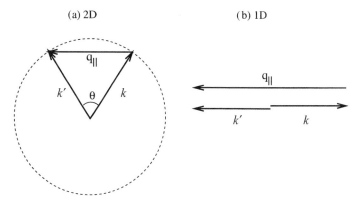

(a) 2D   (b) 1D

Figure 4.7.   Relation between incident and scattered electron wavevector and parallel phonon wavevector component which follow from momentum conservation for (a) 2DEG and (b) 1DEG.

contributes. In low-dimensional systems, the momentum relaxation rate will be substantially smaller than the scattering rate at low temperatures such that the dominant phonons can only cause the electrons to scatter through small angles. The fact that the mobility depends on τ_m and not on τ is physically reasonable, since the electron motion is only expected to be strongly affected by a scattering event which adds a large wavevector component in the direction opposite to the initial motion.

We have seen that there are certain restrictions on the wavevectors of phonons which can strongly scatter electrons. The wavevector component perpendicular to the well or wire must not exceed the inverse transverse dimensions, while the parallel component must be close to $2k_F$. Since the phonon energy is proportional to wavevector magnitude, i.e. $E = \hbar sq$, the phonon energy is similarly restricted. The occurrence of an upper bound on the phonon energy divides the acoustic phonon-dominated temperature range into essentially two well-defined regimes where the mobility has qualitatively different temperature dependences. When $k_B T$ is much larger than this phonon energy bound, the Bose–Einstein distribution (4.28) can be approximately replaced by $N(q) \approx k_B T/\hbar sq$. (For $\hbar sq/k_B T \ll 1$, the exponential in (4.28) can be approximated as $1 + \hbar sq/k_B T$.) Hence all phonon modes which interact strongly with the electrons have *equal* thermal occupation energies $\hbar sq N(q) = k_B T$. This is called the *equipartition* (EP) regime. On the other hand, when $k_B T$ is much less than the upper bound, the thermal occupation number $N(q)$ of the strongly interacting phonon modes is significantly reduced. This is called the *Bloch–Grüneisen* (BG) regime, after a similar regime which occurs for electron-phonon scattering in metals (see, e.g., Chapter 26 of Ashcroft and Mermin, 1976). For a GaAs quantum well with thickness $d = 100\,\text{Å}$, the EP and BG regimes occur respectively above and below the temperature $T \approx \hbar s\pi/k_B d \approx 10\,\text{K}$. In the case of a GaAs quantum wire with width $500\,\text{Å}$ and thickness $100\,\text{Å}$, the EP regime occurs above about $10\,\text{K}$ and the BG regime below about $1\,\text{K}$.

The temperature dependence of the phonon-limited mobility in these two regimes is worked out in Exercises 3 and 4 (see also, e.g., Kawamura and Das Sarma (1992) for quantum wells and Mickevičius and Mitin (1993) for quantum wires). In the EP regime, we find that both the well and wire mobilities are *inversely proportional* to the temperature:

$$\mu \propto T^{-1} \tag{4.31}$$

As is to be expected, phonon scattering causes a *decrease* in the mobility with increasing temperature. In the BG regime and for deformation potential dominated scattering with screening taken into account, the quantum-well mobility is (Price, 1984):

$$\mu \propto T^{-7} \tag{4.32}$$

For piezoelectric interaction dominated scattering with screening, we have:

$$\mu \propto T^{-5} \tag{4.33}$$

Thus, the mobility increases much more rapidly with decreasing temperature than in the EP regime. Comparing the temperature dependences in (4.32) and (4.33), we see that piezoelectric scattering will dominate at sufficiently low temperatures.

The predicted temperature dependence for quantum wires in the BG regime is rather spectacular. Only phonons which have sufficient momentum to backscatter electrons by an amount $2k_F$ will affect the mobility. However, from the form of the Bose–Einstein distribution (4.28), we see that the occupation number of such modes is exponentially suppressed when the temperature drops below $2\hbar s k_F/k_B$, resulting in an *exponential* increase in mobility:

$$\mu \propto \exp(2\hbar s k_F/k_B T) \tag{4.34}$$

It is important to keep in mind that these temperature dependences follow from the assumption that the electron and phonon systems interact to good approximation via single, well-defined electron–phonon scattering events, with energy and momentum being conserved for each event. Such assumptions may not always be valid, however. For example, in sufficiently long quantum wires the Coulomb interaction between electrons can give rise to a phase where the energy eigenstates correspond to plasmon-like, collective excitations, rather than to single electrons (see, e.g., Schulz, 1995). The mobility may then not be such a meaningful measure of the transport properties and we should instead consider the wire conductance. The consequences for the conductance of phonons interacting with these plasmon-like states has yet to be properly understood. Quite apart from modifications originating from Coulomb interactions, it has also been argued that modifications of the phonon system itself due to the electron–phonon interaction may replace the exponential temperature dependence (4.34) with a power law dependence – as we have for the quantum well – hence resulting in much lower mobilities than predicted by (4.34) (Senna and Das Sarma, 1993, 1994).

For quantum wells, the inverse temperature dependence of the mobility in the EP regime is relatively straightforward to observe (see, e.g., Hirakawa and Sakaki, 1986). The BG regime is more difficult to probe because electron-impurity scattering usually dominates in the relevant lower temperature range. The predicted BG temperature dependence (Price, 1984) was first observed in GaAs quantum wells by Stormer *et al.* (1990) (Fig. 4.8). In order to have *quantitative* agreement between theory and experiment for the phonon-limited mobility in both regimes, a larger deformation-potential coupling constant C^{DP} than that measured in bulk GaAs is required. The reason for this discrepancy is as yet unclear.

To our knowledge, there have been no measurements of the phonon-limited mobility of quantum wires. In a typical quantum wire, the temperature dependence of the mobility is determined by scattering from imperfections such as

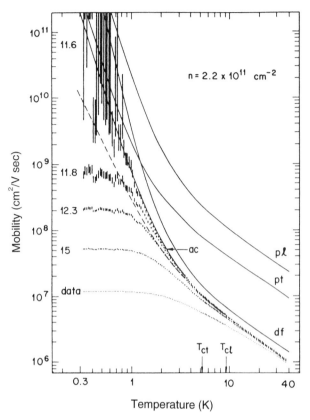

Figure 4.8.   Mobility versus temperature of a 2DEG (Stormer *et al.*, 1990). The trace labelled 'data' is the measured mobility, while the trace identified as 11.6 is the mobility which results after all temperature independent scattering contributions have been removed. This should be the phonon-limited mobility. The solid line labelled 'ac' is the calculated mobility with deformation-potential coupling strength adjusted so as to obtain agreement with the data. The solid curves 'pl', 'pt' and 'df' are the individual longitudinal-piezoelectric, transverse-piezoelectric and deformation-potential contributions to the mobility ($\mu_{ac}^{-1} = \mu_{df}^{-1} + 2\mu_{pt}^{-1} + \mu_{pl}^{-1}$). The BG regime onset temperatures for transverse and longitudinal phonons are T_{ct} and T_{cl}, respectively. (Courtesy H. L. Stomer)

impurities or boundary roughness. Of course, the phonon contribution will dominate at sufficiently high temperatures. However, at the necessary temperatures many subbands are significantly thermally populated, so that the wire loses its one-dimensional character. Smaller wire widths are required in order to further increase the energy separation between subbands. Wires should also be made longer in order to increase the probability for an electron passing through the wire to be scattered by a phonon. Obviously, there should also be fewer imperfections. Considerable effort is being put into fabricating suitable wires, so we can soon expect to be able at least to test for the inverse temperature dependence of the EP regime.

Acoustic phonon pulse spectroscopy provides an alternative, powerful method to probe the electron–phonon interaction in quantum wires. In the experiments of Naylor *et al.* (1996a,b), the change in conductance of GaAs/AlGaAs split-gate wires due to the interaction with incident acoustic phonon pulses is measured. This method gives the added control of being able to vary the phonon distribution with respect to the electron distribution, allowing the possibility to extract the phonon contribution to the wire resistance. Because of the non-equilibrium nature of the distributions, the theoretical analysis is a little more complicated than that given above for the mobility and is worked out in Blencowe and Shik (1996).

In the discussion so far, we have neglected the scattering of electrons by optical phonons. The optical phonon contribution to the scattering rate becomes increasingly important and eventually dominates as we approach room temperature and above. Because the electron Fermi wavelength and transverse dimensions of typical quantum wells and wires are much larger than the interatomic spacing, parallel momentum component conservation and suppression of large normal momentum component (see discussion following equation (4.27)) allow only small-q optical phonons to scatter the electrons. In the preceding section, we saw that small-q GaAs optical modes involve neighbouring Ga and As atoms vibrating almost $180°$ out of phase. The coupling mechanism between the optical phonons and electrons is thus very similar to that of the acoustic piezoelectric interaction, i.e. the electrons are scattered by the long-range electric field produced by the relative displacements of oppositely charged ions. A derivation of the interaction potential, called the *polar optical* (PO) potential, can be found in Section 3.5 of Ridley (1993).

In Fig. 4.2, we see that the small-q GaAs longitudinal optical modes are approximately q-independent and correspond to a temperature of about $400\,\mathrm{K}$. Thus, in typical GaAs quantum wells and wires the scattering of electrons by optical phonons will be highly inelastic and the mobility no longer has a simple expression in terms of a momentum relaxation rate as in the lower temperature acoustic regime (see, e.g., Kawamura and Das Sarma, 1992). Nevertheless, it is still possible to make a few basic observations without having to go through the more complicated analysis for the inelastic case. Because only optical phonons close to zone centre interact with the electrons, we can neglect the q-dependence of the phonon energy and the Bose–Einstein distribution N in the scattering rate τ^{-1} (see equation (4.15)) is

$$N \approx \frac{1}{\exp(\hbar\omega_0/k_\mathrm{B}T) - 1} \tag{4.35}$$

where ω_0 denotes the zone centre frequency. If T is somewhat smaller than the Debye temperature, which is $\hbar\omega_0/k_\mathrm{B} \approx 400\,\mathrm{K}$ (for GaAs), then $\tau^{-1} \propto \exp(-\hbar\omega_0/k_\mathrm{B}T)$ and hence, in the regime where polar optical scattering dominates, we would expect a much sharper decrease in mobility with increasing temperature than in the EP regime. In the experiment of Lin *et al.* (1984) using GaAs quantum wells, the cross-over to optical phonon dominated scattering occurred at around $T = 100$ K and above this cross-over an exponential decrease in mobility with increasing temperature was observed.

At room temperature, the mobility is limited by polar optical scattering. For certain heterostructure device applications, we would like to have as high a mobility as possible (see Chapters 9 and 10) and thus ways must be found to reduce the electron–optical phonon interaction. This might be achieved by modifying in some way the phonon modes and/or electron states in the hetero-structures. As mentioned near the beginning of this section, modifying the phonon modes does not change the mobility very much (Rücker *et al.*, 1992; Tsuchiya and Ando, 1993). This leaves possible modifications of the electron states. Tsuchiya and Ando (1994) showed that an enhancement in quantum-well mobility can be achieved by modulating the electron wavefunction in the direction normal to the well. To understand how this works, consider the electron–phonon matrix element for the PO interaction:

$$\langle \psi_{\mathbf{k}'} | \hat{V}_{\pm}(q) | \psi_{\mathbf{k}} \rangle \sim C^{\mathrm{PO}} q^{-1} \delta_{k'_x, k_x \mp q_x} \delta_{k_y, k_y \mp q_y} \int_0^d \mathrm{d}z \; \psi^*(z) \mathrm{e}^{\mp \mathrm{i} q_z z} \psi(z) \qquad (4.36)$$

where $\psi(z)$ is the part of the wavefunction normal to the well. The matrix element magnitude increases with decreasing q and thus the interaction can be reduced by making $\psi(z)$ rapidly varying within the thickness d, so that the z-integral gives a small contribution for small q. The modulation can be achieved by inserting several thin AlGaAs barrier layers into the well at equally spaced points between $z = 0$ and $z = d$. Another possible way to suppress the electron–optical phonon interaction in quantum wires was proposed by Sakaki (1989). Instead of modulating the electron wavefunction in the transverse y- and z-directions, a periodic, modulating potential is introduced along the length of the wire. This splits the E versus k dispersion relation (4.23) into minibands and minigaps. If the periodic potential is chosen such that the optical phonon energy $\hbar \omega_0$ is larger than the miniband width but smaller than the minigap, then from energy conservation we can have neither intra- nor interminiband scattering: electron–optical phonon scattering is completely suppressed.

4.4   Conclusion

In this chapter we have given a basic description of the phonon modes and electron-phonon interactions in low-dimensional structures. Several topics have not been discussed here. Of considerable interest both at the fundamental physics level and for device applications is the subject of hot electron relaxation in low-dimensional structures. When an electron current flows in a quantum well or wire, electron heating will inevitably occur. In order to find ways to avoid overheating, it is important that we understand the various energy relaxation processes such as phonon emission. Unlike the mobility properties discussed above, this is a non-equilibrium process where the electron temperature (if it can be defined at all) exceeds the phonon temperature. A good source of information on hot electron physics is the proceedings volume edited by Ryan and Maciel (1994).

We also only briefly mentioned the various spectroscopic techniques such as acoustic phonon pulse spectroscopy and Raman spectroscopy. Detailed discussions concerning these techniques as applied to heterostructures can be found in the references listed in the Introduction. An understanding of the material presented in this chapter should enable the reader to explore effectively these other topics.

EXERCISES

1. Consider the following trial decay mode solution to equations (4.3) and (4.4):
$u_j(na, t) = A_j \exp[-i(\omega t - qna)]$, where $q = \alpha + i\beta$ with α and β both real.

(a) Write down the dispersion relation as a function of complex q and show that, when $\beta \neq 0$, the requirement that ω be real gives $\alpha = \pm n\pi/a$, for $n = 0, 1, 2, \ldots$

(b) Show that, for $\alpha = 0$, the dispersion relation is:

$$\omega^2 = \frac{\kappa}{m_1 m_2} \{m_1 + m_2 + [m_1^2 + m_2^2 + 2m_1 m_2 \cosh(\beta a)]^{1/2}\} \qquad (4.37)$$

(c) Show that, for $\alpha = \pi/a$, the dispersion relation is:

$$\omega^2 = \frac{\kappa}{m_1 m_2} \{m_1 + m_2 \pm [m_1^2 + m_2^2 - 2m_1 m_2 \cosh(\beta a)]^{1/2}\} \qquad (4.38)$$

2. Use equations (4.37) and (4.38) to answer the following.

(a) Plot the ω versus βa dispersion curves for GaAs and AlAs. (Note that $m_{Ga} = 70 m_p$, $m_{Al} = 27 m_p$ and $m_{As} = 75 m_p$, where m_p is the proton mass: $m_p = 1.7 \times 10^{-27}$ kg. The force constants are given below equation (4.5).)

(b) Match the GaAs decay mode frequency to the AlAs optical frequency obtained from equation (4.5) to show that the minimum and maximum β values for GaAs are $\beta_{\min} a \approx 2.4$ and $\beta_{\max} a \approx 3.4$.

(c) Match the AlAs decay mode frequency to the GaAs optical frequency obtained from equation (4.5) to show that the minimum and maximum β values for AlAs are $\beta_{\min} a \approx 0.22$ and $\beta_{\max} a \approx 1.0$.

3. The thermally averaged momentum relaxation rate for quantum-well electrons is:

$$\overline{\tau_{\mathrm{m}}^{-1}} \sim \frac{1}{T} \int_0^\infty \mathrm{d}E \int_0^\infty \mathrm{d}q_z \int_0^\pi \mathrm{d}\theta (1 - \cos\theta) S(q_\parallel)^2$$

$$\times (|C^{\mathrm{DP}}|^2 q + |C^{\mathrm{PE}}|^2 q^{-1}) q_z^{-2} [(2\pi/d)^2 - q_z^2]^{-2} \sin^2(q_z d/2)$$

$$\times \{[N(q) + 1]F(E)[1 - F(E - \hbar s q)] + N(q)F(E)[1 - F(E + \hbar s q)]\} \quad (4.39)$$

where $q^2 = q_z^2 + q_\parallel^2$ with $q_\parallel = k_{\mathrm{F}}\sqrt{2(1 - \cos\theta)}$ and where S is the screening function.

(a) Using the approximation

$$\int_0^\infty \mathrm{d}E F(E)[1 - F(E - \epsilon)] \approx \epsilon N(\epsilon)$$

show that equation (39) can be written as:

$$\overline{\tau_{\mathrm{m}}^{-1}} \sim \frac{1}{T} \int_0^\infty \mathrm{d}q_z \int_0^\pi \mathrm{d}\theta (1 - \cos\theta) S(q_\parallel)^2 (|C^{\mathrm{DP}}|^2 q^2 + |C^{\mathrm{PE}}|^2)$$

$$\times q_z^{-2}[(2\pi/d)^2 - q_z^2]^{-2} \sin^2(q_z d/2) N(q)[N(q) + 1] \quad (4.40)$$

(b) By expanding $N(q)$ with respect to $\hbar s q / k_{\mathrm{B}} T$, show that $\overline{\tau_{\mathrm{m}}^{-1}} \propto T$ when $k_{\mathrm{B}} T \gg \max(\hbar s \pi / d, \hbar s k_{\mathrm{F}})$.

(c) When $k_{\mathrm{B}} T \ll \min(\hbar s \pi / d, \hbar s k_{\mathrm{F}})$, we have only small-angle scattering. By making the approximation $\cos\theta \approx 1 - \theta^2/2$ and changing integration variables to $\theta' = \theta \hbar s k_{\mathrm{F}} / k_{\mathrm{B}} T$ and $q_z' = q_z \hbar s / k_{\mathrm{B}} T$, show that $\overline{\tau_{\mathrm{m}}^{-1}} \propto T^5$ for piezoelectric-dominated scattering and $\overline{\tau_{\mathrm{m}}^{-1}} \propto T^7$ for deformation-dominated scattering. (Note that, for a two-dimensional electron gas, $S(q_\parallel) \propto q_\parallel$ for small $q_\parallel$.)

4. For a quantum wire, the thermally averaged momentum relaxation rate is:

$$\overline{\tau_{\mathrm{m}}^{-1}} \sim \frac{1}{T} \int_0^\infty \mathrm{d}q_y \mathrm{d}q_z (|C^{\mathrm{DP}}|^2 q^2 + |C^{\mathrm{PE}}|^2) q_y^{-2}[(2\pi/w)^2 - q_y^2]^{-2}$$

$$\times \sin^2(q_y w/2) q_z^{-2}[(2\pi/d)^2 - q_z^2]^{-2} \sin^2(q_z d/2) N(q)[N(q) + 1] \quad (4.41)$$

where $q^2 = q_y^2 + q_z^2 + q_\parallel^2$ with $q_\parallel = 2k_{\mathrm{F}}$ and where screening has been neglected. Using analogous approximations to those of parts (b) and (c) of Exercise 3, derive the temperature dependences of the momentum relaxation rate in the EP and BG regimes.

References

A. V. Akimov, E. S. Moskalenko, L. J. Challis and A. A. Kaplyanskii, *Physica B* **219–220**, 9 (1996).

N. W. Ashcroft and N. D. Mermin, *Solid State Physics* (Holt, Rinehart and Winston, New York, 1976).

N. Bannov, V. Mitin and M. Stroscio, *Phys. Stat. Sol. (b* **183**, 131 (1994).

A. S. Barker, J. L. Merz and A. C. Gossard, *Phys. Rev. B* **17**, 3181 (1978).

M. P. Blencowe, *J. Phys.: Condens. Matter* **7**, 5177 (1995).

M. P. Blencowe and A. Y. Shik, *Phys. Rev. B* **54**, 13899 (1996).

M. Cardona, in *Light Scattering in Semiconductor Structures and Superlattices*, D. J. Lockwood and J. F. Young, eds. (Plenum Press, New York, 1991 pp. 19–38.

M. Cardona and G. Güntherodt, eds., *Light Scattering in Solids V* (Springer-Verlag, Heidelberg, 1989).

M. P. Chamberlain, M. Cardona and B. K. Ridley, *Phys. Rev. B* **48**, 14 356 (1993).

C. Colvard, T. A. Grant, M. V. Klein, R. Merlin, R. Fischer, H. Morkoc and A. C. Gossard, *Phys. Rev. B* **31**, 2080 (1985).

N. C. Constantinou and B. K. Ridley, *Phys. Rev. B* **49**, 17065 (1994).

J. E. Digby, U. Zeitler, C. J. Mellor, A. J. Kent, K. A. Benedict, M. Henini, C. T. Foxon and J. J. Harris, *Physica Scripta* **T66**, 163 (1996).

D. S. Greywall, B. Yurke, P. A. Busch and S. C. Arney, *Europhys. Lett.* **34**, 37 (1996).

K. Hirakawa and H. Sakaki, *Phys. Rev. B* **33**, 8291 (1986).

J. Jäckle, *Solid State Commun.* **39**, 1261 (1981).

T. Kawamura and S. Das Sarma, *Phys. Rev. B* **45**, 3612 (1992).

M. J. Kelly, *J. Phys. C* **15**, L969 (1982).

A. J. Kent, G. A. Hardy, P. Hawker, V. W. Rampton, M. I. Newton, P. A. Russel and L. J. Challis, *Phys. Rev. Lett.* **61**, 180 (1988).

A. G. Kozorezov, T. Miyasato and J. K. Wigmore, *J. Phys.: Condens. Matter* **8**, 1 (1996).

L. D. Landau and E. M. Lifshitz, *Theory of Elasticity*, 3rd edn (Pergamon Press, Oxford, 1986).

L. D. Landau and E. M. Lifshitz, *Quantum Mechanics* 3rd edn (Pergamon Press, Oxford, 1977).

B. J. F. Lin, D. C. Tsui, M. A. Paalanen and A. C. Gossard, *Appl. Phys. Lett.* **45**, 695 (1984).

D. J. McKitterick, A. Shik, A. J. Kent and M. Henini, *Phys. Rev. B* **49**, 2585 (1994).

A. Mickevičius and V. Mitin, *Phys. Rev. B* **48**, 17194 (1993).

E. Molinari, S. Baroni, P. Giannozzi and S. de Gironcoli, *Phys. Rev. B* **45**, 4280 (1992).

A. J. Naylor, K. R. Strickland, A. J. Kent and M. Henini, *Surface Science* **361–2**, 660 (1996a).

A. J. Naylor, A. J. Kent, P. Hawker, M. Henini and B. Bracher, in *23rd International Conference on the Physics of Semiconductors (Berlin 1996)*, M. Scheffler and R. Zimmermann, eds. (World Scientific, Singapore, 1996b) pp. 1249–52.

N. Nishiguchi, *Jpn. J. Appl. Phys.* **33**, 2852 (1994).

E. G. S. Paige, *Prog. Semicond.* **8**, 1 (1964).

N. Perrin, *Phys. Rev B.* **48**, 12151 (1993).

A. Potts, D. G. Hasko, J. R. A. Cleaver, C. G. Smith, H. Ahmed, M. J. Kelly, J. E. F. Frost, G. A. C. Jones, D. C. Peacock and D. A. Ritchie, *J. Phys.: Condens. Matter* **2**, 1807 (1990a).

A. Potts, M. J. Kelly, C. G. Smith, D. G. Hasko, J. R. A. Cleaver, H. Ahmed, D. C. Peacock, D. A. Ritchie, J. E. F. Frost and G. A. C. Jones, *J. Phys.: Condens. Matter* **2**, 1817 (1990b).

A. Potts, M. J. Kelly, D. G. Hasko, J. R. A. Cleaver, H. Ahmed, D. A. Ritchie, J. E. F. Frost and G. A. C. Jones, *Semicond. Sci. Technol.* **7**, B231 (1992).

P. J. Price *Solid State Commun.* **51**, 607 (1984).

B. K. Ridley, *Quantum Processes in Semiconductors* 3rd edn (Oxford University Press, Oxford, 1993).

F. Rossi, C. Bungaro, L. Rota, P.Lugli and E. Molinari, *Solid-State Electronics* **37**, 761 (1994).

M. Rothenfusser, L. Köster and W. Dietsche, *Phys. Rev. B* **34**, 5518 (1986).

H. Rücker, E. Molinari and P. Lugli, *Phys. Rev. B* **45**, 6747 (1992).

J. F. Ryan and A. C. Maciel, eds., *Semicond. Sci. Technol.*, **9** 411 (1994).

H. Sakaki, *Jpn. J. Appl. Phys.* **28**, L314 (1989).

H. J. Schulz, in *Mesoscopic Quantum Physics (Les Houches 1994)*, E. Akkermans *et al.*, eds., (North Holland, Amsterdam, 1995) pp. 533–603.

J. R. Senna and S. Das Sarma, *Phys. Rev. Lett.* **70**, 2593 (1993).

J. R. Senna and S. Das Sarma, *Phys. Rev. Lett.* **72**, 2811 (1994).

J. Seyler and M. N. Wybourne, *J. Phys.: Condens. Matter* **2**, 8853 (1990).

G. P. Srivastava, *The Physics of Phonons* (Adam Hilger, Bristol, 1990).

H. L. Stormer, L. N. Pfeiffer, K. W. Baldwin and K. W. West, *Phys. Rev. B* **41**, 1278 (1990).

T. S. Tighe, J. M. Worlock and M. L. Roukes, *Appl. Phys. Lett.* **70**, 2687 (1997).

J. Travis, *Science* **263**, 1702 (1994).

M. M. J. Treacy, T. W. Ebbesen and J. M. Gibson, *Nature* **381**, 678 (1996).

T. Tsuchiya and T. Ando, *Phys. Rev.B* **47**, 7240 (1993).

T. Tsuchiya and T. Ando, *Surface Science* **305**, 312 (1994).

K. Yoh, A. Nishida, H. Kawahara, S. Izumiya and M. Inoue, *Semicond. Sci. Technol.* **9**, 961 (1994).

S. G. Yu, K. W. Kim, M. A. Stroscio, G. J. Iafrate and A. Ballato, *Phys. Rev. B* **50**, 1733 (1994).

5 Localization and Quantum Transport

A. MacKinnon

5.1 Introduction

Traditional solid-state physics is based on the concept of the perfect crystalline solid, sometimes with a relatively low density of defects. This perfect crystallinity has played a crucial role in the development of the subject, with Bloch's theorem providing the central conceptual base. Concepts that arise from this theorem, such as bands, Brillouin zones, vertical transitions, effective mass and heavy and light holes, are really only well-defined in a perfect infinite crystal. In the absence of crystallinity none of these concepts is strictly valid, though in some cases it provides a useful starting point. In general, however, a new approach is required to characterize electrons and phonons in disordered solids.

When we consider low-dimensional structures Bloch's theorem may or may not be valid. There is nothing intrinsic to low dimensionality which invalidates it. Many of the simple examples in quantum mechanics and solid-state physics text-books are, in fact, one-dimensional (e.g. the particle in a box, the Kronig–Penney model). Indeed, in a quantum well prepared by any of the standard growth methods (Chapter 1), much of the physics can be understood by using first-year undergraduate quantum mechanics and the effective mass approximation (Chapter 2). This is because a region of adjacent GaAs layers in $Al_xGa_{1-x}As$ can, for many purposes, be regarded as a perfect potential well. By doping the AlGaAs, the electrons in the well can be spatially separated from the scattering due to the ionized donor atoms (Chapter 3). Thus, in many respects, the electrons in this system can be treated as particles in a one-dimensional box.

The quantum well is, however, a very special quasi-two-dimensional system, albeit a very important one. As discussed in Chapter 1, it is very difficult to prepare low-dimensional samples of high quality for other than lattice-matched planar heterostructures. Thus, most heterojunctions, such as those with a significant lattice mismatch, metal-oxide-semiconductor field-effect transistors (MOSFETs), narrow quantum wells and quantum wires, etc., are in practice highly disordered with an effective density of scatterers which can approach the density of atoms. Clearly, in such systems, it cannot be valid to treat the effect of scatterers with perturbation theory using the perfect crystalline case as a starting point.

As we shall see later, there is one sense in which low-dimensional systems are intrinsically different from three-dimensional systems. The amount of scattering

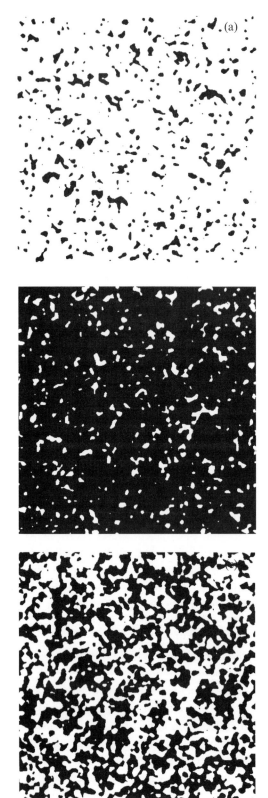

Figure 5.1. Percolation diagrams, with (a) low water level with a few lakes, (b) high water level with a few islands, (c) intermediate (critical) water level.

required to produce dramatic changes in the behaviour can sometimes be so small that perturbation theory may *never* be valid.

5.2   Localization

5.2.1   Percolation

Let us start with a simple classical problem. How does a fluid flow through a random medium? This is a problem of considerable practical importance in its own right: the extraction of oil from porous rock strata.

Consider a random landscape which is being slowly filled with water. At first there will be a continuous land mass with a few lakes (Fig. 5.1(a)). When the water level is very high we have islands in a sea (Fig. 5.1(b)). Let us now suppose there is a dam at the edge of the area which requires large quantities of water to drive a power station. When the water level is low only the lake next to the dam can be used and it will soon run out. As the level is raised this lake becomes larger but still finite. The power station will run longer but will still eventually drain the lake and have to stop. At a critical water level (Fig. 5.1(c)) the system changes from a lake district to an archipelago. This is analogous to the *percolation transition* (Stauffer and Aharony, 1994), where the water first forms a continuous network through the landscape. After this the power station can run indefinitely without fear of running out of water.

This phenomenon has much in common with more conventional phase transitions. There is a characteristic length scale which diverges at the transition: the size of the lakes or islands. There is a well-defined critical water level, rather like the critical temperature of the freezing transition or the ferromagnetic-to-paramagnetic transition in iron. If we think in terms of the density of blockages rather than the water level we see that there is a critical density above which the flow of water stops.

The one-dimensional version of this problem is special. Any blockage of the channel is enough to prevent the flow of water. The critical density is zero. This is an example of a problem which cannot be solved by perturbation theory. There is a discontinuous jump in the behaviour between a system with no blockages and one with a single blockage. In higher dimensions, in contrast, water can flow *around* the blockage.

5.2.2   The Anderson Transition and the Mobility Edge

The concept of the localization of electrons caused by disorder is due to Anderson (1958). He argued that an electron which starts at a particular site cannot completely diffuse away from that site if the disorder is greater than some critical value. Anderson thus introduced the concept of *localized* and *extended* states. The

(a)   (b)

Figure 5.2.   Schematic diagrams of (a) extended and (b) localized states, showing the correlation length, ℓ, and the localization length, ξ.

characteristics of these states can be summarized as follows (Fig. 5.2):

(a) extended   (i)   spread over the entire sample
                    (ii)   not normalizable
                    (iii) contributes to transport
(b) localized   (i)   confined to a finite region
                    (ii)   normalizable
                    (iii) does not contribute to transport

It is worth noting at this point that the phenomenon of localization is not confined to electrons, but can also be observed in other wave phenomena in random media, such as acoustic and optical waves (Section 4.2.2), as well as water waves (Fig. 5.3).

Mott (1968) later introduced the concept of a *mobility edge* (Fig. 5.4). He argued that it is meaningless to consider localized and extended states which are degenerate since any linear combination of a localized and an extended state must be extended. Thus, the concept of localization can only be meaningful if there are separate energy regions of localized and extended states, rather like bands and gaps. These regions are separated by a mobility edge. Mott further argued that the states close to a band edge are more likely to be localized than those in the middle of a band. Since the localized states do not take part in conduction, electrons in a disordered semiconductor must be activated to beyond the mobility edge rather than simply to the band edge to contribute to the conductivity. This activated process would be manifested in a conductivity σ of the form

$$\sigma = \sigma_0 \exp\left(-\frac{E_\mu - E_F}{k_{\mathrm{B}} T}\right) \tag{5.1}$$

where E_μ and E_F are the mobility edge and Fermi energy, respectively. This form should reveal itself as the slope in an *Arrhenius plot* of the conductivity, i.e. a plot of $\ln \sigma$ vs. $1/T$:

$$\ln \sigma = \ln \sigma_0 - \frac{E_\mu - E_F}{k_B T} \tag{5.2}$$

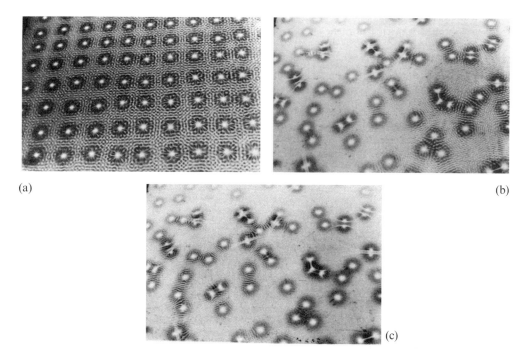

(a)   (b)

(c)

Figure 5.3.   Three photographs of a water bath exposed to an audio-frequency oscillation (not stroboscopic). (a) shows a situation where the obstacles sit in a regular quadratic lattice (frequency 76 Hz). We see strong Bragg reflection corresponding to standing waves. (b) and (c) show randomly spaced obstacles exposed to two different audio frequencies (105 Hz and 76 Hz). Both (b) and (c) show standing wave patterns, but localized in different areas. (With the permission of the authors from Lindelof *et al.* (1986).)

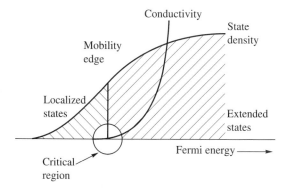

Figure 5.4.   Schematic illustration of the density of states and the conductivity near a band edge in a disordered system. The regions of localized states (no conductivity) and extended states (finite conductivity) are indicated.

5.2.3 Variable Range Hopping

At very low temperatures, when activated conductivity is not significant and the Fermi level is in a region of localized states (e.g. an amorphous semiconductor), transport takes place by *hopping* between localized states. The electrons can gain or lose energy of order $k_B T$ from interactions with phonons and other excitations. In more than one dimension the electron is more likely to find a state in this energy range the further it hops.

On the other hand, the exponential envelopes of the localized states must overlap for the phonon to couple them. In this way Mott (Mott and Davis, 1979) found the famous $T^{-1/4}$ law. More precisely, Mott's law is written as

$$\sigma = \sigma_0 \exp\left[-\left(\frac{T_0}{T}\right)^{1/(d+1)} \right] \tag{5.3}$$

where d is the number of spatial dimensions. This result has been verified many times in different systems. Or has it? To make an accurate measurement of the exponent, $1/(d+1)$, the conductivity must be measured over several decades of temperature while still remaining below the onset of activated transport. Thus, the exponent cannot be determined very precisely. In addition, the measured value of the pre-exponential factor, σ_0, often disagrees with that obtained from theory by several orders of magnitude.

5.2.4 Minimum Metallic Conductivity

Yet another idea from Nevill Mott: the semi-classical conductivity can be written in the form

$$\sigma = \frac{ne^2\tau}{m} = \frac{ne^2\ell}{mv_F} = \frac{ne^2\ell}{\hbar k_F} \tag{5.4}$$

where n is the density of conduction electrons, m and e are the electron mass and charge, respectively, τ is a scattering time, ℓ is the mean free path, and v_F and k_F are the Fermi velocity and wave vector. The density n of electrons is proportional to k_F^d (Ashcroft and Mermin, 1976). The Ioffe–Regel (Ioffe and Regel, 1960) criterion states that the de Broglie wavelength of an electron cannot be greater than the mean free path ℓ (essentially, this is the Heisenberg uncertainty principle) and, in any case, neither can be less than the interatomic distance, a. Hence, the conductivity cannot be less than

$$\sigma_{\min} \propto \frac{e^2}{h} k_F^{d-1}\ell \geq \frac{e^2}{h} k_F^{d-2} \geq \frac{e^2}{h} a^{2-d} \tag{5.5}$$

In two dimensions, the material-dependent quantities k_F and a do not appear and $\sigma_{\min}$ may be a *universal* quantity: $\sigma_{\min} = e^2/h$. There have been many experiments which purported to measure a value for $\sigma_{\min}$ corresponding to $25\,813\ \Omega$ in

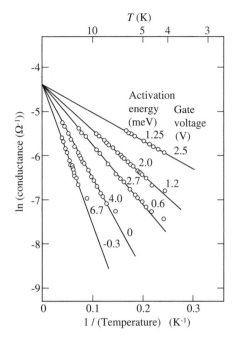

Figure 5.5.   Arrhenius plot on a Silicon MOSFET, for various gate voltages V_g/V. Note the common intercept of the lines (Pepper, 1978). (Courtesy SUSSP Publications)

two-dimensional (2D) systems. By plotting equation (5.1) for several different systems there appears to be a common value of $\ln \sigma_0$ (see, e.g. Fig. 5.5). This is now believed to be characteristic of an intermediate regime and not a proof of the existence of $\sigma_{\min}$.

5.3   Scaling Theory and Quantum Interference

5.3.1   The Gang of Four

A decisive breakthrough in the theory of the metal-insulator transition was made when Abrahams, Anderson, Licciardello and Ramakrishnan (1979) – the 'gang of four' – published their scaling theory. The essence of their idea is that the transport properties of a disordered system can be expressed in terms of a single extensive variable, which can be chosen to be the conductance G (i.e. the inverse of the resistance, *not* the conductivity, which is the inverse of the resistivity). In particular, consider a block of disordered d-dimensional material. What is the change in the conductance when we join 2^d such blocks together to form a new block with all d of its dimensions doubled? The assumption (which can be justified) is that we can write

$$G(2L) = f(G(L)) \tag{5.6}$$

or, as a differential equation,

$$\frac{\mathrm{d}\ln g}{\mathrm{d}\ln L} = \beta(\ln g) \tag{5.7}$$

where $g = (\hbar/e^2)G$ is the dimensionless conductance, L is the length of an edge of a d-dimensional hypercube and β is some function of $\ln g$.

What are the properties of the function $\beta(\ln g)$? For strong disorder, the states are highly localized and we expect g to fall exponentially with the size of the system:

$$g = g_0 \exp(-\alpha L) \tag{5.8}$$

where α is an inverse decay length. Substituting this relation into (5.7) yields

$$\beta(\ln g) = \ln g - \ln g_0 \tag{5.9}$$

Thus, for $g \ll 1$, $\beta(\ln g)$ is always negative.

In the case of weak disorder the classical behaviour of the conductance (Exercise 7) should be valid:

$$g = \frac{\hbar}{e^2}\sigma L^{d-2} \tag{5.10}$$

and we obtain

$$\beta(\ln g) = d - 2 \tag{5.11}$$

Thus, for $g \gg 1$, $\beta(\ln g)$ is positive for $d = 3$, negative for $d = 1$ and zero for $d = 2$. Since g and β should be smooth functions of disorder, energy, etc., β must change sign for $d = 3$, may change sign for $d = 2$ and probably does not change sign for $d = 1$ (Fig. 5.6).

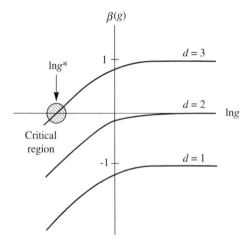

Figure 5.6.   The β function (5.7) for the conductance g in 1, 2 and 3 dimensions.

What then is the meaning of the sign of β? If β is negative, then according to (5.7), g is decreasing with increasing L, so β becomes even more negative, and so on until g vanishes. That is, there is no conductance. When β is positive, by contrast, g is increasing with increasing L and will eventually approach the classical behaviour in (5.10) and (5.11). Thus, there is no metal-insulator transition for $d = 1$. All states are localized, just as in the percolation example discussed in Section 5.2.1. For $d = 3$, there is always a *metal-insulator transition*, which corresponds to the point $\beta = 0$.

What about $d = 2$? This is the marginal case. The existence of a transition depends on whether $\beta(\ln g)$ approaches zero from above or from below as $g \to \infty$. Abrahams *et al.* (1979) were able to show that the leading term in an expansion of (5.7) in $1/g$ has the form

$$\beta(\ln g) = -\frac{a}{g} \tag{5.12}$$

Thus, β is always negative and all states are localized. In other words, there is no true metallic conductivity in two dimensions.

While it is true in a strict mathematical sense that all states are localized in two dimensions, this result requires some interpretation. After all, there is no shortage of experimental evidence that there is considerable conductivity in some 2D systems (e.g. HEMTs, as discussed in Chapter 10). Moreover, there have been very few attempts to calculate the actual numerical value of the localization length for a real 2D system. Results of computer simulations (MacKinnon and Kramer, 1983a) suggest that it can be of the order of *centimetres*, even when the fluctuations in the potential are of the same order of magnitude as the band width. If we ask for the localization length in a very pure sample, then numbers larger than the length scale of the universe (e.g. $10^{10^{30}}$) tend to emerge (MacKinnon and Kramer, 1983b). Is it meaningful then to talk about localization in this case, when the localization length is often much larger than the sample size?

5.3.2   Experiments on Weak Localization

In practice, although localization often cannot be observed directly there are various precursor effects which can be observed fairly easily. These are collectively referred to as *weak localization*. By substituting (5.12) into (5.7) and integrating, we obtain a formula for the conductivity in two dimensions:

$$\sigma = \sigma_0 - a \ln L \tag{5.13}$$

This is still not of much use to us. The sample size is not an easily varied quantity. However, the discussion so far has been only in terms of disorder effects or elastic scattering. Inelastic effects such as scattering by phonons and by other electrons must also be considered. There is one important distinction between elastic and inelastic scattering. In elastic scattering there is a well-defined phase relationship between an incident and a scattered wave, whereas inelastic scattering destroys such phase coherence. Since localization is really an interference phenomenon, it can be destroyed by inelastic scattering. This can be built into our picture in a simple way. Equation (5.13) is valid until the electron is scattered inelastically, so we can identify the length L with the inelastic scattering length L_{inel}. In general, the inelastic scattering length scales with the temperature as $L_{inel} \propto T^{-\alpha}$, so that by substituting this into (5.13) we obtain

$$\sigma = \sigma_0 + \alpha a \ln T \qquad (5.14)$$

This logarithmic temperature dependence of the conductivity has been observed in a number of systems (MOSFETs, thin films, etc.) and was considered a confirmation of the concept of weak localization in two dimensions.

5.3.3   Quantum Interference

What is the origin of the negative coefficient in equation (5.12)? In order to understand this we first consider a simpler problem: that of quantum interference between two waves travelling in opposite directions around a ring. Our bulk system will then be treated as an ensemble of such rings.

Consider a disordered ring with contacts at two points diagonally opposite each other. We assume that the circumference of the ring is large compared to the electronic mean free path but small compared to the localization length. There are many scattering events but no localization. An electron which starts at one contact can travel to the other contact by one of two routes (Fig. 5.7(a)). The two waves which arrive at the other side have been scattered differently. There is, therefore, no particular phase relationship between them. On the other hand, if we follow the two waves all the way around the ring and consider the effect back at the origin, then the two waves have been scattered *identically* (Fig. 5.7(b)). One is the time-reversed case of the other. They thus arrive back at the origin with the same phase. The probability of returning to the origin is twice what it would have

(a)   (b)

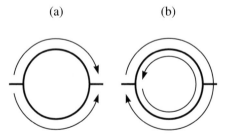

Figure 5.7.   Scattering paths round a ring. (a) Waves round opposite sides of the ring interfere randomly on the other side as the paths travelled are different. (b) Waves round the whole ring but in opposite directions interfere constructively back at the origin as the paths travelled are the same.

been had we ignored the interference between the waves and simply added their intensities.

If we consider the probability that a particle in a general disordered system will eventually return to the origin, then the expression we obtain will contain terms which refer to pairs of waves which go around rings in opposite directions, in addition to lots of other terms. These other terms, at least for weak scattering, do not produce any non-classical effects, but the ring interference terms still give an enhanced probability of returning to the origin, and thus a tendency towards localization.

The same interference phenomenon can also be described in **k**-space. In this case a wave which starts with a wavevector **k** has a higher probability of being scattered to −**k** than to any other direction. This *correlated backscattering* has given rise to a number of experiments that have looked for optical analogies of weak localization. To avoid any misunderstanding it should perhaps be pointed out that backscattering is different from specular reflection. In the latter case only the component of **k** perpendicular to the surface is reversed, whereas in the weak localization effect all the components of the wavevector are reversed.

5.3.4   Negative Magnetoresistance

What happens when we introduce a magnetic field into this system? Consider again a single ring of radius R. When a magnetic flux, $\Phi = \pi R^2 B$, is fed through this ring the momentum term in the Schrödinger equation[†] is changed in the usual way (Goldstein, 1950) according to the replacement

$$p^2 \rightarrow (\mathbf{p} - e\mathbf{A})^2$$

$$= (\mathbf{p} - \tfrac{1}{2}eBR\hat{\theta})^2 \qquad (5.15)$$

where $\mathbf{A} = \tfrac{1}{2}BR\hat{\theta}$ is the magnetic vector potential. This breaks the symmetry between +**k** and −**k**. Thus, the two opposite paths around the ring are no longer equivalent, the probability of a particle returning to the origin is no longer enhanced and there is no enhanced backscattering. Translated into the language of a solid rather than a single ring, we see that the chief mechanism which leads to weak localization is no longer active. Hence, the resistance decreases in a magnetic field, i.e. negative magnetoresistance (Fig. 5.8).

Returning though to the single ring, it so happens that it is now possible to make small rings or cylinders with dimensions such that these phenomena can be observed directly. Firstly, consider again the interference between two waves which cross to the opposite side of the ring by different routes. The phase difference between the two waves will be

$$\theta = \theta_0 + \left(2\pi R \times \frac{eBR}{2\hbar}\right) = \theta_0 + \frac{e}{\hbar}\Phi \qquad (5.16)$$

[†] Strictly speaking, we are referring here to the tangential component of the momentum, which is the only important component in a simple ring.

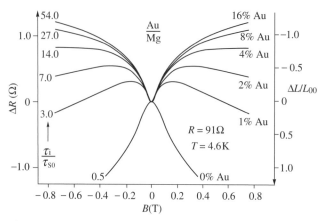

Figure 5.8.   The magnetoresistance ΔR of thin Mg–films. The clean film shows a negative magnetoresistance indicating localization. When the film is covered with a small number of gold atoms the magnetoresistance becomes positive due to increasing spin-orbit scattering. The right scale shows the magnetoconductance ΔL. On the left, the ratio of the inelastic scattering time and the spin-orbit scattering time is indicated (Adapted from G. Bergmann (1984), *Phys. Rep.* **107**, with permission of Elsevier Science.

This has the value $\theta = \theta_0 + 2n\pi$ whenever $\Phi = n(h/e)$. Therefore, we expect the current through the ring to vary with a period of one flux quantum.[†] This is one variant of the well known Aharonov–Bohm effect (Aharonov and Bohm, 1959). In fact, when Sharvin and Sharvin (1981) performed the experiment on a hollow magnesium cylinder the period was found to be $2h/e$ (Fig. 5.9). Why? It certainly does not indicate pairing of electrons as might be suspected by anyone familiar with flux quantization in superconducting rings.

In fact, Altshuler, Aronov and Spivak (1981) showed that there is an alternative interference process with precisely this period. As before, we must consider interference back at the origin. In this case the total flux enclosed is doubled and the period is, therefore, $2h/e$. However, since the two paths are identical in the absence of a magnetic field, the term θ_0 vanishes. By contrast, in the Aharonov–Bohm effect, θ_0 is non-zero with a random value from sample to sample. Since Sharvin and Sharvin's cylinder may be considered as an ensemble of such rings, the phase is randomized between samples and no oscillation of the resistance with period h/e is observed.

5.3.5   Single Rings and Non-local Transport

More recently, it has become possible to etch very fine patterns on metal films, leaving very fine wires. An example is shown in Fig. 5.10. Note the periodic

---

[†] The flux quantum is h/e rather than $h/2e$, as we are dealing with a single electron effect rather than one due to pairs of electrons, such as superconductivity.

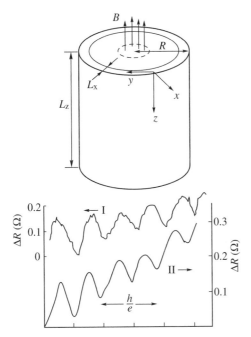

Figure 5.9. Aharonov–Bohm like magnetoconductance oscillations observed in normally conducting Mg cylinders of diameter 1.5 μm by Sharvin and Sharvin (1981). Left and right resistance scales correspond to samples 1 and 2, respectively. The periodicity of the oscillations corresponds to $\Delta\Phi = h/2e$. (Courtesy V. Gantmakher)

oscillations of the current as a function of the magnetic fields with the period of a single flux quantum (h/e), in contrast with the Sharvin and Sharvin experiment discussed in the preceding section.

An even more dramatic example of quantum interference effects on transport in microstructures can be seen in Fig. 5.11. Here the figure without the ring (or head) shows universal conductance fluctuations (Section 9.2.3) whereas, in the second figure, with a ring, a periodic oscillation is superimposed. Note that, classically, the ring is irrelevant, as it constitutes a dead end for the current, but that quantum interference between different paths around the ring can still contribute, leading to the oscillations. This is the first of a range of phenomena involving *non-local* transport, in which classically irrelevant interference paths can contribute to transport.

Consider a sample with several different leads in which a current is sent between leads k and l and a potential difference is measured between leads i and j. The result may be defined in terms of a generalized resistance $R_{ij;kl}$ such that

$$V_{ij} = R_{ij;kl}I_{kl} \tag{5.17}$$

This is a very general notation for describing most common transport measurements, and is often used to represent non-local effects.

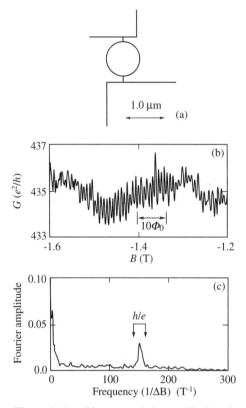

Figure 5.10. Aharonov–Bohm oscillations in a small ring. The period of the oscillations is one flux quantum (i.e. $\Delta\Phi = h/e$) (Umbach *et al.*, 1987). (Courtesy C. Van Haesendonck)

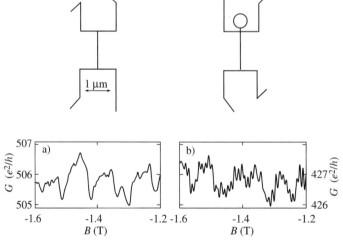

Figure 5.11. Non-local transport in thin wires. In (a) only random fluctuations are observed. In (b), however, interference round the 'head' contributes a periodic oscillation (Umbach *et al.*, 1987). (Courtesy C. Van Haesendonck)

5.3.6   Spin-orbit Coupling, Magnetic Impurities, etc.

Soon after the discovery of weak localization the quantum Hall effect (Section 5.5) was discovered. It rapidly became clear that equation (5.12) is not universally valid. In fact, there are three important exceptions. Besides high magnetic fields (see below), spin-orbit coupling and magnetic impurities will also give deviations.

In the case of spin-orbit coupling the deviation from the classical behaviour is positive. In fact, it has not been shown analytically that there can be any localized states for purely spin-orbit scattering. In a beautiful set of experiments on Au doped Mg, Bergmann (1984) was able to demonstrate the validity of the perturbation theory for such systems (Fig. 5.8) (note that spin-orbit scattering rises as Z^4). The spin-orbit effect is sometimes termed *weak anti-localization*.

Magnetic impurities destroy the weak localization effect by destroying the time reversal symmetry. The β-function has a leading term of $-a/g^2$, and the localization is even weaker than before. Magnetic impurities can also give rise to a divergence of resistance at low temperatures known as the Kondo (1964) effect.

5.3.7   Universal Conductance Fluctuations

So far in our discussion of the conductivity of disordered systems, we have implicitly assumed that the transport properties are *self-averaging*. As is usual in statistical mechanics, we have assumed that the sample is so big that the distribution of possible values of the resistance is very narrow, essentially a δ-function in the case of a large enough sample. In fact, this assumption is invalid. As long as we are working in a regime where the inelastic scattering length is larger than the sample size, the sample cannot be considered as made up of a large number of statistically independent systems. A small change in one place may have consequences for the entire sample. In the regime discussed earlier in this section, the conductance is a number of order e^2/h. In fact, as $T \rightarrow 0$ the standard deviation $(\delta g)^2$ is also of order e^2/h. Note that this behaviour is exactly what would be expected if the conductance takes the values e^2/h or zero randomly.

Experimentally this can be measured not by comparing different samples, but by looking at the way the conductance depends on such quantities as gate voltage or magnetic field. In the former case the Fermi level is moved through the spectrum, alternately seeing regions of allowed states and gaps. In the latter the spectrum is moved in a systematic way. The results look like noise (Fig. 5.11(a)). Unlike noise, however, the structure is reproducible. Different samples behave qualitatively similarly but differ in details. These *universal conductance fluctuations* are discussed at length in Section 9.2.3.

5.3.8   Ballistic Transport

By using a so-called *split gate* it is possible to study the transition from 2D to one-dimensional behaviour. If the scattering is weak and the one-dimensional channel short enough, it is possible to measure the remarkable phenomenon discussed in Section 3.4.1. When the conductance of such a channel is measured as a function of electron density (i.e. conventional gate voltage) it is found to be quantized as (Section 3.4.1)

$$G = \frac{2e^2}{h} i_{\max}$$

5.4   Interaction Effects

5.4.1   The ln T Correction

Although the ln T term in equation (5.14) seems to constitute a proof of weak localization, there is unfortunately another effect which gives rise to a similar term. If we consider interacting free electrons and treat the disorder with perturbation theory, the term α in (5.14) becomes $1 - F$, where F depends on the details of the particle–particle scattering. F is difficult to estimate, but is probably of order unity. Given the uncertainties in the coefficients it is impossible to distinguish between the two effects on the simple basis of resistance measurements alone.

However, the interaction effect leads to a conventional *positive* magnetoresistance. Thus the two effects can be distinguished by studying the influence of a magnetic field on the resistance. In fact, the interaction effect depends on the electron density, such that Uren *et al.* (1980) were able to measure a change in the sign of the magnetoresistance as the gate voltage is varied on a MOSFET.

5.4.2   Wigner Crystallization

A gas of electrons behaves very differently from a gas composed of neutral weakly interacting particles. One of the most striking differences is the behaviour of these two types of gases as a function of the density. At large densities, interactions between the particles in atomic and polyatomic gases become increasingly important. But for an electron gas, the phenomenon of *screening* leads to behaviour that for many purposes may be regarded as that of free electrons. Thus, a high-density electron gas behaves essentially like an ideal gas of fermions. As the density of an atomic or polyatomic gas is lowered, the interactions diminish in importance and the gas approaches ideal behaviour. For an electron gas, however, decreasing the density *increases* the effect of the Coulomb potential because the screening effect becomes much less effective.

These observations led Eugene Wigner (Wigner, 1934, 1938) to propose the existence of a lattice of electrons as the ground state of an interacting gas – what is now called a *Wigner crystal* (Mellor, 1992). Wigner argued that below a certain critical density the kinetic energy will be negligible in comparison to the potential energy. Thus, at low enough temperatures the energy of a system of electrons would be dominated by the pair-wise Coulomb potential between the particles and the behaviour of the gas will be determined by the configuration that minimizes this potential energy. Since the potential of a random array is higher than that of an ordered array, electrons in this regime will form a *crystal*. In three dimensions, the case that Wigner considered, the lowest potential energy is obtained for a body-centred cubic crystal.

In two dimensions there are two regimes to consider: the *quantum* regime, where $k_{\mathrm{B}}T \ll E_{\mathrm{F}}$, and the *classical* regime, where $k_{\mathrm{B}}T \gg E_{\mathrm{F}}$. The classical regime of Wigner crystallization is relatively easy to achieve when the density n_s of electrons is small, since $E_{\mathrm{F}} \propto n_s$. The potential energy V per electron can then be estimated by $V \approx e^2/4\pi\varepsilon_0 r \propto n_s^{1/2}$ The average kinetic energy can be obtained from the equipartition theorem, so the cross-over temperature where the kinetic and potential energies are of comparable magnitude is $T \propto n_s^{1/2}$. The first observation of a Wigner crystal was, in fact, in the classical regime for electrons on the surface of liquid helium (Grimes and Adams, 1979).

The higher densities n_s (and lower effective masses) of 2DEGs in semiconductors means that $E_{\mathrm{F}} \gg k_{\mathrm{B}}T$. In this (quantum) regime, the kinetic energy of the electrons remains non-zero down to the lowest temperatures, being of order E_{F}, which leads to the kinetic and potential energies being of comparable magnitude. Thus, electrons in most semiconductors remain in a 'liquid' state even at the lowest temperatures. Achieving lower densities is technically very demanding, so an alternative approach has been to apply a large ($\sim 10\,\mathrm{T}$) magnetic field perpendicular to the 2DEG which has the effect of confining electrons to small ($\sim 5\,\mathrm{nm}$) orbits. This makes the 2DEG easier to solidify and there have been a number of experiments carried out that support the notion that 2DEGs in GaAs crystallize in very high magnetic fields and low temperatures (Goldman *et al.*, 1990).

5.5   The Quantum Hall Effect

5.5.1   General

The classical picture of carrier mobility in a magnetic field is modified substantially for the corresponding measurement on a 2DEG. For a 2DEG in the (x, y)-plane in a uniform magnetic field along the z-direction, $\mathbf{B} = B\mathbf{k}$, the Schrödinger equation may be written in the form

$$\frac{1}{2m}\left[p_x^2 + (p_y - eBx)^2\right]\psi = E\psi \tag{5.18}$$

where $\mathbf{A} = Bx\,\mathbf{j}$ is the magnetic vector potential. By using the substitution

$$\psi(x, y) = \varphi(x)\exp(\mathrm{i}ky) \tag{5.19}$$

this equation can be transformed into the Schrödinger equation for a harmonic oscillator,

$$\left[\frac{p_x^2}{2m} + \frac{e^2 B^2}{2m}\left(x - \frac{\hbar k}{eB}\right)^2\right]\varphi = E\varphi \tag{5.20}$$

where $\omega_c = eB/m$ is the cyclotron frequency. If we focus only on the orbital motion of the electrons, then the allowed energies are those of a quantum harmonic oscillator:

$$E_n = (n + \tfrac{1}{2})\hbar\omega_c \tag{5.21}$$

The states corresponding to different n are called *Landau levels*. The transformation (5.19) shows that the effect of the magnetic field is to change only the motion along the x-direction; the motion along the y-direction corresponds to free electrons. The solutions φ of the Schrödinger equation (5.20) are harmonic oscillator wavefunctions centred at x_0, which is given by

$$x_0 = \frac{\hbar k}{eB} \tag{5.22}$$

where k is the wavevector associated with the motion of the electron along the y-direction. Thus, x_0 is seen to be a good quantum number.

The energy spectrum of this system is thus a regularly spaced sequence of Landau levels, each separated by an energy $\hbar\omega_c$. To distribute the original density of states among these discrete levels requires each level to have an enormous degeneracy. This degeneracy can be calculated from the number of centre co-ordinates x_0 that can be accommodated within the sample subject to the Pauli exclusion principle. For a sample of dimensions $L_x \times L_y$, the centre coordinates are separated along the y-direction by

$$\Delta x_0 = \frac{\hbar}{eB}\Delta k = \frac{\hbar}{eB} \times \frac{2\pi}{L_y} = \frac{h}{eBL_y} \tag{5.23}$$

which corresponds to a degeneracy N_0 of

$$N_0 = \frac{L_x}{\Delta x_0} = \frac{L_x L_y eB}{h} \tag{5.24}$$

The degeneracy N per unit area is therefore given by

$$N = \frac{N_0}{L_x L_y} = \frac{eB}{h} \tag{5.25}$$

Note that this degeneracy depends on two fundamental constants (e and h) and on an experimentally controllable quantity (B). In particular, there is no dependence on any parameters associated with the particular material, such as the effective mass of the electrons. When disorder or impurities are included this degeneracy is broken, but as long as the cyclotron energy is large compared with the potential fluctuations the basic structure remains.

Let us now consider the behaviour of the conductivity as the electron density is varied (e.g. in a MOSFET, as discussed in Chapter 10). Whenever the 2D electron density n_s is varied by more than the degeneracy of a Landau level, the Fermi level jumps from one Landau level to another. When the Landau level is broadened we expect the conductivity roughly to follow the density of states, with a maximum in the longitudinal conductivity σ_{xx} when each level is half-filled. The conductivity will therefore vary periodically as the density is varied. The Fermi level E_F is in a Landau level when

$$E_F = (n + \tfrac{1}{2})\hbar\omega_c = (n + \tfrac{1}{2})\hbar\frac{eB}{m} \qquad (5.26)$$

In the middle of the Landau level the index n can be related to n_s by using the degeneracy N. To do so we must ignore the contribution to the energy due to the interaction of electron spin with the magnetic field, in effect doubling the degeneracy. Thus

$$n_s = 2N(n + \tfrac{1}{2}) = 2\frac{eB}{h}(n + \tfrac{1}{2}) \qquad (5.27)$$

Alternatively we can use the filling factor $i = 2n + 1$ to obtain

$$n_s = iN = i\frac{eB}{h} \qquad (5.28)$$

It is instructive to relate this to the Fermi energy E_F by using (5.26) and (5.27):

$$n_s = 2\frac{eB}{h} \times \frac{E_F}{\hbar eB/m} = \frac{E_F m}{\pi\hbar^2} \qquad (5.29)$$

which is exactly the relationship one obtains from the density of states in (2.14) in the absence of the magnetic field. If the density of states at the Fermi energy of the 2DEG is zero, i.e. if the Fermi lies between filled and unfilled Landau levels, the carriers cannot be scattered and the cyclotron orbit drifts in a direction perpendicular to the electric and magnetic fields. In this case, the conductivity σ_{xx} (current flow in the direction of the electric field) becomes zero, since the electrons are moving like free particles perpendicular to the electric field with no diffusion (originating from scattering) in the direction of the field. Experimentally, σ_{xx} is never precisely zero, but becomes unmeasurably small at high magnetic fields and low temperatures.

In the Shubnikov–de Haas effect we measure the longitudinal resistivity ϱ_{xx} of the sample as the magnetic field is varied. At low temperatures, $k_B T \ll \hbar\omega_c$, where $\hbar\omega_c$ is the Landau level separation, $\sigma_{xx} \to 0$ and $\varrho_{xx} \to 0$ whenever the Landau levels are full, i.e. equation (5.28) holds. By plotting ϱ_{xx} against $1/B$ and looking for the periodicity it is possible to measure the electron density. Complications occur in real systems due to the additional spin splitting and to valley effects in semiconductors. The reason why both σ_{xx} and ϱ_{xx} vanish for full Landau levels can be appreciated from Exercise 9.

5.5.2   The Quantum Hall Effect Measurements

Probably the most remarkable effect observed in low-dimensional systems was first discovered by von Klitzing, Dorda and Pepper in 1979 (von Klitzing *et al.*, 1980). Their results are illustrated in Fig. 5.12. A constant current I_x is imposed on the 2DEG between source and drain (see inset Fig. 5.12), the longitudinal voltage U_{pp} measured between two probes along the sample and the Hall voltage U_H measured between two probes across the sample, as the magnetic field B perpendicular to the 2DEG is varied. The voltage measurements are usually interpreted in terms of the component of a resistivity tensor ϱ with

$$\varrho_{xy} = \frac{U_H}{I_x} \quad \text{and} \quad \varrho_{xx} = f\,\frac{U_{pp}}{I_x} \tag{5.30}$$

where f is a geometrical factor. The components of the conductivity tensor $\sigma = \varrho^{-1}$ are then related to those of ϱ by (Exercise 9)

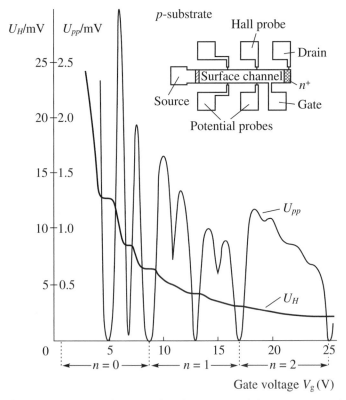

Figure 5.12.   Normal U_{pp} and Hall U_H potential versus gate voltage, V_g, in a silicon MOSFET at $T = 1.5$ K with a magnetic field of 18 T and a source drain current of 1 μA (von Klitzing *et al.*, 1980). The inset shows a diagram of the sample. (Courtesy K. von Klitzing)

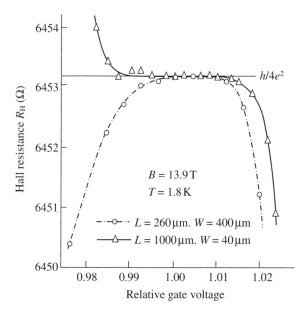

Figure 5.13.   Close up of a single plateau for two samples of different geometry. Note the scale and the similarity between the plateau values for the two samples (von Klitzing *et al.*, 1980). (Courtesy K. von Klitzing)

$$\varrho_{xx} = \varrho_{yy} = \frac{\sigma_{xx}}{\sigma_{xx}^2 + \sigma_{xy}^2}, \quad \varrho_{xy} = -\varrho_{yx} = -\frac{\sigma_{xy}}{\sigma_{xx}^2 + \sigma_{xy}^2} \qquad (5.31)$$

For a 2DEG with electron density per unit area n_s given by (5.28), one expects (Exercise 10)

$$\varrho_{xy} = \frac{h}{ie^2} \qquad (5.32)$$

But the real surprise comes on closer inspection of the plateau, as in Fig. 5.13. The value of ρ_{xy}, the Hall resistivity, for this plateau is $h/4e^2 \, (1 \pm 10^{-6})$. The measurement can now be almost routinely carried out with an accuracy of around 1 part in 10^8. This accuracy is astonishing. Typically, agreement between theory and experiment of 1% or even 10% is considered good in condensed matter physics. The only other phenomenon which comes anywhere close to this accuracy is the AC Josephson effect in superconductors. In fact, the accuracy is such that the quantum Hall effect has now been internationally adopted as the standard of resistance.

5.5.3   The Semiclassical Theory

In order to gain some understanding of the effect we consider a simple approximate picture. It should be borne in mind, however, that a full explanation may not contain any significant approximations as the result appears to be exact.

The equation of motion for a classical particle in a magnetic field is

$$m \frac{\partial \mathbf{v}}{\partial t} = e\mathbf{B} \times \mathbf{v} + e\mathbf{E} \tag{5.33}$$

where $\mathbf{v}$ is the velocity of the particle and $\mathbf{B}$ and $\mathbf{E}$ the magnetic and electric fields, respectively. The solution of this equation has the form

$$\mathbf{v} = \omega_c \left(A_x \cos \omega_c t, A_y \sin \omega_c t \right) + \frac{1}{B^2} \mathbf{B} \times \mathbf{E} \tag{5.34a}$$

$$\mathbf{r} = \mathbf{r}_0 + \left(A_x \sin \omega_c t, -A_y \cos \omega_c t \right) + \frac{t}{B^2} \mathbf{B} \times \mathbf{E} \tag{5.34b}$$

This behaviour consists of two parts: a circular motion with radius A and frequency $\omega_c = eB/m$ (the cyclotron frequency), and a drift velocity perpendicular to both fields with magnitude $v_d = E/B$:

$$v_d = \frac{1}{B^2} \mathbf{B} \times \mathbf{E} \tag{5.34c}$$

In our case, where we define the magnetic field to be perpendicular to the 2D plane and the electric field to be in this plane, it is important that v_d is perpendicular to the electric field.

If we return to the general problem of an electron in a disordered system in a magnetic field and assume now that the radius A is small compared with the rate at which the potential changes, then the electron sees a constant local potential gradient. This is of course indistinguishable from an electric field. With this idea in mind we substitute the local potential gradient for the electric field in the drift velocity to obtain

$$v_d = \frac{1}{B^2} \mathbf{B} \times \nabla \phi \tag{5.35}$$

Hence, the drift velocity is always perpendicular to the local potential gradient, $\nabla \phi$.

Let us return now to our analogy with a random landscape. The drift velocity follows the contour lines rather than going up or down the mountains. It is as though it had misread its map and mistaken the contour lines for paths. Using Fig. 5.1 again, we see that most contours are around the tops of mountains or the bottoms of valleys. In fact, for a random landscape there is only one level where a contour crosses the whole system. This is the situation depicted in Fig. 5.1(c). Using our previous picture of varying water levels, the electron is now constrained to follow the shoreline. It is only at this one level that it can manage to cross.

Translating this result into the language of localized and extended states we have the result that all states are localized except for those at a single level. Thus, almost none of the states can contribute to the current. Clearly this gives us a simple mechanism by which it is possible to vary the number of electrons in the system while the current remains unchanged.

Now let us consider the current through a cross-section of the system, as illustrated in Fig. 5.14. Firstly, consider a system with a uniform charge density, $n_s e$ corresponding to a full Landau level. This condition will be relaxed later. The current density through the cross-section is

$$J_x = n_s e v_d = \frac{n_s e}{B} \frac{\partial \phi}{\partial x} \tag{5.36}$$

and the total current is given by the integral from one side to the other

$$I_x = \frac{n_s e}{B} \int \frac{\partial \phi}{\partial x} \, dx = \frac{n_s e}{B} \Delta \phi \tag{5.37}$$

Hence, the total current is independent of the details of the potential but only depends on the potential difference, $\Delta \phi$, across the sample. Since, however, we already know that only the electrons at a particular level can contribute to the current, it must be possible to remove the electrons from the localized states at the tops of mountains or the bottoms of valleys without altering the total current.

Since we have a well-defined current which does not change when electrons are added or removed, we have a mechanism for the plateaux in the quantum Hall effect. Note, however, that we do not yet have a value for the Hall conductivity associated with each plateau. For that we need quantum mechanics. The result e^2/h contains Planck's constant, after all.

We return therefore to (5.34) and look at the part describing circular motion. The root mean square (rms) deviation of the electron from the orbit centre is

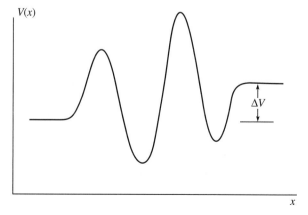

Figure 5.14.   Cross-section through a random potential with a net potential difference ΔV between the two sides.

$\Delta x = A/\sqrt{2}$. Since the motion is circular, this immediately gives us the rms deviation of the momentum, namely $\Delta p = m\omega_c \Delta x$, where $\omega_c = eB/m$, the cyclotron frequency. Using the Heisenberg uncertainty principle, $\Delta x \Delta p_x = \hbar$, gives us a formula for the radius of the orbit:

$$A^2 = \frac{2\hbar}{m\omega} = \frac{2\hbar}{eB} \qquad (5.38)$$

The Pauli exclusion principle reserves the area of a circle of radius A for each electron, so that the maximum uniform charge density must be

$$n_s e = e\frac{1}{\pi A^2} = \frac{e^2 B}{2\pi\hbar} = \frac{e^2}{h}B \qquad (5.39)$$

Substituting this in (5.37) gives us the result we have been seeking:

$$I_y = \frac{e^2}{h}\Delta\phi \qquad (5.40a)$$

or

$$J_y = \sigma_{xy} E_x \qquad (5.40b)$$

which serves as a definition of the Hall conductivity, σ_{xy}. Our analysis in terms of a semiclassical picture of electron motion in a magnetic field and a slowly varying potential illustrates several important aspects of the physics of the quantum Hall effect.

• The presence of disorder and, hence, of localized states is vital to the explanation of the effect and its precision.
• The density corresponding to a full Landau level and, hence, the quantized state corresponds to that of an incompressible fluid. The Pauli exclusion principle forbids us to compress the system.
• Higher Landau levels contain the same density of electrons, so a full Landau level corresponds to a Hall conductivity of $\sigma_{xy} = ie^2/h$.
• The number of open contours in the above semiclassical argument is proportional to the potential difference across the sample. In fact, that number is independent of the disorder. Hence, there is no reason to worry that a macroscopic current is being carried by a single state.

5.5.4   The Fractional Quantum Hall Effect

Just as the quantum Hall effect seemed to be understood, at least at a basic level, Tsui, Störmer and Gossard (1982) of Bell Laboratories in Murray Hill, New Jersey published some remarkable experiments on GaAs/$Al_{0.3}Ga_{0.7}$As prepared by MBE. The experiment produced the startling result that the Hall conductivity not only has steps in integer multiples of $\sigma_{xy} = n(e^2/h)$, where n is an integer, but they also observed steps at $n = \frac{1}{3}$ and $n = \frac{2}{3}$. Since then features have been observed at $n = \frac{1}{3}, \frac{2}{3}, \frac{2}{5}, \frac{3}{5}, \frac{3}{7}, \frac{4}{7}, \frac{4}{9}, \frac{5}{9}, \frac{4}{3}, \frac{5}{3}$ and many more rational fractions, all with odd denominators (Chang *et al.*, 1984). Typically, these have an accuracy of 10^{-5} for the $\frac{1}{3}$ feature and lower accuracy on the others. For some fractions the feature is only observed as a peak in σ_{xx}. Fig. 5.15 shows a typical spectrum.

There have been many attempts to explain these results. Some of the theories rely on some very exotic mathematics and many are still controversial. Here, we will try to give a simple explanation based on those aspects on which there is general agreement.

Consider first the wavefunction for 2D electrons in a perpendicular magnetic field expressed in the so-called symmetric gauge, where the magnetic vector potential in (5.19) takes the form $\mathbf{A} = \frac{1}{2}(-By, Bx, 0)$. The wavefunctions in the lowest Landau level are

$$\psi(z) \propto z^m \exp\left(-\frac{|z|^2}{4l_c^2}\right) \tag{5.41}$$

where $z = x + iy$ and l_c is the cyclotron radius. Laughlin (1983) proposed the following generalization of this to N electrons

$$\Psi_m(z_1, z_2, \ldots, z_N) \propto \prod_{j<k}(z_j - z_k)^m \prod_i \exp\left(-\frac{|z_i|^2}{4l_c^2}\right) \tag{5.42}$$

A useful insight into the meaning of this wavefunction can be obtained by thinking of $|\Psi|^2$ as a probability distribution of a classical plasma, obeying a partition

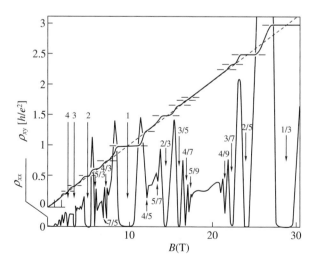

Figure 5.15. A recent example of the Hall resistivity, ρ_{xy}, and normal resistivity, ρ_{xx}, as a function of magnetic field, in Tesla (Willet *et al.*, 1987). Note the large number of fractional features marked. (Courtesy A. C. Gossard)

function $Z = \exp(-\beta\Phi)$ in which $\beta = 1/m$ and

$$\Phi = -2m^2 \sum_{jk} \ln(z_j - z_k) + \tfrac{1}{2}m \sum_i \frac{|z_i|^2}{l_c^2} \qquad (5.43)$$

This corresponds to a system of particles which repel each other logarithmically and a uniform background density $\rho_m \propto 1/m$.[†]

Laughlin's wavefunction has several properties which are useful for understanding the effect.

- Ψ_m corresponds to $1/m$th occupation of the lowest Landau level.
- When m is odd, the wavefunction is odd under exchange of electrons.
- This restriction to odd m explains why fractional densities with odd denominators are special.
- At the special densities it can be shown that the system has an energy gap (Laughlin, 1983): more energy is required to add an electron than is gained by removing one. As with the gap in a semiconductor, or, more accurately in this case, a superconductor, the state is stable unless excitations are possible across the gap, due to finite temperature, radiation, etc.

Interesting though Laughlin's wavefunction certainly is, in its simplest form it is restricted to fractions of the form $1/m$. How can we understand the occurrence of other fractions and the order in which they appear? A very simple semiclassical picture is illustrated in Fig. 5.16. Here, we see a ring of states with one-third of the states occupied (Fig. 5.16(a)). When one particle is added we find a situation as in Fig. 5.16(b). This is not the ground state, however. If the particles mutually repel one another the ground state will be one in which their density is most uniform, as shown in Fig. 5.16(c). Note the peculiar feature of the relaxed system: instead of a single *defect* due to the additional particle there are now three defects which are as far apart as possible. Thus, the additional particle is behaving as if it is three *quasi-particles*, each with charge of $\tfrac{1}{3}$. A similar result, with one particle removed, is shown in Figs. 5.16(d,e). Again there are three *defects*.

To a semiconductor physicist there should be no difficulty in understanding the concept of quasi-particles illustrated here. We are all well used to the concept of a *hole*, which is used to describe an almost full band of electrons. Indeed we often tend to forget that it is not a true particle.

We now require a leap of the imagination. It has been shown that a $\tfrac{1}{3}$ charged Landau level behaves as if it is *full* of particles of charge $\tfrac{1}{3}$. Similarly for all $1/m$th full levels. Consider now a level which is $1/m$th full of $\tfrac{1}{3}$ charged quasi-particles. These could now form a special state in which a new generation of *quasi*-quasi-particles is formed. These can then condense into another special state and form *quasi*-quasi-

---

[†] The solution of Poisson's equation in two dimensions is logarithmic, so that *truly* 2D particles would repel each other logarithmically. Alternatively, a system of charged rods in three-dimensional space also interacts logarithmically.

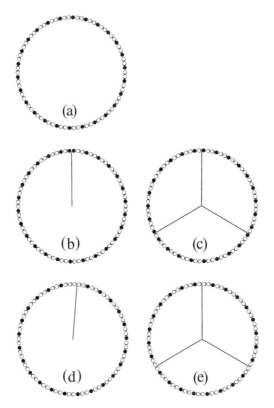

Figure 5.16.   Mutually repulsive classical particles restricted to specific ordered sites on a ring. (a) Density is $\frac{1}{3}$; (b) density $\frac{1}{3}$ with one additional particle (unrelaxed); (c) as (b) (relaxed); (d) density $\frac{1}{3}$ with one particle removed (unrelaxed); (e) as (d) (relaxed).

quasi-particles, and so on. A whole hierarchy of different states involving different generations of quasi-particles can arise, each corresponding to a different fractional density and each giving rise to a feature in the transport measurements.

EXERCISES

1.   The quantity e^2/h is sometimes called the *quantum of conductance* because it arises as the unit of quantization in, e.g., the quantum Hall effect, ballistic transport and universal conductance fluctuations.

  (a) Show that the quantum conductance has the correct units for a conductance (i.e. inverse ohms).
  (b) Show that a single quasi-1D channel in a crystalline system carries a current $I = (e^2/h)\Delta V$, where ΔV is the voltage drop along the length of the channel.

(c) Use the Ioffe–Regel criterion to show that in two dimensions the minimum metallic conductivity is $\sigma_{\min} \propto e^2/h$.

2.   Define the terms 'localized,' 'extended' and 'mobility edge' as used to describe the behaviour of electrons in a non-crystalline solid.

3.   Describe briefly the two main modes of electronic transport in a disordered system at low temperatures when the chemical potential lies in an energy range in which the states are localized.

4.   A measurement of the conductivity of a silicon MOSFET gives a straight line when the conductivity is plotted against the logarithm of the temperature. Describe the physical process which gives rise to the logarithmic temperature dependence and explain how this is related to the behaviour of electrons in small rings.

5.   The result of a 4-probe measurement of resistivity is often written in the form

$$V_{ij} = R_{ij,kl}I_{kl}$$

where the voltage difference, V_{ij}, between probes i and j is associated with the current, I_{kl}, between probes k and l.

(a) Two samples of wires of submicron dimensions are prepared as in Fig. 5.17. Explain (quantitatively where possible) the behaviour you would expect for the temperature and magnetic field dependence of $R_{12,34}$ and $R_{13,24}$ in these samples.
(b) A third sample is prepared in which many rings, like the one in Fig. 5.17(a) and nominally all of the same diameter, are arranged in parallel between the two horizontal wires. What differences would you expect to observe in this sample compared with the sample in Fig. 5.17(b)?

6.   The metal-insulator transition is often described. in terms of a β function

$$\frac{d\ln g}{d\ln L} = \beta(\ln g)$$

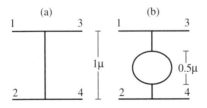

Figure 5.17.   Figure for Exercise 5.

where g is the dimensionless conductance, $g = (e^2/h)G$, and L is the length of a side of a cubic sample.

(a) By assuming that g is a smooth, i.e. differentiable, function of the Fermi energy, E_F for all finite L, show that g depends on L as $g(L/\xi)$, where ξ is independent of L.

(b) Close to the transition g can be written in the form

$$\ln g = \ln g_c + A(E_F - E_c)L^\alpha$$

where $\alpha = d\beta/d\ln L$ when $\beta = 0$. Hence, show that $\xi \propto |E_F - E_c|^{-\nu}$.

7. Verify equation (5.10) in the form

$$G = \sigma L^{d-2}$$

for the particular cases $d = 3$ and $d = 2$ starting from Ohm's law:

$$\mathbf{J} = \sigma \mathbf{E}$$

(a) In $d = 3$, consider a conductor of rectangular cross-section ac with the current flow parallel to a side of length b.

(b) In $d = 2$, consider a planar conductor of width a and current flow parallel to a side of length b.

8. The Schrödinger equation for electrons in the (x, y)-plane subject to a magnetic field B in the z-direction and an electric field E in the x-direction may be written, in the Landau gauge, as

$$\frac{\hbar^2}{2m}\frac{\partial^2 \psi}{\partial x^2} + \frac{1}{2m}\left(\frac{\hbar}{i}\frac{\partial}{\partial y} - eBx\right)^2 \psi + eEx\psi = \epsilon\psi$$

(a) Using the substitution $\psi \to \phi(x)\exp(iky)$, show that the Schrödinger equation may be transformed into that of a harmonic oscillator centred around

$$X = \frac{\hbar k}{eB} - \frac{meE}{e^2 B^2}$$

with a spring constant $K = e^2 B^2/m$, plus some additional constants.

(b) Using your knowledge of the harmonic oscillator, confirm that the eigen-energies are given by

$$\epsilon_n = (n + \tfrac{1}{2})\hbar\frac{eB}{m} + \text{additional terms}$$

and give an interpretation of the additional terms.

9. The relationship between the conductivity tensor σ and the resistivity tensor ϱ is $\sigma = \varrho^{-1}$.

(a) Derive the expressions for ϱ_{xx} and ϱ_{xy} in equation (5.30) starting from the conductivity tensor

$$\sigma = \begin{pmatrix} \sigma_{xx} & \sigma_{xy} \\ -\sigma_{xy} & \sigma_{xx} \end{pmatrix}$$

(b) At high B-field in a high-mobility 2DEG, $\sigma_{xy} \gg \sigma_{xx}$. In this limit, what is the relationship between ϱ_{xy} and σ_{xy}?

(c) In the limit in (b), what happens to ϱ_{xx} if $\sigma_{xx} \to 0$?

10. Assuming that the Hall bar in the inset to Fig. 5.12 has width a and that the distance between the longitudinal probes is b, show that if

$$\varrho_{xx} = f \frac{U_{pp}}{I_x} = fR_L$$

in the Hall condition ($J_y = 0$), then $f = a/b$. Furthermore, show that if

$$\varrho_{xy} = \frac{U_{\mathrm{H}}}{I_x}$$

in the Hall condition, then

$$R_{\mathrm{H}} = \frac{E_y}{J_x B_z} = \frac{\varrho_{xy}}{B_z}$$

Hence, using equations (5.34c) and (5.36), show that if (5.39) holds, then

$$\varrho_{xy} = \frac{h}{ie^2}$$

11. Shubnikov–de Haas oscillations were studied in a high-mobility 2DEG in a heterostructure at 4.2 K. Minima were observed in the longitudinal resistance at certain values of the perpendicular **B**-field and the filling factors i at the various minima were identified. A graph of i against $1/B$ was plotted and the data gave a straight line of gradient 8 T. Use this information to find n_s, the electron density per unit area in the 2DEG.

In the same experiment the longitudinal resistance at zero **B**-field was measured to be 100 Ω. The distance between the potential probes along the length of the Hall bar (b) was found to be 7.5 times the width of the bar (a). Use this information and the value of n_s calculated above to estimate the mobility of the 2DEG.

References

E. Abrahams, P. W. Anderson, D. C. Licciardello and T. V. Ramakrishnan, *Phys. Rev. Lett.* **42**, 673 (1979).

Y. Aharonov and D. Bohm, *Phys. Rev.* **115**, 485 (1959).

B. L. Altshuler, A. G. Aronov and B. Z. Spivak, *Sov. Phys. JETP Lett.* **33**, 94 (1981).

P. W. Anderson, *Phys. Rev.* **109**, 1492 (1958).

N. W. Ashcroft and N. D. Mermin, *Solid State Physics* (Holt, Rinehart and Winston, New York, 1976)

G. Bergmann, *Phys. Rep.* **107**, 1 (1984).

A. M. Chang, P. Berglund, D. C, Tsui, H. L. Störmer and J. C. M. Hwang, *Phys. Rev. Lett.* **53**, 997 (1984).

V. J. Goldman, M. Santos, M. Shayegan and J. E. Cunningham, *Phys. Rev. Lett.* **65**, 2189 (1990).

H. Goldstein, *Classical Mechanics* (Addison-Wesley, Reading, MA, 1950).

C. C. Grimes and G. Adams, *Phys. Rev. Lett.* **42**, 795 (1979).

A. F. Ioffe and A. R. Regel, *Prog. Semicond.* **4**, 237 (1960).

S. Kawaji, T. Igarashi and J. Wakabayashi, *Prog. Theor. Phys. Suppl.* **57**, 176 (1975).

K. von Klitzing, G. Dorda and M. Pepper, *Phys. Rev. Lett.* **45**, 494 (1980).

J. Kondo, *Prog. Theoret. Phys.* **32**, 37 (1964).

R. B. Laughlin, *Phys. Rev. Lett.* **50**, 1395 (1983).

P. E. Lindelof, J. Nørregaard and J. Hanberg, *Phys. Scr.* **T14**, 17 (1986).

A. MacKinnon and B. Kramer, *Z. Physik* **B53**, 1 (1983a).

A. MacKinnon and B. Kramer, in *High Magnetic Fields in Semiconductors*, G. Landwehr, ed., Lecture Notes in Physics, Vol. 177 (Springer, Berlin, 1983b) p. 74.

C. Mellor, *New Scientist* **135**(1833), 36 (1992).

N. F. Mott, *J. Non–Cryst. Sol.* **1**, 1 (1968).

N. F. Mott and E. A. Davis, *Electronic Processes in Non-Crystalline Materials*, 2nd edn (Clarendon Press, Oxford, 1979).

M. Pepper, in *The Metal Non-Metal Transition in Disordered Solids*, L.R. Friedman and D.P. Tunstall, eds. (SUSSP, Edinburgh, 1978) pp. 285.

D. Y. Sharvin and Y. V. Sharvin, *Sov. Phys. JETP Lett.* **34**, 272 (1981).

D. Stauffer and A. Aharony, *Introduction to Percolation Theory* (Taylor and Francis, London, 1994).

D. C. Tsui, H. L. Störmer and A. C. Gossard, *Phys. Rev. Lett.* **48**, 1559 (1982).

C. P. Umbach, P. Santhanam, C. Van Haesendonck and R. A. Webb, *Appl. Phys. Lett.* **50**, 1289 (1987).

M. J. Uren, R. A. Davies and M. Pepper, *J. Phys. C* **13**, L985 (1980).

E. Wigner, *Phys. Rev.* **46**, 1004 (1934).

E. Wigner, *Trans. Farad. Soc.* **34**, 678 (1938).

R. Willet, J. P. Eisenstein, H. L. Stormer, D. C. Tsui, A. C. Gossard and J. H. English, *Phys. Rev. Lett.* **59**, 1776 (1987).

6 Electronic States and Optical Properties of Quantum Wells

J. Nelson

6.1 Introduction

Recent progess in epitaxial growth techniques has promoted the use of semiconductor heterostructures in optoelectronic devices. The physics of these materials relies upon the similarity between the electronic band structures of the different semiconductors. If the bulk band structures are sufficiently similar then changes in composition can be represented primarily as changes in the band splittings and other bulk parameters. In a direct band-gap semiconductor an abrupt change in composition from wide to narrow band gap results in a discontinuity in the conduction and valence band profiles in the growth direction. The heterointerface so formed is Type I or Type II, depending on the band-gap alignments, determined by the conduction band offset (Fig. 6.1).

A quantum well (QW) is made by growing a thin layer – typically a few nanometres (nm) or 10s of nm – of narrower gap material within a wider-gap semiconductor, where the inserted layer is thin enough to cause quantum confinement of the carriers. QWs are similarly classified as Type I or II in direct-gap materials (Fig. 6.2).

In indirect gap materials we need to consider the band-edge discontinuities at different points in the band structure. The AlAs/GaAs heterointerface, for example, is Type I at the Γ point but Type II at the X point (Fig. 6.3). The overall band structure of an AlAs/GaAs QW system thus depends on the relative well and barrier widths.

In a multi-quantum well (MQW) system a series of QWs is separated by layers of wider-gap (barrier) material thick enough to isolate carriers in the wells (Fig. 6.4). A regular MQW with a barrier thin enough to allow carrier communication between the wells is known as a superlattice (SL).

The choice of materials for heterostructures is influenced by lattice spacings as well as by the symmetry of the crystal band structures and differences in the effective band gaps (Fig. 6.5). Strain forces at the interface between materials of very different lattice constant may produce imperfections which degrade the material and interface quality. However, in a small-period SL the thin layers may accommodate the alternating strain forces to produce useful heterostructures where the potential wells in different crystal bands are distorted by strain (Fig. 6.6).

180

Type I   Type II

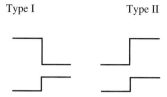

Figure 6.1.   Type I and Type II heterointerfaces.

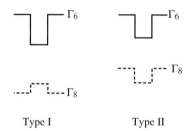

Type I   Type II

Figure 6.2.   Type I and Type II quantum wells.

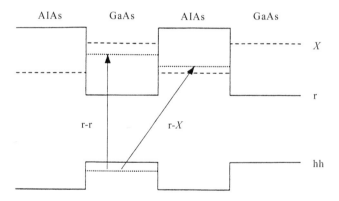

Figure 6.3.   Conduction and valence band profiles in AlAs/GaAs at the Γ point and the X points.

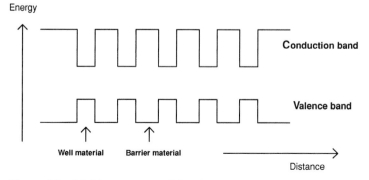

Figure 6.4.   Multi-quantum well band profiles.

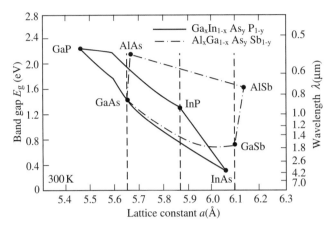

Figure 6.5. Room temperature band gap E_g versus lattice constant for the main binary III–V semiconductors. Tertiary alloys are represented by the lines joining relevant binaries.

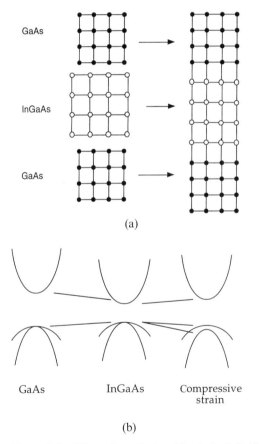

(a)

(b)

Figure 6.6. Dispersion relationships in bulk InGaAs and in strained GaAs/InGaAs.

The focus of this chapter will be the theoretical modelling of the electronic and optical features of quantum wells. We will restrict attention to the case of Type I QWs in direct-band-gap materials in the limit of low carrier concentrations. The basic theoretical framework, the envelope function approximation (EFA) as applied to heterostructures, is introduced in Section 6.2 and then applied to the electronic structure of a QW in the simple parabolic band model in Section 6.3. We discuss refinements to the electronic structure due to band mixing in Section 6.4, interwell coupling in Section 6.5, Coulomb interaction in Section 6.6, and applied electric fields in Section 6.7. Finally we calculate the optical absorption of a QW in Section 6.8 and discuss optical characterization and the features of QW optical spectra in Section 6.9.

6.2   The Envelope Function Scheme

Approaches to modelling the electronic band structures of bulk crystals range from detailed linear combination of atomic orbital methods to more approximate effective potential methods. In a semiconductor, features at energies close to the band edge are well described by the effective mass or $\mathbf{k} \cdot \mathbf{p}$ theory, which assumes a parabolic carrier dispersion of the form $E = \hbar^2 k^2 / 2m^*$, exactly as for a free electron but with the crystal potential accounted for through the effective mass m^*. In this picture the carrier wavefunction ψ has the Bloch form

$$\psi_{\mathbf{k}}(\mathbf{r}) = e^{i\mathbf{k}\cdot\mathbf{r}} u_{\mathbf{k}}(\mathbf{r}) \tag{6.1}$$

where the rapidly varying periodic (crystal) part of the wavefunction $u_{\mathbf{k}}(\mathbf{r})$ is modulated by a slowly-varying plane wave or 'envelope function' (Fig. 6.7).

Most common III–V semiconductor materials occur in the zinc-blende crystal structure. The band structure is due to the overlap of the atomic valence orbitals

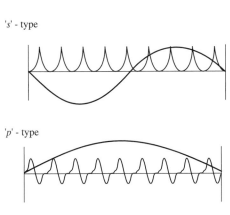

's' - type

'p' - type

Figure 6.7.   Schematic one-dimensional Bloch wave functions for bands of 's' and 'p' like symmetry showing the localized, atomic function and slowly varying envelope function separately.

into three valence bands of p-like atomic symmetry (which are completely filled at absolute zero) and one conduction band of s-like atomic symmetry (which is unoccupied at absolute zero). In quantum mechanical terms, the bands are identified by their total angular momentum J and its projection J_z. One valence band $(J = \frac{1}{2}, \ m_J = \pm\frac{1}{2})$ is split away by the spin–orbit interaction; the remaining two $(J = \frac{3}{2})$ are degenerate at the band edge but have different dispersions, and so are known as 'heavy' $(m_J = \pm\frac{3}{2})$ and 'light' $(m_J = \pm\frac{1}{2})$ valence bands through their different effective masses (Fig. 6.8). These four bands are generally close in energy compared with other valence and conduction bands (remote bands), and so in many materials it is a good approximation to treat the crystal electronic structure in the basis of eight Bloch states formed by these four spin-degenerate bands. Applying $\mathbf{k} \cdot \mathbf{p}$ theory to this basis – the Kane model (Kane, 1957) – yields a set of dispersion relations describing the non-parabolicity of the bands in terms of the band splittings and a material dependent dipole matrix element. The Kane model works well in bulk materials typically to within $100 \, \mathrm{meV}$ of the band edge.

In a heterostructure, for an abrupt interface between two electronically similar and lattice-matched materials A and B of different band gap grown with the same crystal orientation, we may represent the heterointerface as a step in the conduction and valence band edges (Fig. 6.9). The heterointerface has effectively added an external one-dimensional potential $V(z)$ to the crystal potential $V_c(\mathbf{r})$ which

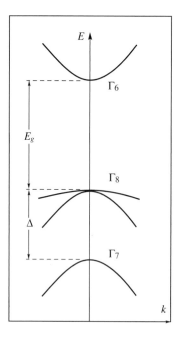

Figure 6.8.   Schematic band dispersion for a bulk semiconductor in the Kane model, showing the conduction (Γ_6) band, heavy and light valence bands (Γ_8) and the spin-orbit split off band (Γ_7).

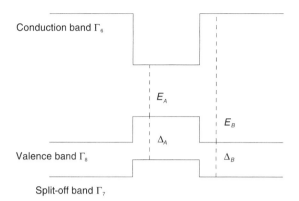

Figure 6.9.   Conduction and valence band profiles for a quantum well in the Kane model.

determines the electronic states in the bulk. Thus, the Schrödinger equation for this heterostructure is

$$\mathcal{H}\psi(\mathbf{r}) = \left[-\frac{\hbar^2}{2m^*}\nabla^2 + V_c(\mathbf{r}) + V(z) \right]\psi(\mathbf{r}) = E\psi(\mathbf{r}) \tag{6.2}$$

where E is the energy of the state $\psi(\mathbf{r})$. Recognizing that the one-dimensional heterostructure potential will not influence the plane wave motion normal to the growth direction, we can express the wavefunction $\psi(\mathbf{r})$ in either material as a sum over the crystal bands, $u_n(\mathbf{r})$, in the basis

$$\psi(\mathbf{r}) = \sum_n u_n(\mathbf{r})\, e^{i\mathbf{k}_\parallel \cdot \mathbf{r}} F_n(z) \tag{6.3}$$

where $\mathbf{k}_\parallel$ is the wavevector in the plane of the interface and $F_n(z)$ is known as the *envelope function* (for the nth band). This formalism becomes useful when we make the approximation which is central to the EFA in heterostrucures: that the atomic structure of the two materials, A and B, is sufficiently similar that the functions $u_n^A(\mathbf{r})$ and $u_n^B(\mathbf{r})$ may be equated, and that bulk behaviour is maintained up to the interface in either material. Then substituting for $\psi(\mathbf{r})$ and applying the $\mathbf{k} \cdot \mathbf{p}$ expansion to $\mathcal{H}$ we obtain a set of equations for the envelope functions $F_n(z)$ in terms of the carrier wavevector $\mathbf{k}$ and a matrix element of the bulk band structure. The four-band Kane model yields a Schrödinger equation of the form

$$\mathcal{H}\psi = E\psi \tag{6.4a}$$

The wavefunction ψ is a column vector comprising the pertinent atomic basis states, which are shown in (6.4b) on page 186. The Hamiltonian $\mathcal{H}$ is an 8×8 matrix, also shown in equation (6.4b), where k_i $(i = +, -, z)$ is the momentum operator $(k_i = -i\hbar d/dr_i)$ and $\mathbf{P}$ is the dipole matrix element linking an s-like to a p-like atomic Bloch function $\mathbf{P} = \langle s|x|p_x \rangle$. All other matrix elements vanish by symmetry. $\mathbf{P}$ is usually expressed in terms of a parameter known as the Kane

$$
\begin{array}{c|cccccccc}
 & |iS\uparrow\rangle & |\tfrac{3}{2},\tfrac{3}{2}\rangle & |\tfrac{3}{2},-\tfrac{1}{2}\rangle & |\tfrac{1}{2},\tfrac{1}{2}\rangle & |iS\downarrow\rangle & |\tfrac{3}{2},\tfrac{1}{2}\rangle & |\tfrac{3}{2},-\tfrac{3}{2}\rangle & |\tfrac{1}{2},-\tfrac{1}{2}\rangle
\end{array}
$$

	$|iS\uparrow\rangle$	$|\tfrac{3}{2},\tfrac{3}{2}\rangle$	$|\tfrac{3}{2},-\tfrac{1}{2}\rangle$	$|\tfrac{1}{2},\tfrac{1}{2}\rangle$	$|iS\downarrow\rangle$	$|\tfrac{3}{2},\tfrac{1}{2}\rangle$	$|\tfrac{3}{2},-\tfrac{3}{2}\rangle$	$|\tfrac{1}{2},-\tfrac{1}{2}\rangle$
$|iS\uparrow\rangle$	$\epsilon_0 + V_s(z) + V_{\text{ext}}(z)$	Phk_-	$-Phk_+/\sqrt{3}$	$Phk_z/\sqrt{3}$	0	$-\sqrt{2/3}\,Phk_z$	0	$-\sqrt{2/3}\,Phk_+$
$|\tfrac{3}{2},\tfrac{3}{2}\rangle$	Phk_+	$-\epsilon_A + V_p(z) + \epsilon_0 + V_{\text{ext}}(z)$	0	0	0	0	0	0
$|\tfrac{3}{2},-\tfrac{1}{2}\rangle$	$-Phk_-/\sqrt{3}$	0	$-\epsilon_A + V_p(z) + \epsilon_0 + V_{\text{ext}}(z)$	0	$-\sqrt{2/3}\,Phk_z$	0	0	0
$|\tfrac{1}{2},\tfrac{1}{2}\rangle$	$Phk_z/\sqrt{3}$	0	0	$-\epsilon_A - \Delta_A + V_\delta(z) + \epsilon_0 + V_{\text{ext}}(z)$	$\sqrt{2/3}\,Phk_-$	0	0	0
$|iS\downarrow\rangle$	0	0	$-\sqrt{2/3}\,Phk_z$	$\sqrt{2/3}\,Phk_+$	$\epsilon_0 + V_s(z) + V_{\text{ext}}(z)$	$Phk_+/\sqrt{3}$	Phk_-	$Phk_z/\sqrt{3}$
$|\tfrac{3}{2},\tfrac{1}{2}\rangle$	$-\sqrt{2/3}\,Phk_z$	0	0	0	$Phk_-/\sqrt{3}$	$-\epsilon_A + V_p(z) + \epsilon_0 + V_{\text{ext}}(z)$	0	0
$|\tfrac{3}{2},-\tfrac{3}{2}\rangle$	0	0	0	0	Phk_+	0	$-\epsilon_A + V_p(z) + \epsilon_0 + V_{\text{ext}}(z)$	0
$|\tfrac{1}{2},-\tfrac{1}{2}\rangle$	$-\sqrt{2/3}\,Phk_+$	0	0	0	$Phk_z/\sqrt{3}$	0	0	$-\epsilon_A - \Delta_A + V_\delta(z) + \epsilon_0 + V_{\text{ext}}(z)$

(6.4b)

energy, E_k, through $\mathbf{P}^2 = 2m_e E_k$, where m_e is the electron rest mass. Kane energies are typically 20 eV for III–V semiconductors (Bastard, 1988).

In equation (6.4b), ϵ_0 represents the quantity $\hbar^2 k^2 / 2m_e$; ϵ_A the band gap of material A; Δ_A the spin-orbit split-off band of material A; $V_s(z)$, $V_p(z)$ and $V_\delta(z)$ the potentials due to the heterostructure in the conduction, valence and spin-orbit split-off bands respectively; $V_{ext}(z)$ an external applied potential; and $k_\pm = (1/\sqrt{2})(k_x \pm ik_y)$.

The solution of (6.4a,b) is complicated in comparison to the bulk case by the loss of symmetry in the crystal. The fixing of a direction for J_z means that it is no longer possible to diagonalize $\mathcal{H}$ by appropriate choice of $\mathbf{k}$ and to obtain analytic dispersion relations for each band. In general the problem must be solved numerically.

6.3   The Parabolic Band Model

It can be seen from (6.4) that in the Kane model the different crystal bands are mixed together by the proximity of the band edges in energy and by the carrier's in-plane momentum. If we focus on states very close to the band edge, where the carrier kinetic energy is small, we may allow the band splittings E_g and Δ to tend to infinity and, if we further restrict attention to the band edges where $\mathbf{k}_\parallel = (k_x, k_y) = 0$, then (6.4) reduces to a set of four doubly-degenerate, one-dimensional Schrödinger-like equations for the envelope functions of each band. These are known as *effective mass equations*:

$$\left[-\frac{\hbar^2}{2m^*(z)}\frac{d^2}{dz^2} + V(z) \right] F_n(z) = EF_n(z) \tag{6.5}$$

where E is the energy due to motion in the z-direction and the effective mass $m^*(z)$ describes the parabolicity of the band n in the z-direction. The central problem is to find the envelope functions $F(z)$ and the related energies E.

Since the barriers are of finite height there is always a finite probability of finding the carrier in the barrier material, even at energies below the barrier band edge. At either interface the solution is required to satisfy continuity of the wavefunction, and therefore of $F(z)$, and of some quantity related to its derivative. While a range of possible conditions ensure that the solutions for E will be real, the Ben–Daniel–Duke condition that $(1/m^*)dF/dz$ be continuous (Bastard, 1985) is favoured because it also ensures continuity of the probability current and, in the case of AlGaAs/GaAs QWs, is consistent with experimental spectra (Duggan *et al.*, 1985b). (See also Section 2.3.3.)

For a quantum well, the heterostructure potential $V(z)$ in the conduction band has the form

$$V(z) = \begin{cases} 0, & |z| \leqslant \tfrac{1}{2}L \quad \text{(material A)} \\ V_e, & |z| > \tfrac{1}{2}L \quad \text{(material B)} \end{cases} \tag{6.6}$$

if we choose the band edge of the well material as our zero of energy. For a piecewise constant potential the solution for $F(z)$ has the form $A\,e^{ikz} + B\,e^{-ikz}$ in either material, where $k^2 = 2m^*E/\hbar^2$. Applying the final boundary condition that $F(z) \to 0$ as $z \to \infty$ for $E < V_e$ we obtain the usual quantum mechanical solution for a finite square well: a discrete set of confined states below V_e and a continuum of propagating states above V_e. The symmetry of the well imparts a definite parity to the confined state envelope functions, so that

$$F_n(z) = \begin{cases} A\,e^{\kappa_n z}, & z < -\tfrac{1}{2}L \\ A\cos(k_n z), & -\tfrac{1}{2}L \leqslant z \leqslant \tfrac{1}{2}L \\ A\,e^{-\kappa_n}, & z > \tfrac{1}{2}L \end{cases} \tag{6.7a}$$

for even states (odd n), and

$$F_n(z) = \begin{cases} -A\,e^{\kappa_n z}, & z < -\tfrac{1}{2}L \\ A\sin(k_n z), & -\tfrac{1}{2}L \leqslant z \leqslant \tfrac{1}{2}L \\ A\,e^{-\kappa_n}, & z > \tfrac{1}{2}L \end{cases} \tag{6.7b}$$

for odd states (even n). The confinement energy is given by

$$E_n = \frac{\hbar^2 k_n^2}{2m_A^*} \tag{6.8}$$

where

$$\tan(\tfrac{1}{2}k_n L) = \frac{m_A^* \kappa_n}{m_B^* k_n}, \qquad n = 1, 3, \ldots \tag{6.9a}$$

for even states and

$$\cot(\tfrac{1}{2}k_n L) = -\frac{m_A^* \kappa_n}{m_B^* k_n}, \qquad n = 2, 4, \ldots \tag{6.9b}$$

for odd states, and κ_n, the 'imaginary' wavevector for the evanescent barrier solution, satisfies

$$V_e - E_n = \frac{\hbar^2 \kappa_n^2}{2m_B^*} \tag{6.10}$$

where m_A^*, m_B^* are the electron effective masses in materials A and B. We solve for the E_n numerically. Typical solutions are shown in Fig. 6.10.

Since in this simple approximation the carrier's in-plane motion is decoupled from the effects of $V(z)$, the electron's total wavefunction in the nth confined state (or subband) in the well is simply

$$\psi(\mathbf{r}) = A u_n(\mathbf{r})\,e^{i\mathbf{k}_\parallel \cdot \mathbf{r}} \times \begin{cases} \cos(k_n z), & n \text{ odd} \\ \sin(k_n z), & n \text{ even} \end{cases} \tag{6.11}$$

with energy

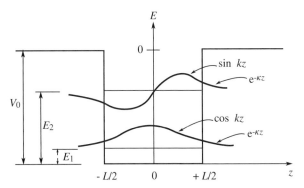

Figure 6.10.   The first two confined state envelope functions in a finite quantum well.

$$E = E_n + \frac{\hbar^2 k_\parallel^2}{2m_\parallel{}^*} \tag{6.12}$$

where $m_\parallel^*$ is the in-plane effective mass. From (6.12) we recover for the density of states $g(E)$ the staircase-like function characteristic of a quasi-two dimensional system as discussed in Chapter 2:

$$g(E) = \frac{m_\parallel^*}{\pi\hbar^2} \sum_n \theta(E - E_n) \tag{6.13}$$

where $\theta(E)$ is the Heaviside function. This crosses the usual parabolic function for the well material in the bulk at energies close to the subband edges (Fig. 6.11). An exactly analogous analysis applies for holes in each of the valence bands.

The number N of confined states admitted by the QW,

$$N = \mathrm{Int}\left(\frac{\sqrt{2m_A^* V_e}L}{\pi\hbar}\right) + 1 \tag{6.14}$$

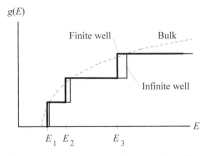

Figure 6.11.   Density of states function $g(E)$ for a finite quantum well (full thick line), an infinitely deep quantum well (full thin line) compared with that of the bulk material (dashed line). The confinement energies $E_1, E_2, \ldots$ mark the onset of a new subband, each of which is like a new two-dimensional electron gas. Notice how the function for the infinite quantum well touches the parabola.

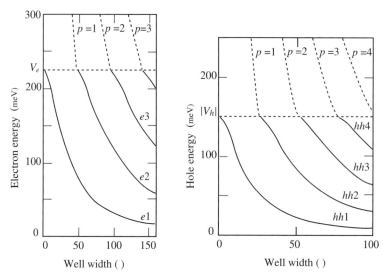

Figure 6.12.   Confinement energies for electrons and heavy holes as a function of well width in an Al$_{0.3}$Ga$_{0.7}$As/GaAs quantum well. Quasi-confined states above the well are represented by dashed lines.

where 'Int' means the integer part of the enclosed expression, increases with the well width, the well depth and carrier effective mass. Consequently, more heavy than light hole states are contained in the same well. In the limit of a wide well, subband spacings become smaller as the QW density of states converges on the bulk limit, while in the limit of a narrow well the energy of the lowest confined state (which always exists) tends to V_e (Fig. 6.12). Since the first available energy level is always shifted to a higher energy with respect to the bulk, the lowest energy for a valence-conduction band transition is always increased over the bulk band gap. Above the barrier band edge, where propagating states are allowed, the effect of the confining potential of the QW persists as a series of quasi-bound states. These are manifested as resonances in the transmission coefficient of the QW and may act as efficient states for carrier capture (Fig. 6.13).

   Carrier penetration into the barrier material increases for light particles and for states at higher energy in the well and, hence, also for very narrow wells, meaning that greatest carrier confinement is achieved at some intermediate well width. Thus, knowledge of only the QW width L, depths V_e and V_h, and the well and barrier effective masses is needed to find the band structure of a QW in this simple approximation. For example, a 100 Å GaAs QW in Al$_{0.3}$Ga$_{0.7}$As should admit three electron, four heavy hole (Fig. 6.14) and two light hole subbands. This calculation was made by taking a conduction band offset Q_e of 0.67, well and barrier effective masses m_A^* and m_B^* of 0.067 and 0.092 for the electron, 0.34 and 0.46 for the heavy hole and 0.094 and 0.11 for the light hole, values which are fairly well established. However, knowledge of the band offset and carrier effective

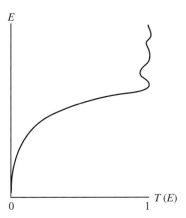

Figure 6.13.   Transmission coefficient $T(E)$ through a quantum well double barrier structure as a function of electron kinetic energy E. Quasi-confined states above the well produce oscillations in $T(E)$.

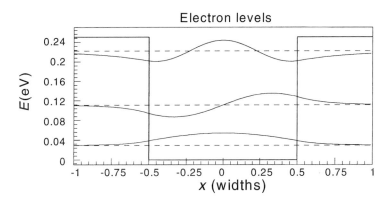

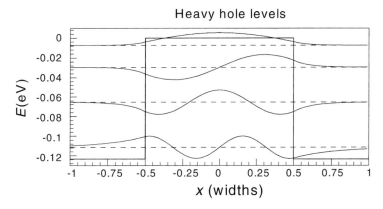

Figure 6.14.   Confined state envelope functions for electrons and heavy holes in a 100 Å $Al_{0.3}Ga_{0.7}As/GaAs$ quantum well.

masses, particularly in the valence band, where they are non-isotropic and differ strongly from bulk values, is often poor. Indeed, for less well-understood materials systems it may be necessary to use theoretical calculations of the sub-band energies in order to derive these parameters from spectroscopic data.

6.4   Effects of Band Mixing

If we relax either of the approximations made in the simple analysis above, the heterostructure Hamiltonian mixes the crystal bands and destroys the parabolicity of the band structure. Finite band splittings upset the band dispersions, as in the bulk, but in QWs the principal effect of this is to shift the positions of the subband edges. Admitting coupling of the carrier's in-plane and z motions through finite $k_\parallel$, in contrast, influences not the subband edges, for which $k_\parallel = 0$ by definition, but the magnitude of the density of states away from the step edges.

6.4.1   Light Particle Band Non-parabolicity

If we set $k_\parallel = 0$ in (6.4), then the Hamiltonian reduces to a doubly degenerate 3×3 matrix which couples together the envelope functions for the three light $(m_J = \pm\frac{1}{2})$ particle bands and an uncoupled effective mass equation for the heavy holes, exactly as in bulk materials (Bastard, 1988). This yields an effective mass equation for each light particle but with an energy-dependent m^* derived from the non-parabolic dispersion for k_z. For the electron

$$\frac{1}{m^*(E)} = \frac{2}{3}P^2\left[\frac{2}{E + E_g} + \frac{1}{E + E_g + \Delta}\right] \tag{6.15}$$

where E_g and Δ are the band splittings for that material. In the well, $m^*(E)$ obeys

$$[k_A(E)]^2 = \frac{2m_A^*(E)}{\hbar^2} \tag{6.16}$$

Numerical solution of the transcendental equations (6.8)–(6.10) using the variable $m^*(E)$ yields a series of energies E_n which are now depressed in energy relative to the parabolic case since the effective masses increase with E (Fig. 6.14). The effect is stronger for narrower wells and for higher states in the well, and may introduce extra subbands into the QW density of states which are not predicted by the parabolic model. For example, for a 100 Å GaAs QW in $Al_{0.3}Ga_{0.7}As$, this light particle band mixing depresses the $n = 1$ electron state by ≈ 1 meV but the $n = 3$ state by ≈ 30 meV.

In principle the dispersion relation (6.15) above should permit calculation of carrier band-edge effective masses from knowledge of the Kane matrix element **P** and the band splittings. However, actual differences in **P** between materials, i.e. failure of the EFA, and the influence of remote bands, i.e. failure of the Kane model, make this a poor approximation, particularly for the hole masses. In

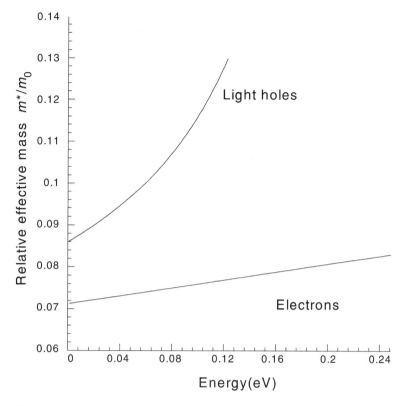

Figure 6.15.   Energy dependence of the light particle effective masses in GaAs using the Kane model. In each case the dipole matrix element **P** is chosen to produce the observed band-edge mass.

calculations it may be preferable to define a **P** for either material from experimentally derived values for the effective masses (Fig. 6.15).

6.4.2   Valence Band Non-parabolicity

We can see from (6.4) that carriers' in-plane motion immediately couples together the heavy and light hole motions. This effect is specific to heterostructures and arises from the loss of crystal symmetry in comparison with the bulk material. $\mathcal{H}$ can no longer be made diagonal in the basis of the (J, J_z) states. Exact solution of the 8×8 system of differential equations (6.4) is very time-consuming and moreover would not be expected to describe the heavy hole dispersion accurately since it does not explain the heavy hole effective mass even at $k_\parallel = 0$. The heavy hole effective mass is influenced by second-order $\mathbf{k} \cdot \mathbf{p}$ coupling to remote bands not included in the Kane model. The approximate features of the heavy and light hole dispersions can, however, be described by solving a 4×4 system, the Luttinger Hamiltonian, which couples the valence bands at finite $k_\parallel$ while neglecting

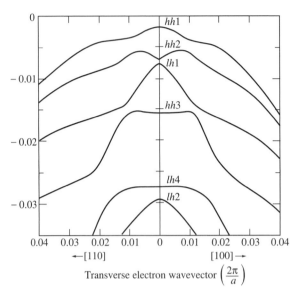

Figure 6.16. Calculated in-plane dispersion for heavy and light holes in a GaAs/ Al$_{0.25}$Ga$_{0.75}$As superlattice showing strongly non-parabolic behaviour. After Chang and Schulman (1985) (Courtesy Y. C. Chang)

conduction and spin-orbit split-off bands (Luttinger, 1956). Band mixing strength at $k_\parallel$ is determined by material-dependent Luttinger parameters.

Calculations of the valence band mixing effects in this approximation show a significant deviation of the band structure from the parabolic case (Fig. 6.16). Band distortion around the band edge can cause dramatic changes of mass – usually for the light hole – where its in-plane effective mass becomes negative. Further from $k_\parallel = 0$, unphysical level crossings predicted by the parabolic model are replaced with anti-crossings which strongly distort the band curvature and confuse the character of the light and heavy holes. Relative to the parabolic case the subband dispersion curves are shifted typically by several meV. The effects are in general stronger for the light hole bands.

Coupling between bands of different parity n means that the QW wavefunctions become linear combinations of functions of different symmetry and so lose their well-defined parity. This effect is apparent in optical spectra and represents a major failing of the Kane model. For an asymmetric QW, valence band coupling further lifts the bands' spin degeneracy at $k_\parallel = 0$.

6.5   Multiple Well Effects

Increasing the number of identical quantum wells when sufficiently well separated in the growth direction – a MQW system – serves principally to multiply the

Figure 6.17.   Splitting of levels in a double quantum well.

degeneracy of the electronic structure and the magnitude of the density of states. Reducing the separation between a pair of wells mixes the envelope functions into symmetric and antisymmetric mixed states and lifts the degeneracy of their energy levels, much as atomic orbitals are mixed in a diatomic molecule (Fig. 6.17). In the same way, coupling between envelope functions in an infinite periodic array of closely spaced QWs or a superlattice (SL) broadens discrete energy levels into mini-bands. This imposes an SL miniband structure of smaller zone width onto the existing crystal band structure, as discussed in Section 3.5.

To solve for the SL band structure within the EFA the long range QW boundary condition of vanishing ψ, which ensures confinement in the QW, is replaced by the periodic boundary condition

$$F(z + nd) = F(z)\, e^{inqd} \tag{6.17}$$

where d is the SL period and q the SL wavevector. Taking the usual forms for the envelope function in a piecewise constant potential and applying the interface boundary conditions yields a dispersion equation for the SL wavevector:

$$\cos(qd) = \cos(k_A L_A)\cos(K_B L_B) - (\xi^{-1} + \xi)\sin(k_A L_A)\sin(k_B L_B) \tag{6.18}$$

with

$$\xi = \frac{k_A m_B^*}{k_B m_A^*} \tag{6.19}$$

where L_A and L_B are the well and barrier widths and k_A and k_B the well and barrier z-wavevectors, such that

$$k_A^2 = \frac{1}{\hbar^2} 2 m_A^* E, \qquad k_B^2 = \frac{1}{\hbar^2} 2 m_B^* (E - V) \tag{6.20}$$

as before.

Equation (6.18) has solutions for real q over a set of bands of energies, rather than at discrete values (Fig. 6.18). Physically, these bands represent the energies at which the carrier may propagate through the SL. On account of the finite barrier thickness the barrier can support a propagating state, even when $E < V_e$ and the solution in the barrier material is evanescent.

In contrast to the single QW the pattern of allowed and forbidden energies continues above the well, although miniband gaps contract to eventually leave a continuum of allowed states. SL bandwidths, which should vary like $\exp[-L_B k_B(E)]$ (see Section 3.5.2) according to tight-binding calculations, increase when barrier penetration is stronger, that is, for light particles, for higher energy states and for narrower barriers. In such cases the carrier is more likely to be found in the barrier and can travel more easily through the SL.

Interwell coupling affects the density of states function by broadening the step edges, particularly for higher order states, and reducing the effective subband–subband transition energy (see Fig. 3.28). The choice of d thus offers another variable in band structure design.

The SL envelope functions have lost the spatial parity of the QW solutions except at band extrema. There the envelope function has the parity of the corresponding QW state and is either repeated ($q = 0$) or inverted ($q = \pi/d$) in the next well. Isolated QW solutions are recovered in the limit of infinite d (Fig. 6.19).

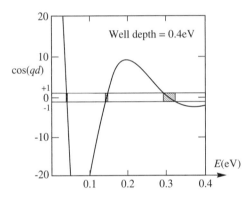

Figure 6.18.   Schematic plot of $\cos(qd)$ as a function of electron kinetic energy E in a superlattice of period d. Solutions exist for real q over bands of energies (shaded) which become broader at higher E.

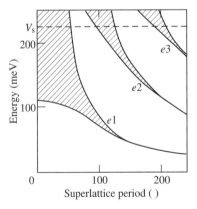

Figure 6.19.   GaAs/Al$_{0.3}$Ga$_{0.7}$As superlattice bands (shaded) as a function of period d for fixed well and barrier widths. Note how the bands narrow on to discrete energy levels as d increases.

While the EFA provides an adequate description of SL band structure for direct-gap materials with moderate SL period d, it may break down in the small period limit where atomic scale effects appear (Bastard, Brum and Ferreira, 1991).

6.6   Effects of the Coulomb Interaction

Until now we have neglected the Coulomb interaction between the electrons and holes in the semiconductor. This has allowed us to solve effective mass equations such as (6.5) separately for the electron and hole band structure. The electrostatic interaction immediately compounds the two problems into a single Schrödinger equation for the coupled electron–hole state χ:

$$\left[-\frac{\hbar^2}{2m_e}\nabla_e^2 + V(\mathbf{r}_e) - \frac{\hbar^2}{2m_h}\nabla_h^2 + V(\mathbf{r}_h) - \frac{e^2}{4\pi\epsilon_r\epsilon_0|\mathbf{r}_e - \mathbf{r}_h|} \right]\chi = E\chi \qquad (6.21)$$

where $\chi = \chi(\mathbf{r}_e, \mathbf{r}_h)$ and $\mathbf{r}_e$ and $\mathbf{r}_h$ are the electron and hole coordinates, respectively. If we expand χ in the basis of the independent wavefunctions of an electron and hole with relative momentum $\mathbf{k}$,

$$\chi(\mathbf{r}_e, \mathbf{r}_h) = \sum_{\mathbf{k}} A_{\mathbf{k}}\psi_{\mathbf{k}}(\mathbf{r}_e)\psi_{\mathbf{k}}(\mathbf{r}_h) \qquad (6.22)$$

then we may define an envelope funtion for the combined state $F_{\text{ex}}(\mathbf{r})$ as

$$F_{\text{ex}}(\mathbf{r}) = \sum_{\mathbf{k}} A_{\mathbf{k}}\, e^{i\mathbf{k}\cdot\mathbf{r}} \qquad (6.23)$$

where $\mathbf{r}$ is the electron-hole separation.

6.6.1   Excitons in Bulk Semiconductors

In a bulk material we exploit the three-dimensional symmetry to recover a hydrogenic effective mass equation for F_{ex}:

$$\left[-\frac{\hbar^2}{2\mu^*}\nabla^2 - \frac{e^2}{4\pi\epsilon_r\epsilon_0 r} \right]F_{\text{ex}}(\mathbf{r}) = E_B F_{\text{ex}}(\mathbf{r}) \qquad (6.24)$$

Here μ^* is the reduced effective mass of the electron–hole pair,

$$\frac{1}{\mu^*} = \frac{1}{m_e^*} + \frac{1}{m_h^*} \qquad (6.25)$$

and E_B is the binding energy of the electron–hole pair – the amount by which the joint energies of the electron and hole are reduced by their interaction. By analogy with the hydrogen atom, equation (6.24) yields a series of bound states or *excitons* of binding energy

$$E_B = -\frac{R^*}{n^2}, \qquad n = 1, 2, \ldots \qquad (6.26)$$

where R^* is the effective Rydberg:

$$R^* = \frac{\mu^*}{m_0 \epsilon_r^2} R \qquad (6.27)$$

which differs from the hydrogenic constant $R = 13.6\,\text{eV}$ through the reduced effective mass and increased permittivity ϵ_r of the semiconductor. The oscillator or line strength f_{osc} of the s-symmetric states (the only optically allowed excitons) varies with n as

$$f_{osc} \sim \frac{1}{n^3} \qquad (6.28)$$

so that the ground state exciton is by far the strongest.

Above $E_B = 0$ there exists a continuum of 'free carrier' excitons where the interaction enhances the joint electron–hole density of states by a factor known as the Sommerfeld factor, which is large near to the band edge but tends to unity at higher energies. Together the excitonic and continuum effects modify the joint density of states, as shown in Fig. 6.21. In a typical semiconductor $R^* \approx 5\,\text{meV}$, which means that excitons are nearly always ionized at room temperature, and so are not important in bulk materials.

6.6.2 Excitons in Quantum Wells

In a quantum well the analysis is immediately complicated by the loss of symmetry and approximate methods must be used. Some insight is gained from the case of a perfectly two-dimensional system, where a two-dimensional hydrogenic effective mass equation is recovered for F_{ex} (Elliot, 1984). This is exactly soluble and yields bound states of energy

$$E_B^{2D} = -\frac{R^*}{\left(n - \frac{1}{2}\right)^2}, \qquad n = 1, 2, \ldots \qquad (6.29)$$

and oscillator strength

$$f_{osc} \sim \frac{1}{\left(n - \frac{1}{2}\right)^3} \qquad (6.30)$$

The two-dimensional ground state exciton is thus eight times as strong and four times as well bound as in the corresponding bulk material, which suggests that excitonic effects will be important in quasi-two-dimensional systems such as QWs. This would be expected intuitively since quantum confinement would impede excitonic ionization, at least through motion in the z-direction.

Quantum wells differ from perfect geometries in two and three dimensions in that the condition for bound state excitons – that $k_\parallel$ vanish identically for electron and hole – occurs not once but many times, at the subband edge for every electron–hole transition. QW excitons could then, in principle, be found for every electron–heavy-hole and electron–light-hole subband pair.

If we decouple the in-plane and z-motion of the individual carriers to define an effective Coulomb potential $V(\rho)$,

$$V(\rho) = \frac{e^2}{4\pi\epsilon_r\epsilon_0} \int\int \frac{|F_e(z_e)|^2 |F_h(z_h)|^2}{\sqrt{\rho^2 + (z_e - z_h)^2}} \, dz_e \, dz_h \qquad (6.31)$$

we obtain an effective mass equation for F_{ex} as a function of the in-plane separation ρ:

$$\left[\frac{\hbar^2}{2\mu^*}\nabla^2 - V(\rho) \right] F_{ex}(\rho) = E_B F_{ex}(\rho) \qquad (6.32)$$

Here μ^* is the reduced in-plane electron-hole effective mass, and $F_e(z_e)$ and $F_h(z_h)$ are the envelope functions for the electron and hole in their respective subbands. The interaction potential is reduced relative to the perfect two-dimensional value by the delocalization, or smearing, of the electron and hole across the QW width, and hence smaller binding energies might be expected.

Equation (6.32) may be solved variationally for E_B with a trial wave function such as

$$F(\rho) = \left(\frac{2}{\pi\lambda^2} \right)^{1/2} e^{-\rho/\lambda} \qquad (6.33)$$

which works fairly well as a first approximation. Better estimates of the binding energy may be obtained by integrating (6.32) numerically. Either method produces binding energies of typically ≈ 10 meV and a strong QW width dependence.

At large L, E_B tends to its bulk value as quantum confinement diminishes and the exciton gains mobility in the z-direction. In the narrow well limit, however, E_B again falls rapidly as the individual electron and hole states are pushed up in the well and leak more strongly into the barrier. An optimum E_B exists at some intermediate L. In $Al_{0.3}Ga_{0.7}As/GaAs$, for example, E_B peaks for the first electron–heavy-hole transition at ≈ 10 meV for a $40\,\text{Å}$ well (Fig. 6.20). Some calculations find that E_B for light holes is higher than for heavy holes. However, these calculations depend upon the values used for the in-plane hole masses which are not well-known, even in widely studied materials, on account of valence band mixing.

It may be convenient to parametrize the QW excitonic states in terms of the QW two-dimensionality. If we define

$$E_B = \frac{R^*}{(s - \nu)^2}, \qquad s = 1, 2, \ldots \qquad (6.34)$$

where ν is a parameter between 0 (the three-dimensional limit) and 0.5 (the two-dimensional limit), the binding energies obtained numerically may be fitted to this form to fix ν. The parametrization fits fairly well, giving $\nu = 0.3$ for a $100\,\text{Å}$

AlGaAs/GaAs QW and offers an estimate of oscillator strength from the parallel parametric form (Klipstein and Apsley, 1986):

$$f_{osc} \sim \frac{1}{(s-\nu)^3} \tag{6.35}$$

As in the bulk material, the Coulomb interaction also couples the motion of QW electrons and holes above the ground state and enhances the joint density of states by a Sommerfeld-like factor (Fig. 6.21). Normally only the $1s$ exciton is visible for

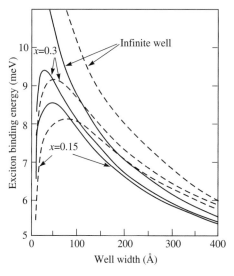

Figure 6.20. Calculations of exciton binding energy as a function of $Al_xGa_{1-x}As/GaAs$ quantum well width. Full lines are for first electron–heavy-hole exciton and broken lines are for the first electron–light-hole exciton. (From Greene *et al.*, 1984), (Courtesy R. L. Greene)

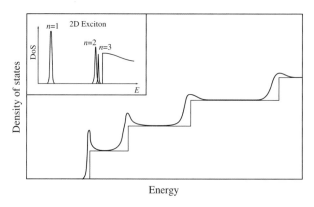

Figure 6.21. Joint density of states function for a quantum well. The coulombic interaction produces an excitonic series at every subband edge.

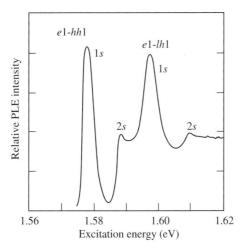

Figure 6.22. Photoluminescence excitation (PLE) spectrum of a $GaAs/Al_{0.33}Ga_{0.67}As$ multi-quantum well showing the $1s$ and $2s$ excitons for the first electron–heavy-hole and electron–light-hole transitions (Dawson *et al.*, 1986). (Courtesy P. Dawson)

each transition. In very good QWs it is, however, sometimes possible to distinguish the $2s$ exciton (Fig. 6.22). The relative positions offer a very good guide to the actual binding energy.

In practice, the most important effect of the Coulomb interaction on QW band structure is the appearance of strong, sharp spectral resonances just below the band edge for the first electron–heavy-hole and electron–light-hole transition. In real QWs the excitons are broadened by processes such as phonon interactions, well-width fluctuations, variations in impurity levels and electric field effects. Nonetheless, in good material the peaks are quite distinct at room temperature and by deliberate tuning of the peak position, for example through an applied bias, can act as effective optical switches. The strength of QW excitons is largely responsible for the interest in QWs in the device field.

6.7   Effects of Applied Bias

A static electric field $\mathbf{E}$ modifies the crystal potential from its flat band value $V_c(\mathbf{r})$ to $V_c(\mathbf{r}) - e\mathbf{E} \cdot \mathbf{r}$. This tilts the conduction and valence band edges in a semiconductor so that where the electron's potential energy is increased by $e\mathbf{E} \cdot \mathbf{r}$ the hole's energy is reduced by the same amount. The net effect is to pull electron and hole in opposite directions (as would be expected!). In a bulk material this is the Franz–Keldysh effect which reduces the effective band gap and reduces the (already small) electron–hole interaction.

In a QW the effect depends critically upon the direction of the applied field with respect to the growth direction. An electric field applied in the well plane acts as in the bulk. Distortion of the in-plane potential accelerates electrons and holes in opposite directions. This destroys the strong quasi-two-dimensional excitons until at moderate field of a few $V\mu m^{-1}$ the exciton is effectively washed out and the band structure begins to represent the bulk (Fig. 6.23).

An applied bias in the growth direction of the QW may be treated within the effective mass approximation (Fig. 6.24). The piecewise constant potential $V(z)$ in equation (6.5) is modified by a linear term $e\mathbf{E}z$ so that for electrons, for instance, the envelope function now satisfies

$$\left[-\frac{\hbar^2}{2m^*}\frac{d^2}{dz^2} + V(z) + e\mathbf{E}z\right]F_e(z) = EF_e(z) \tag{6.36}$$

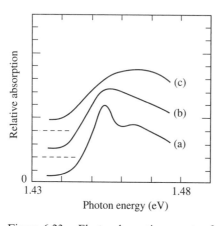

Figure 6.23.   Electroabsorption spectra for an AlGaAs/GaAs quantum well with electric field $\mathbf{E}$ applied in the plane of the well. (a) $\mathbf{E} = 0$, (b) $\mathbf{E} = 2\,V\,\mu m^{-1}$, (c) $\mathbf{E} = 5\,V\,\mu m^{-1}$ (Miller et al., 1985). (Courtesy D. S. Chemla)

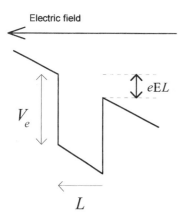

Figure 6.24.   Band profile of a quantum well with bias applied in the growth direction.

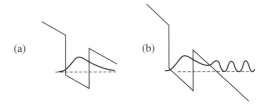

Figure 6.25.   Envelope functions for a quantum well 'confined' state under (a) low and (b) high applied bias. The delocalized form of the envelope function at high bias shows the high probability of tunnelling out of the quantum well.

The solutions for $F_e(z)$ are Airy functions in both well and barrier (Fig. 6.25). Strictly there are no bound solutions to this problem. All QW energies are allowed for $F_e(z)$ and all solutions are time-dependent, that is, the carrier will eventually tunnel out of the QW. Clearly this effect may be expected to destroy the features of confinement, and at high biases the discrete nature of the QW band structure is lost.

At low biases an exact solution for $F_e(z)$ using Airy functions and a phase shift method of calculating the density of states shows that the field broadens discrete QW confined states into resonances which are narrow enough to qualify as quasi-bound states. This implies a broadening of the step edges in the basic QW density of states and a finite lifetime τ for the carrier in that subband:

$$\tau \sim \frac{h}{\Delta E} \tag{6.37}$$

where ΔE is the resonance width.

Subband edges are shifted as well as broadened by the field. This is known as the quantum confined Stark effect (QCSE). In the quasi-bound state limit variational, perturbative or phase-shift methods may be used to locate the resonances and estimate their widths. The $n = 1$ subband is always shifted down in energy in an isolated QW, by an amount which varies approximately like F^2 (Fig. 6.26). The shifts increase with QW width.

Electron and hole shifts combine to reduce the subband transition energy and hence the effective band gap. Shifts depend strongly on well width, increasing for wide wells. Higher bands are also shifted up or down, depending on the band symmetry and other factors. The envelope functions are distorted so as to lose the well-defined parity of the flat band QW, and electron and hole accumulate on opposite sides of the well (Fig. 6.27). The strength of excitonic transitions is necessarily reduced by the spatial separation of the carriers. Yet the envelope functions are still confined at moderate fields and allow excitons to persist to much higher fields than in bulk material. Excitonic binding energies are reduced, but by a small degree compared with the ground state Stark shift. The net result for a single QW is still a shift to a smaller effective band gap (Fig. 6.28).

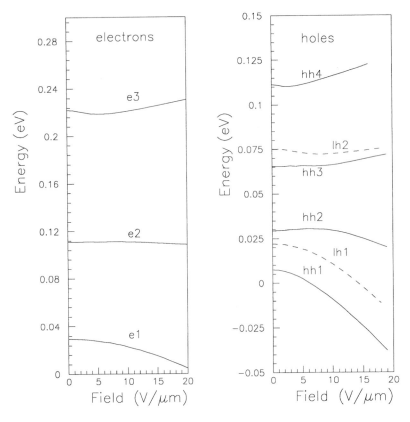

Figure 6.26. Calculated confinement energies for all confined states in a 100 Å GaAs/Al$_{0.3}$Ga$_{0.7}$As QW as a function of transverse electric field. Note how the lowest state of each type is always depressed in energy, and how the highest heavy-hole state is lost at about 15 V μm^{-1}.

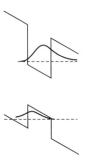

Figure 6.27. Distortion of electron and heavy-hole envelope functions in a biassed quantum well.

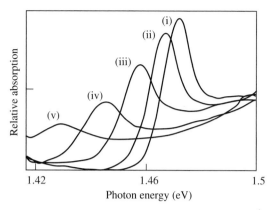

Figure 6.28.   Electroabsorption spectra for a 94 Å AlGaAs/GaAs quantum well under a transverse electric field **E** of (i) $0\,\mathrm{V}\,\mu\mathrm{m}^{-1}$, (ii) $6\,\mathrm{V}\,\mu\mathrm{m}^{-1}$, (iii) $11\,\mathrm{V}\,\mu\mathrm{m}^{-1}$, (iv) $15\,\mathrm{V}\,\mu\mathrm{m}^{-1}$ and (v) $20\,\mathrm{V}\,\mu\mathrm{m}^{-1}$ (Schmitt-Rink *et al.*, 1989). (Courtesy D. S. Chemla) Note how the exciton persists to higher fields than for a parallel electric field (Fig. 6.23).

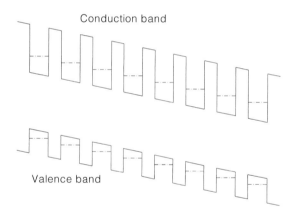

Figure 6.29.   Superlattice in a transverse electric field.

In an SL the electric field also impedes communication between wells and breaks up the SL band into a set of localized states (Fig. 6.29). Thus by narrowing the SL band into discrete states the applied bias may first cause an increase in the effective band gap, or blueshift, before the usual Stark shift takes over.

6.8   Optical Absorption in a Quantum Well

Optical absorption and emission processes in a semiconductor depend directly on the form of the electronic density of states. Hence optical methods are central in the study of QW electronic structure. Here we will use our theory of electronic structure to calculate the optical absorption of a QW and look at how the various effects descibed above appear in spectra.

By Fermi's golden rule, the rate of electronic transitions W from an initial state $|i\rangle$ to a final state $|f\rangle$ under the influence of an electromagnetic field of frequency ω and polarisation $\mathbf{e}$ is given by

$$W = \frac{2\pi}{h}|\langle f|\mathcal{H}'|i\rangle|^2 \Delta(E_f - E_i - \hbar\omega) f(E_i)[1 - f(E_f)] \tag{6.38}$$

where $f(E)$ is the Fermi–Dirac probability that the state at E is occupied and $\mathcal{H}'$ the interaction Hamiltonian which varies like $\mathbf{p}\cdot\mathbf{e}$ in the electric dipole approximation. Defining the absorption coefficient of our semiconductor $\alpha(\hbar\omega)$ as the rate of energy loss in the propagation direction so that light intensity falls like (Fig. 6.30)

$$I(x) = I(0)\,e^{-\alpha(\hbar\omega)x} \tag{6.39}$$

and, substituting for $\mathcal{H}'$, we find

$$\alpha(\hbar\omega) = \frac{\beta}{\omega}\sum_{i,f}|\langle f|\mathbf{p}\cdot\mathbf{e}|i\rangle|^2 \Delta(E_f - E_i - \hbar\omega)[f(E_i) - f(E_f)] \tag{6.40}$$

where β is a sample-dependent optical constant. Thus knowledge of all states $|i\rangle$ and $|f\rangle$ through the electronic band structure yields the absorption spectrum.

The central feature of (6.40) is the optical or dipole matrix element. The Fermi–Dirac occupancy factors in a semiconductor merely mean that absorption is likely to be strong for transitions from the valence band (initial states which are normally full) to the conduction band (final states which are normally empty), at least in the low intensity limit. Calculation of $\alpha(\hbar\omega)$ consists of summing the dipole matrix elements for all electron–hole transitions, both subband–subband and excitonic contributions.

In a QW, within the decoupled band EFA, the independent hole (initial) and electron (final) state wavefunctions have the form

$$\langle\mathbf{r}|i\rangle = \exp(i\mathbf{k}_\parallel\cdot\mathbf{r}_\parallel)u_v(\mathbf{r})F_h(z) \tag{6.41a}$$

and

$$\langle\mathbf{r}|f\rangle = \exp(i\mathbf{k}_\parallel\cdot\mathbf{r}_\parallel)u_c(\mathbf{r})F_e(z) \tag{6.41b}$$

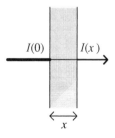

$I(0)$ $I(x)$

x

Figure 6.30.   Absorption in a slab of width x.

with $u_v(\mathbf{r})$ $(u_c(\mathbf{r}))$ the valence (conduction) band atomic basis state and $F_h(z)$ $(F_e(z))$ the hole (electron) envelope function. Expanding the matrix element as a product we find

$$\langle f|\mathbf{p}\cdot\mathbf{e}|i\rangle = [\langle u_c|\mathbf{p}\cdot\mathbf{e}|u_v\rangle\langle F_e|F_h\rangle + \langle u_c|u_v\rangle\langle F_e|\mathbf{p}\cdot\mathbf{e}|F_h\rangle]\delta(\mathbf{k}_{\|,e} - \mathbf{k}_{\|,h}) \qquad (6.42)$$

which immediately yields the selection rules for optical transitions. Conservation of in-plane wavevector $\mathbf{k}_\|$ is expected for optical transitions, in analogy with total $\mathbf{k}$ conservation in the bulk.

For electron–hole transitions the orthogonality of the atomic basis states means that the second term on the right-hand side of (6.42) vanishes and the matrix element depends upon the envelope function overlap. Since in a QW these functions have a well-defined parity, this means that optical transitions should only be observed between conduction and valence subbands of the same parity (Fig. 6.31); moreover they will be weak between near-orthogonal subbands of the same parity but different n. The relative strength of transitions involving the different valence bands is given by the bulk optical matrix element $\langle u_c|\mathbf{p}\cdot\mathbf{e}|u_v\rangle$ and depends upon polarization, so that for $\mathbf{e}$ in the well plane (the usual case of interest) heavy-hole transitions are three times as strong as light-hole transitions:

$$|\langle u_e|\mathbf{p}\cdot\mathbf{e}|u_{hh}\rangle|^2 = 3|\langle u_e|\mathbf{p}\cdot\mathbf{e}|u_{lh}\rangle|^2 = \tfrac{1}{2}\mathbf{P}^2 \qquad (6.43)$$

with $\mathbf{P}$ the Kane optical matrix element. For $\mathbf{e}$ parallel to z, light hole transitions are forbidden.

Different selection rules apply for intraband excitations, i.e. excitations between different subbands within the same crystal valence or conduction band. Here the first term in (6.42) vanishes and optical transitions for in-plane polarization are allowed only between subbands of different parity.

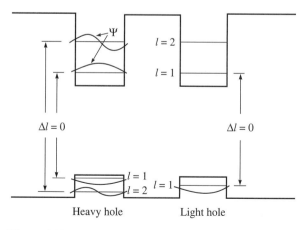

Figure 6.31. Promotion of an electron in a quantum well. Only optically strong transitions (between subbands of the same order) are shown.

The staircase form of the density of states in equation (6.13) means that for a QW we expect to see an abrupt threshold for absorption at the smallest electron–hole subband gap and a staircase-like variation thereafter. This reflects the form of the joint density of states for both electron–heavy-hole and electron–light-hole transitions. Substituting for the matrix element into the expression for the absorption coefficient in equation (6.40) and summing over $\mathbf{k}$ states we find, in the absence of excitonic interaction,

$$\alpha(\hbar\omega) = \frac{\beta}{\omega} \sum_{m,n} \gamma_h |\langle F_{e,n} | F_{h,m} \rangle|^2 \theta(\hbar\omega - E_g - E_n - E_m) \tag{6.44}$$

where γ_h is the relative strength of the transition,

$$\gamma_{hh} = \tfrac{3}{4}\mu_{\parallel,hh} \qquad \text{(electron–heavy-hole)}$$
$$\gamma_{lh} = \tfrac{1}{4}\mu_{\parallel,lh} \qquad \text{(electron–light-hole)} \tag{6.45}$$

and μ the respective reduced effective masses.

For excitonic absorption, the dipole matrix element couples the ground and excitonic excited state and adds the extra selection rule that only s-symmetric excitons may be observed. Excitonic optical transitions appear as peaks just below the absorption edge for each subband-subband transition. Including, for

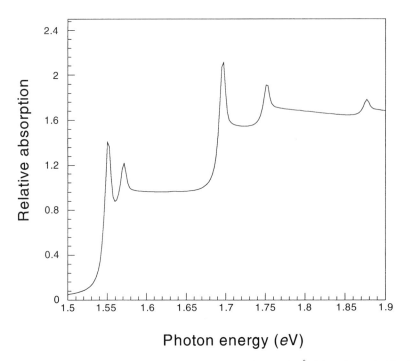

Figure 6.32. Calculated absorption spectrum for an 80 Å $Al_{0.34}Ga_{0.66}As/GaAs$ quantum well at 77 K.

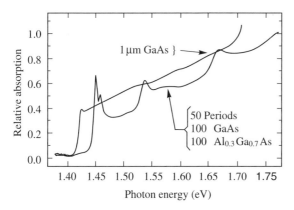

Figure 6.33.   Absorption spectrum for a GaAs/AlGaAs multi-quantum well in comparison with the equivalent thickness of GaAs. The features of quantum confinement are shown: the redshift in the absorption edge, the strong excitons and the step-like density of states (Schmitt-Rink *et al.*, 1989). (Courtesy D. S. Chemla)

simplicity, only the first exciton in each series, we finally obtain for the QW absorption

$$\alpha(\hbar\omega) = \frac{\beta}{\omega} \sum_{m,n} \gamma_h |\langle F_{e,n} | F_{h,m} \rangle|^2$$

$$\times \left[\theta(\hbar\omega - E_g - E_n - E_m) + r_n \delta(\hbar\omega - E_g - E_n - E_m + E_B) \right] \tag{6.46}$$

where r_n represents the exciton oscillator strength relative to the continuum (Stevens *et al.*, 1988) (Fig. 6.32). In practice exciton line widths will be broadened by inhomogenous broadening mechanisms and the step height enhanced near each subband edge by the unbound excitonic interaction (Fig. 6.33).

6.9   Optical Characterization

6.9.1   Measurement of Absorption

The absolute magnitude of the absorption for a single QW is low: approximately 1% of the light is absorbed on the step above the first light-hole and first heavy-hole transitions. This means that direct absorption measurements are difficult to perform except for systems with many QWs. Alternative methods of probing the absorption spectrum are photocurrent (PC) and photoluminescence excitation (PLE) spectroscopy.

In a photoconductivity measurement the QW sample must be contacted (typically as a *p-i-n* device) so that a bias may be applied. The QW is then illuminated with monochromatic light of frequency ω and a small bias is applied to remove carriers photogenerated in the QW (Fig. 6.34). The photocurrent collected at the contacts, when measured as a function of ω, reflects the QW absorption, but only to within a factor representing the quantum efficiency for escape from the well. At low temperatures, or in deep wells, this factor may vary significantly with the excitation energy and produce an unreliable picture of the relative absorption strength, though not of the positions of features. Another difficulty with this technique, particularly when dealing with wide wells, arises from the Stark shifting of the subband edges by the applied bias.

Photoluminescence excitation spectroscopy may be used to study uncontacted samples. In a PLE measurement, the QW is illuminated by light of frequency ω and a detector is fixed at an energy just below the effective absorption edge (the first electron–heavy-hole exciton). As the energy of the incident light is varied, electrons are promoted from the valence band to the conduction band according to the absorption strength. Carriers first thermalize rapidly to their respective band edges and then recombine slowly by radiative recombination across the band gap (Fig. 6.35). The PL signal will thus vary with ω in harmony with the strength of the absorption (Fig. 6.36). PLE will be more reliable for states close to the absorption edge, where carriers are less vulnerable to removal by other mechanisms.

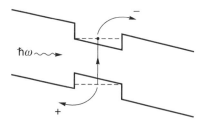

Figure 6.34.   Measurement of photoconductivity.

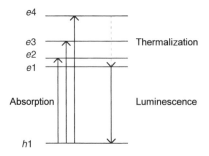

Figure 6.35.   Photogeneration and recombination of carriers in a quantum well in photo-luminescence excitation.

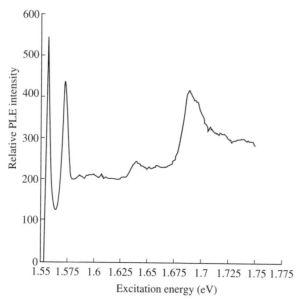

Figure 6.36. Photoluminescence excitation spectrum of an $84\,\text{Å}$ $Al_{0.33}Ga_{0.67}As/GaAs$ multi-quantum well at $11\,\text{K}$. As well as the main transitions between subbands of the same order ($e1$–$hh1$, $e1$–$lh1$ and $e2$–$hh2$) the exciton for the optically weak $e1$–$hh3$ transition can be seen at $1.65\,\text{eV}$. (Courtesy R. Murray)

6.9.2   Features of Optical Spectra

The simplest picture of optical absorption predicts a staircase-like spectrum for the QW, modified by a set of excitonic spikes below the step edges. Refinements to the electronic band structure due to band mixing, interwell coupling and applied electric field may be expected to alter this simple picture. In this section, each of these effects will be discussed.

6.9.2.1 Band Non-parabolicity

We have seen in Section 6.4 that non-parabolicity of the light particle bands in the Kane model is expected to affect the positions of the subband edges. This effect is modelled for a typical QW in Fig. 6.37. In optical characterization of real QWs we may contour the features of the optical spectrum to assess the importance of such effects (Duggan *et al.*, 1985b). In this method, a theoretical model is used to determine which values of QW parameters, such as well width L and band offset Q, would predict each excitonic feature at the energy observed experimentally. The results of these calculations are then plotted as contours of the QW parameters. If the contours for the various subband–subband transitions overlap at a point, this implies that the model, using the values for L and Q at that point, is appropriate for that sample. In the sample shown in Fig. 6.38, contour plot (b)

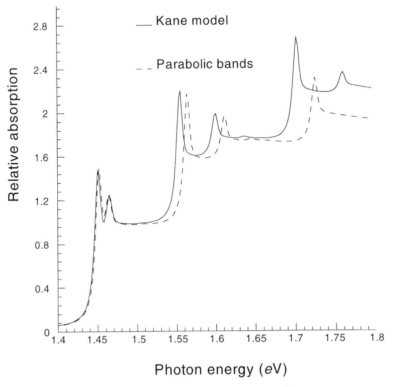

Figure 6.37. Calculated absorption spectra for a 100 Å AlGaAs/GaAs quantum well in the four-band Kane model (full line) and in the parabolic approximation (dashed line).

was produced using a parabolic band model and plot (a) using the Kane model. The much better overlap in (a) implies that non-parabolicity is important.

But a word of caution. The multiplicity of uncertain parameters in QW modelling (effective masses, well width, band offsets, exciton binding energies) means that it is dangerous to assess theoretical models on the basis of a single spectroscopic feature or, indeed, upon a single sample.

6.9.2.2 Valence Band Mixing

In Section 6.4 we saw how mixing of the light- and heavy-hole envelope functions at finite $k_{\parallel}$ will break the well-defined parity of the band-edge functions. Thus, away from subband edges the overlap integral $\langle F_{e,n}|F_{h,m}\rangle$ for n and m of different parity may become non-zero and permit nominally 'forbidden' transitions. The strongest such transition in AlGaAs/GaAs QWs is often the $e2$–$lh1$ transition which occurs at well widths where the $lh1$ band is close in energy to the $hh2$ band and so has strong $hh2$ character. As a result the strengths of the two transitions readjust so that they may appear equally important (Fig. 6.39).

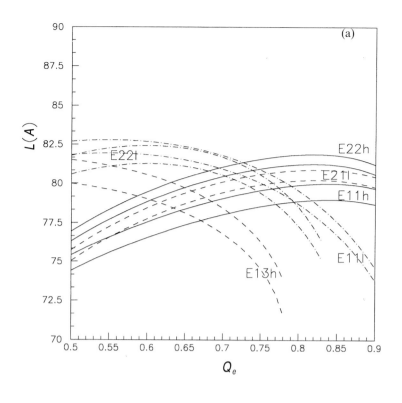

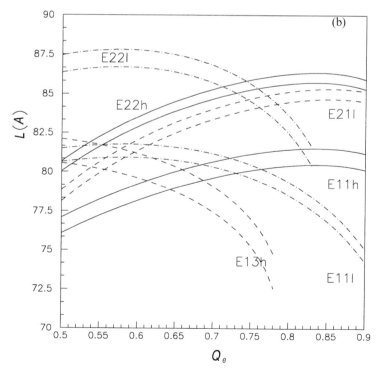

Figure 6.38. Contour plots of the observed exciton positions for a GaAs/AlGaAs quantum well of nominal width 84 Å for (a) the Kane model, and (b) parabolic bands.

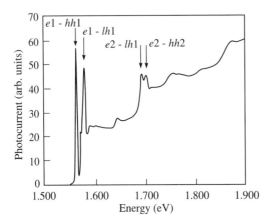

Figure 6.39.   Photocurrent spectrum of an 85 Å GaAs/AlGaAs quantum well at 30 K. The exciton for the forbidden *e2–lh*1 exciton can be seen at 1.70 eV just to the left of the *e2–hh*2 exciton.

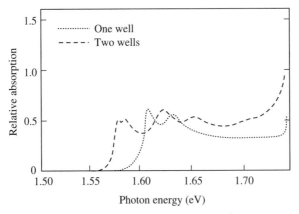

Figure 6.40.   Experimental spectra of a 50 Å single quantum well in comparison with a coupled double quantum of same width, showing the redshift in the absorption edge (After R. Dingle *et al.*, 1975). (Courtesy A. C. Gossard)

6.9.2.3 Interwell Coupling

In Section 6.5 we saw how interwell coupling of the envelope functions broadens the QW confined states into SL bands and so spreads the step edges. This is seen in SL spectra as a broadening of the spectral features – particularly for higher transitions – and a redshift with respect to the single QW spectrum in the absorption edge for each transition (Fig. 6.40).

6.9.2.4 Electric Field

As described in Section 6.7, a transverse electric field breaks the parity of the electron and hole envelope functions so that $\langle F_{e,n}|F_{h,m}\rangle$ becomes non-zero for all n

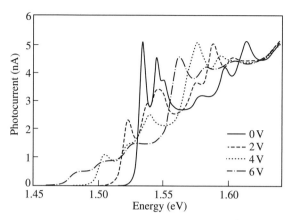

Figure 6.41.   Photocurrent spectra for a 110 Å GaAs/AlGaAs quantum well under reverse bias at 80 K showing the Stark shift of the absorption edge, exciton washout and the appearance of forbidden transitions.

and m and as bias increases, becomes progressively stronger for 'forbidden' transitions and weaker for 'allowed' transitions with $n = m$. At the same time the QCSE shifts the absorption edge to lower energies, and spatial separation of the electron and hole reduce the strength of the excitonic interaction (Fig. 6.41), although much more slowly than if the bias were applied in the well plane.

6.10   Quantum-well Solar Cells

Quantum-well structures are important in optoelectronics because they offer the joint benefits of highly confined electron and hole populations and a tunable band gap. Their advantages for lasers, photodetectors and other nonlinear optical devices are discussed in Chapters 7 and 8. One rather less obvious application is to high efficiency photovoltaics: the quantum-well solar cell (QWSC). This section is concerned with how the theoretical apparatus described in this chapter can be used to characterize and design QW structures for solar energy conversion.

6.10.1 Photoconversion

A simple, single band gap photoconverter works by absorbing incident photons of energy greater than the band gap and separating the charges so produced to deliver an electric current to an external circuit. In a semiconductor solar cell the charges are separated by the built-in electric field of a p-n (or p-i-n) junction. Not all the light energy absorbed can be converted into electrical energy since photogenerated carriers quickly decay to their ground state, losing any excess energy as heat. A photon of energy $\hbar\omega$ absorbed in a semiconductor of band gap E_g can thus deliver no more than E_g of electrical potential energy to the external circuit. Increasing the band gap increases the potential – and hence the cell voltage – but decreases the photocurrent since only high energy photons can be absorbed. Therefore, for any broad band incident spectrum, optimum power conversion is achieved at some intermediate band gap E_{opt}.

These considerations place a limit on the efficiency available from a perfect single threshhold photoconverter. In the standard air mass 1.5 solar spectrum a limit of 31% is reached at a band gap of 1.35 eV (Henry, 1980). In principle, such limits may be surpassed in a multi-band-gap or tandem system, where different parts of the spectrum are preferentially absorbed in materials of different band gap. Independent optimization of band gaps and photocurrents increases the theoeretical efficiency to about 45% for the standard solar spectrum or 50% under a concentration of 1000 (Henry, 1980). Practical improvements fall far short of these limits, usually through electrical losses.

Quantum well structures are interesting for photoconversion firstly as an alternative multi-band-gap approach to the tandem cell. Quantum wells added to the space charge region of a single band gap *p-i-n* solar cell extend its spectral response to longer wavelengths and so increase photocurrent (Fig. 6.42). At the same time the QWs act as centres for enhanced recombination and reduce the cell's operating voltage. If the improvement in current outweighs the loss in voltage, a net increase in power conversion efficiency is achieved. An increase in the limiting efficiency of a monolithic solar cell requires that such enhancements be available even using a host cell of optimum band gap. This situation essentially requires that the quasi-thermal equilibrium between the QW and host material is

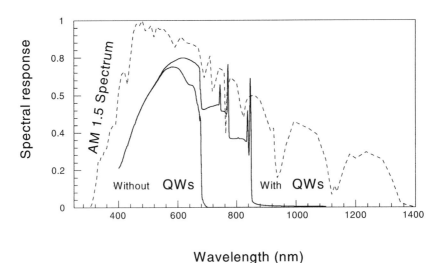

Figure 6.42.   Calculated spectral response of an $Al_{0.33}Ga_{0.67}As$ *p-i-n* solar cell with and without 50 87 Å GaAs quantum wells in the 0.8 μm intrinsic region. Quantum wells extend the spectral response to longer wavelengths where the solar spectral irradiance is still high. (AM air mass)

broken, so that recombination proceeds more slowly than expected. This suggestion and related experimental studies are discussed in detail by Araujo *et al.* (1994) and by Nelson (1995).

Less radically, QW structures are interesting in the possibility of superior practical performance to the homogenous cell of equivalent band gap. By careful choice of materials for well and barrier it may be possible to design a QWSC where the net recombination current is smaller than in the equivalent bulk alloy. This has been experimentally confirmed in $InP/In_{0.52}Ga_{0.48}As$ QW structures.

QWSC behaviour depends upon the QW electronic structure in several ways. The joint density of states function determines the QW absorption spectrum (Section 6.8) which, in turn, determines the photocurrent enhancement. Density of electron states in the conduction band and hole states in the valence band determine the spatial distribution of electrons and holes and, hence, the recombination current. Finally, the mixing of states between QW and host material or between neighbouring QWs can affect the efficiency of carrier transport.

6.10.2 Basic Principles

The net current density $J(V)$ which a solar cell at bias V delivers to the external circuit is the sum of the photocurrent J_{photo} and the opposing diode or darkcurrent, J_{dark},

$$J(V) = J_{photo} - J_{dark} \tag{6.47}$$

where J_{photo} is determined mainly by the absorptive properties of the material and J_{dark} is determined mainly by the carrier recombination rates. Maximum power conversion efficiency is achieved when the product $J \times V$ is a maximum. We therefore seek to maximize J_{photo} and minimize J_{dark} at as high a forward bias as possible.

6.10.2.1 Photocurrent

The photocurrent depends on both the incident spectrum and the spectral response – or external quantum efficiency – of the cell. The spectral response $SR(\hbar\omega)$ is the probability that, for each photon of energy $\hbar\omega$ incident on the sample, one electron will be delivered to the external circuit. For an incident photon flux density b_{in}

$$J_{photo} = \int_0^\infty b_{in} SR(\hbar\omega)\, d(\hbar\omega) \tag{6.48}$$

In a *p-i-n* structure $SR(\hbar\omega)$ may be expressed as the sum of contributions from the *p*-doped, the *n*-doped and the depleted or 'space charge' region (SCR):

$$SR(\hbar\omega) = SR_p(\hbar\omega) + SR_n(\hbar\omega) + SR_{SCR}(\hbar\omega) \tag{6.49}$$

SR_p and SR_n are calculated from semiconductor transport equations, and SR_{SCR} is obtained simply from the photon flux absorbed in the SCR, assuming efficient majority carrier transport. When N QWs of width L are present in the SCR,

$$SR_{\mathrm{SCR}}(\hbar\omega) = [1 - R(\hbar\omega)]\, e^{-x_p \alpha_b(\hbar\omega)}$$

$$\times\, [1 - e^{-\alpha_b(\hbar\omega)(W-NL)-N\alpha_{\mathrm{QW}}(\hbar\omega)L}]\eta_{\mathrm{esc}}(\hbar\omega) \qquad (6.50)$$

where W is the total width of the SCR including depleted p- and n-doped layers, x_p is the undepleted p-region width, $R(\hbar\omega)$ is the reflectivity of the device at $\hbar\omega$, $\alpha_b(\hbar\omega)$ is the absorption of the host material and $\alpha_{\mathrm{QW}}(\hbar\omega)$ is the absorption of the QW material which is calculated by the method of Section 6.8. Although in a p-i-n structure the QW is subject to the built-in electric field of the junction, in practice the Stark shift is negligible at the small forward bias where a solar cell operates,

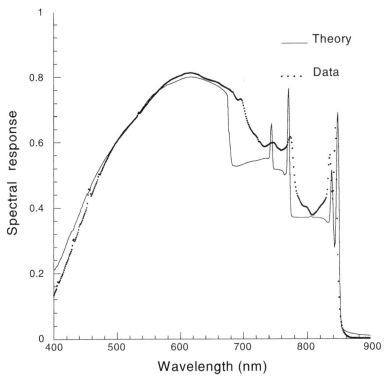

Figure 6.43.   Modelled and measured SR for an $\mathrm{Al_{0.33}Ga_{0.67}As}$ p-i-n solar cell with 50 85 Å GaAs QWs in the undoped region relative to an identical AlGaAs p-i-n solar cell without QWs. The addition of QWs enhances photocurrent by about 150% in standard solar spectrum and reduces the open circuit voltage by about 10%. The efficiency is approximately doubled. It should be stressed that the band gap of the host cell is already well above the optimum, so improved efficiency would be expected simply by the reduction in absorption edge, whether or not this is achieved using QWs.

and $\alpha_{QW}(\hbar\omega)$ may be calculated for flat band. The quantity η_{esc} represents the quantum efficiency for escape from the QW, which in most practical cases is unity at room temperature.

At energies between the effective band gap of the QW, E_{llh}, and the band gap of the host material, E_g, the SR is due solely to the QW absorption. Above E_g the SR is due primarily to the host material but is usually enhanced by the QW layers which absorb more strongly than the equivalent thickness of wide gap material. The photocurrent enhancement is therefore due to the QWs through the non-zero SR below E_g and the increased absorption above E_g (Fig. 6.43).

In Fig. 6.44 and 6.45 calculated and measured spectral response curves are compared for QW and homogenous *p-i-n* test cells in the materials systems $Al_xGa_{1-x}As/GaAs$ and $InP/In_{0.52}Ga_{0.48}As$. The discrepancy between measured and modelled SR at energies in the QW is due to several factors. For the calculated SR the absorption of the QW layer is assumed to change abruptly from that of a perfect square potential well (calculated by the methods of Sections 6.4 and

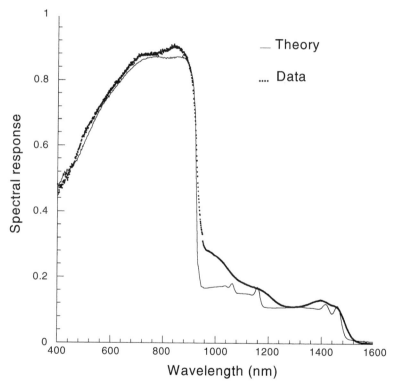

Figure 6.44.   Modelled and measured SR for an InP *p-i-n* solar cell with 15 $In_{0.52}Ga_{0.48}As$ 60 Å QWs in the undoped region. Addition of QWs enhances photocurrent by 190%, reduces the open circuit voltage by 25% and increases efficiency by 105% in a lamp resembling a 3000 K blackbody spectrum. This is an example of how better performance can be achieved using QWs even when the host band gap is near to optimum.

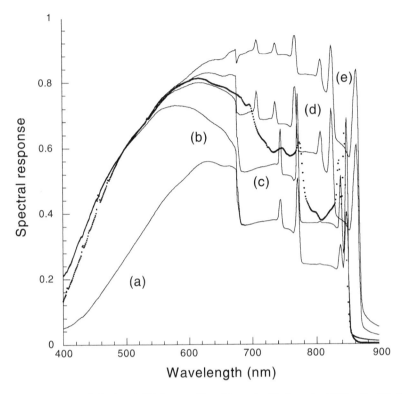

Figure 6.45.   Calculated SR for an $Al_{0.33}Ga_{0.67}As/GaAs$ p-i-n QW cell with the following structures: (a) 0.3 μm p region, 0.48 μm i region, 30 85 Å QWs; (b) 0.15 μm p region, 0.48 μm i region, 30 85 Å QWs; (c) 0.15 μm p region, 0.8 μm i region, 50 85 Å QWs; (d) as (c) with 140 Å QWs; (e) 0.15 μm p region, 1.6 μm i region, 100 140 Å QWs (which may be achieved with a back surface mirror). The thick line is the measured SR for a cell of design (c), which is a 14% efficient solar cell in standard test conditions.

6.8) to that of the QW material in the bulk (taken from published measurements) at E_g. This does not allow for any contribution to the joint density of states from transitions between confined states in one band and continuum states in the other, which tend to enhance the absorption below E_g, nor for the effect of QWs on the density of states of the well material above E_g, for instance through quasi-confined states. Other factors are the neglect of valence band mixing, which was discussed in Section 6.4.2, and the neglect of the Somerfeld enhancement of inter-subband transitions, which was mentioned in Section 6.6.2. Non-abruptness of the QW interfaces through atomic diffusion can also lead to an enhancement of the density of states at energies high in the QW where the effective well width is greater.

Since, unlike other QW devices, the QW solar cell is primarily a broad band absorbing device, these uncertainties in the density of states at energies high in the

QW are important. The theoretical techniques described in this chapter are useful for characterizing QW structures from their primary excitons but do not generally predict the photocurrent enhancement to closer than 10–20%.

Models incorporating the theory of QW electronic structure are, nonetheless, very useful for optimizing the SR of QW solar cells (Renaud *et al.*, 1995; Anderson, 1995; Corkish and Green, 1993). Increased QW photocurrent results from an increase in the number, depth or width of QWs or simply by increasing the amount of light reaching the undoped region. Figure 6.45 shows how SR may be improved by exploiting two of these effects.

6.10.2.2 Recombination Current

Unfortunately, optimizing the peformance of a QW solar cell is not simply a matter of optimizing its SR. Increasing light absorption in the QW by whatever means increases the populations of photogenerated electrons and holes available for recombination. Both radiative and non-radiative recombination currents increase with the electron–hole product, causing a reduction in voltage. The object in QW solar cell design is to maximize the effect of QWs on photocurrent while minimizing their effect on recombination.

Recombination in a QW – both radiative or non-radiative – depends on electron and hole populations as well as on material parameters such as the recombination lifetimes and the efficiency of the competing process of carrier escape. The carrier populations may be calculated from the density of states functions discussed in Sections 6.4 and 6.6. For electrons with a density of states $g(\mathbf{k})$ in phase space,

$$n = \int g(\mathbf{k}) f_{\text{FD}}[E(\mathbf{k}, T, \mu]\, \mathrm{d}\mathbf{k} \tag{6.51}$$

where f_{FD} is the Fermi–Dirac distribution function, μ the electron quasi-Fermi level and T the effective electron temperature. (The integral should be carried out with respect to wavevector, rather than energy, because of the anisotropy of the density of states.) The electron and hole quasi-Fermi levels themselves must be found by self-consistent solution of the semiconductor transport and Poisson's equations across the device structure, and depend on the nature of carrier escape from and capture into the QW. A detailed treatment is beyond the scope of this chapter.

6.10.2.3 Carrier Escape

The escape current density of carriers leaving the QW can be calculated semiclassically from the QW density of states. For electrons,

$$J_{\text{esc}} = \int g(\mathbf{k}) f_{\text{FD}}[E(\mathbf{k}, T, \mu)] T(k_z)(\hbar k_z/m^*)\, \mathrm{d}\mathbf{k} \tag{6.52}$$

where $T(k)$ is the transmission coefficient through the barrier and $\hbar k/m^*$ is the classical velocity of the particle. J_{esc} depends on the width, depth and electric field acting on the QW through g and T. It has been shown by Nelson *et al.* (1993) that photogenerated carriers escape efficiently from QWs at room temperature, even at the low operating bias of a solar cell, and that quantization – particularly through the position of light particle levels in the QW – is important in determining the rate of escape.

Thus, given knowledge of the dominant carrier removal processes and appropriate lifetimes, the theoretical apparatus of Sections 6.4–6.8 can be used to evaluate the recombination and photocurrent of a QW solar cell and hence its current-voltage characteristics. There is some evidence that through carrier escape the recombination currents in the QW are lower than would be expected for a layer of the bulk material of equivalent band gap and optical depth, which indicates that theoretical efficiency enhancements may be possible, in addition to those already observed in practical material (Barnham *et al.*, 1996).

6.11   Concluding Remarks

The AlGaAs/GaAs materials system remains the best understood III–V materials system for heterostructures and has been used for nearly all illustrations in this chapter. The theoretical methods are equally applicable to other direct gap III–V materials, with modifications for strained layer systems. However, their quantitative reliability depends upon the reliability of the important effective materials parameters (band gaps, band offsets, effective masses, optical matrix elements), many of which are not yet well known for other materials. In fact, theoretical techniques such as these, when used interactively with experimental spectra, provide a powerful tool for material characterization as well as for modelling and optimization of new semiconductor devices.

EXERCISES

1.   Using the infinite barrier approximation, derive an expression for the density of states for electrons in a quantum well in terms of the well width L and electron effective mass m^*. Show that it coincides with the bulk density of states at the subband edges.

2.   Show, with the help of sketches, how the joint density of states for electrons and holes in a quantum well may be expected to change as the following effects are included:

(a) finite height of the barrier potential,
(b) electron–hole interactions,
(c) an electric field applied along the growth direction.

3. In an optical experiment on a $100 \, \text{Å} \, \text{Al}_{0.3}\text{Ga}_{0.7}\text{As}/\text{GaAs}$ multi-quantum-well sample, the $n = 1$ and $n = 2$ electron–heavy-hole excitons are observed to lie at $1.456 \, \text{eV}$ and $1.696 \, \text{eV}$.

A student calculates that the same features should appear at $1.457 \, \text{eV}$ and $1.705 \, \text{eV}$, using a finite square model and assuming values for the well width, band offset, effective masses and exciton binding energy. Which of the following is most likely to explain the discrepancy?

(a) Uncertainty in the well width
(b) Uncertainty in the exciton binding energy
(c) Presence of a transverse electric field
(d) Coupling between adjacent quantum wells

Give brief reasons for your answer.

4. A multi-quantum-well system is said to be Type II if the lowest point in the conduction band and the highest point in the valence band occur in different materials. AlAs is an indirect gap semiconductor where the conduction band minimum occurs at a different point in **k** space (the X point) to the valence band maximum (the Γ point), while in GaAs both occur at the Γ point. By tuning the width of the QW in an AlAs/GaAs multi-quantum-well system, we can change the nature of the QW from Type I to Type II.

Using the information below, calculate the height of the electron energy level above the GaAs conduction band minimum when a AlAs/GaAs multi-quantum-well becomes Type II. Estimate the well width at which this occurs, assuming the infinite barrier approximation for the QWs.

The energy difference between Γ minimum and valence band in GaAs is $1.519 \, \text{eV}$ at $4.2 \, \text{K}$. The energy difference between X minimum and valence band in AlAs is $2.243 \, \text{eV}$ at $4.2 \, \text{K}$. The energy difference between Γ minimum and valence band in AlAs is $3.113 \, \text{eV}$ at $4.2 \, \text{K}$. The conduction band offset is 0.67 and the electron effective mass in GaAs is $m^* = 0.0667$.

5. In perturbation theory, the second-order energy shift ΔE_n in the nth energy level of a system induced by a perturbing potential $V(x)$ is given by

$$\Delta E_n = \langle n|V|n \rangle + \sum_{m \neq n} \frac{\langle m|V|n \rangle \langle n|V|m \rangle}{E_n - E_m}$$

where the E_n are the energy levels, $|n\rangle$ are the normalized wavefunctions of the unperturbed system and $\langle n|V|m \rangle$ represents the overlap integral between the nth and mth unperturbed wavefunction and V:

$$\langle n|V|m \rangle = \int \psi_n^*(x) V(x) \psi_m(x) \, dx$$

A small electric field E applied along the growth direction of a quantum well introduces an additional electron potential $V(x) = eEx$. By considering the parity of the confined state wavefunctions, show that

(a) the term $\langle n|V|n \rangle$ is zero for all confined states,
(b) the energy shift is proportional to E^2 and that it is negative for the lowest confined state.

You may take the confined state wavefunctions to be those for an infinite well of width L:

$$\psi_n(x) = \begin{cases} (2/L)^{1/2} \cos k_n x, & \text{for odd } n \\ (2/L)^{1/2} \sin k_n x, & \text{for even } n \end{cases}$$

where $k_n = n\pi/L$.

6. In an optical study of a GaAs/In$_x$Ga$_{1-x}$As single QW under bias, the energy of the lowest exciton shifts *linearly* with the applied electric field. In view of the result obtained in Exercise 5(a), what is a possible explanation?

7. The density of states *per unit area* for electrons confined to two dimensions is (see Section 2.1)

$$g_{2D} = \frac{m}{\pi\hbar^2}$$

(a) How would you modify this formula to describe the joint density of states for allowed optical transitions between a 2D valence band with effective mass m_h^* and a 2D conduction band with effective mass m_e^*?
(b) Fig. 6.46 gives the joint density of states (JDOS) for allowed optical transitions in a single quantum well. What information can you deduce about the well? Give estimates and state any assumptions you make.

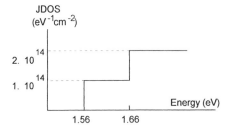

Figure 6.46.   Figure for Exercise 7(b).

(c) What other quantity would you need to predict the optical absorption in the absence of excitonic effects?

8. A GaAs QW of nominal width $100\,\text{Å}$ is grown in Al$_{0.3}$Ga$_{0.7}$As. Assuming an aluminium fraction of 0.3, a conduction band offset of 0.67 and electron and hole effective masses of 0.067 m_0 and 0.34 m_0, respectively, calculate how many electron–heavy-hole transitions we expect to see. (At room temperature, the band gap of GaAs is 1.424 eV and that of $Al_xGa_{1-x}As$ is $1.424 + 0.247x$ eV.)

Suppose that the $n = 3$ electron–heavy-hole transition is, in fact, observed. Which of the following could be an explanation?

(i) Uncertainty in the well width
(ii) Uncertainty in the electron effective mass
(iii) Uncertainty in the conduction band offset

In each case, make a quantitative estimate of the uncertainty and indicate how the suggested explanation could be tested.

References

N. Anderson, *J. Appl. Phys.* **78**, 1850 (1995).

G. L. Araujo, A. Marti, F. W. Ragay and J. H. Wolter, in *Proceedings of the 12th European Photovoltaic Solar Energy Conference*, pp. 1481–84 (1994).

K. W. J. Barnham *et al.*, *J. Appl. Phys.* **80**, 1201 (1996).

G. Bastard, *Wave Mechanics Applied to Semiconductor Heterostructures* (Les Editions de Physique, Paris, 1988).

G. Bastard and J. A. Brum, *IEEE J.Quantum Electron.* **22**, 1625 (1986).

G. Bastard, J. A. Brum and R. Ferreira, *Solid State Physics* **44**, 229 (1991).

Y. C. Chang and J. N. Schulman, *Phys. Rev.* **B31**, 2069 (1985).

R. Corkish and M. A. Green, in *Proceedings of the 23rd IEEE Photovoltaic Specialists Conference*, pp. 675–80 (1993).

P. Dawson, K. J. Moore, G. Duggan, H. I. Ralph and C. T. Foxon, *Phys. Rev. B* **34**, 6007 (1986).

R. Dingle, A. C. Gossard and W. Wiegmann, *Phys. Rev. Lett.* **34**, 1327 (1975).

G. Duggan, H. I. Ralph, K. S. Chan and R. J. Elliott, *Materials Research Society Proceedings* **47**, K. Ploog and N. T. Linh, eds. (Les Editions de Physique, Paris, 1985a).

G. Duggan, H. I. Ralph and K. J. Moore, *Phys. Rev. B* **32**, 8395 (1985b).

R. J. Elliot, in *Polarons and Excitons in Polar Semiconductors and Ionic Crystals*, J.T. Devreese and F. Peeters eds. (Plenum, New York, 1984), pp. 271–92.

R. Eppenga, M. F. H. Schuurmans and S.Colak, *Phys. Rev. B* **36**, 1554 (1987).

R. L. Greene, K. K. Bajaj and D. E. Phelps, *Phys. Rev. B* **29**, 1807 (1984).

C. H. Henry, *J. Appl. Phys.* **51**, 4494 (1980).

M. Jaros, *Physics and Applications of Semiconductor Microstructures* (Clarendon Press, Oxford, 1989).

E. O. Kane, *J. Phys. Chem. Solids* **1**, 249 (1957).

P. C. Klipstein and N. Apsley, *J. Phys. C* **19**, 6461 (1986).

J. M. Luttinger, *Phys. Rev.* **102**, 1030 (1956).

D. A. B. Miller, D. S. Chemla, T. C. Damen, A. C. Gossard, W. Wiegmann, T. H. Wood, and C. A. Burrus, *Phys. Rev. B* **32**, 1043 (1985).

J. Nelson, *Physics of Thin Films*, **21**, 311 (1995).

J. Nelson *et al.*, *IEEE J. Quantum Electron.* **29**, 1460 (1993).

H. I. Ralph, *Solid State Comm.* **3**, 303 (1965).

P. Renaud *et al.*, in *Proceedings of the First World Energy Conference on Photovoltaic Energy Conversion*, pp. 1787–1790 (1995)

S. Schmitt–Rink, D. S. Chemla and D. A. B. Miller, *Adv. Phys.* **38**, 89 (1989).

P. J. Stevens, M. Whitehead, G. Parry and K. Woodbridge, *IEEE J. Quantum Elect.* **24**, 2007 (1988).

C. Weisbuch and B. Vinter, *Quantum Semiconductor Structures: Fundamentals and Applications* (Academic Press, Boston, 1991).

7 Non-linear Optics in Low-dimensional Semiconductors

C. C. Phillips

7.1 Introduction

When an electromagnetic wave $\mathbf{E} = \mathbf{E}_0 \exp[\mathrm{i}(\omega t - \mathbf{k} \cdot \mathbf{x})]$ of optical frequency ω propagates through a solid, its oscillatory field $E(\mathbf{x}, t)$ introduces a polarization $P(\mathbf{x}, t)$. For values of ω in the optical range the E-field oscillates at a frequency much higher than the characteristic lattice vibration frequencies, and the contributions to $\mathbf{P}$ are dominated by the oscillatory response of the electrons in the solid. For the majority of situations the polarization is simply linearly proportional to the instantaneous electric field, and given by

$$P_i(\mathbf{x}, t) = \chi_{ij}\epsilon_0 E_j(\mathbf{x}, t) \tag{7.1}$$

where the linear susceptibility χ is a dimensionless complex constant which depends on the solid and on ω, but is independent of the amplitude of the oscillating electric field. For anisotropic materials the tensor component χ_{ij} relates the polarisation in the ith direction to the electric field in the jth direction, but in isotropic materials χ is independent of the direction of $\mathbf{E}$ and is a simple scalar quantity. χ is related to the complex refractive index, n, through

$$\mathrm{n}^2 = 1 + \chi \tag{7.2}$$

where

$$\mathrm{n} = n(1 + i\kappa) \tag{7.3}$$

Hence, the real (n) and imaginary ($in\kappa$) parts of the refractive index and the absorption coefficient α are also independent of the strength, $\mathbf{E}_0$, of the oscillatory electric field, i.e. independent of the incident light intensity.

This situation generally prevails because, in most practical situations, the E-field associated with the light beams is very much smaller than the typical electric fields surrounding the atoms in the solid; the electrons in the solid move in potential wells which are to a good approximation parabolic and their displacement, z, under the applied E-field obeys an electrostatic equivalent of Hooke's law and is just proportional to $\mathbf{E}$. They produce a polarization $\mathbf{P} = Ne\mathbf{z}$ where N is the number of electrons per unit volume.

The electric fields in such a linear optical material obey the electromagnetic principle of superposition. At all times the polarization $\mathbf{P}(\mathbf{x}, t)$ in the presence of two electric field distributions $\mathbf{E}_1(\mathbf{x}, t)$ and $\mathbf{E}_2(\mathbf{x}, t)$ will simply be the sum of the polarizations which would be induced by $\mathbf{E}_1(\mathbf{x}, t)$ and $\mathbf{E}_2(\mathbf{x}, t)$ independently. If, for example, $\mathbf{E}_1(\mathbf{x}, t)$ and $\mathbf{E}_2(\mathbf{x}, t)$ happen to describe two propagating light beams in a solid, the principle of superposition tells us that the propagation of beam 1 is entirely independent of whether or not the electric field due to beam 2 is present.

If we shine two beams through a block of glass (Fig. 7.1(a)), with optical frequencies ω_1 and ω_2 say, we would normally be very surprised indeed if beam 2 were to change its direction of propagation, its intensity or its wavelength distribution, for example, as beam 1 is turned on and off. Whilst being very convenient for everyday life, the extreme optical linearity of most materials makes it impossible to control or modulate the propagation of one light beam with another within them. In linear optical materials there can be no optical analogue of the electronic switch or transistor (Fig. 7.1(b)).

Recent decades have seen rapid developments in optoelectronics technologies and these have driven a substantial research effort into investigating the regimes where this linear optical response approximation breaks down and all-optical switching and modulating devices become possible. There are many areas of technology where information is either naturally present or is artificially coded (e.g. two-dimensional pattern recognition, optical fibre telecommunications, fibre interconnects, etc.) in optical form, but where any signal processing has to take place electronically. This currently requires comparatively slow and expensive interconversion between the electronic and optical domains.

For tomorrow's computers, computer systems engineers are already looking to the massive parallelism ($>10^6$ individual channels) available in two-dimensional optical image arrays, and to the enormous bandwith in optical signals (e.g. 2×10^{14} Hz at the optical fibre wavelength of 1.5 µm) as a means of producing dramatic increases in computing power. This will only be possible if practical 'photonic computers', i.e. data processing devices where all signal switching takes place in the optical domain (Lee, 1981; Feitelson, 1988; Wherrett and Tooley, 1988), can be developed.

The key to exploiting these ideas lies in the development of materials which exhibit non-linear optical (NLO) responses at usable wavelengths and optical

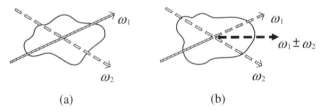

(a)   (b)

Figure 7.1. Schematic representation of light beam propagation in (a) a linear optical material and (b) a non-linear optical material. In (b), modulations in the intensity of beam 2 can produce modulations in the intensity of beam 1 and/or in the intensity of the frequency mixing component at $\omega_1 + \omega_2$.

intensities. The research field of non-linear optics is not in itself particularly young, but the advent of artificial low-dimensional semiconductor materials has led to many important advances in the 1990s, and what follows is an attempt to summarize these.

For the purposes of discussion I have chosen to divide optical non-linearities into two classes, namely, (a) 'non-dissipative' processes, where the fraction of the optical energy absorbed by the solid is negligible and non-linear effects take place over the full optical bandwidth, and (b) 'dissipative' processes, where the primary mechanism is a significant perturbation of the electronic structure of a solid following the absorption of energy from the light field. It is in the latter class of effect that the physics of low-dimensional semiconductors (LDS) has had the most impact in recent years.

7.2   Non-dissipative NLO Processes

If the light frequencies of interest lie in a region well away from any electronic resonances in a solid, the solid will be transparent and have an almost wholly real value of χ. If in this case the light intensity (and hence $\mathbf{E}_0$) is increased, the amplitude of the electron oscillations will ultimately become large enough for higher-order anharmonic terms in the potential function in which the electrons move to become significant (Yariv, 1989). When this happens the 'electrostatic Hooke's law' approximation breaks down (Fig. 7.2) and the electronic

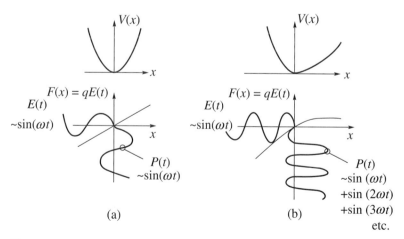

Figure 7.2.   Schematic representation of the polarization induced by electron motion in response to an optical sinusoidal applied electric field (a) in the linear case, when the electron moves in a harmonic (parabolic) electrostatic potential well, where $P(t) \propto \chi E(t)$, and (b) when the electron moves in an anharmonic electrostatic potential well, where $P(t) = \chi E(t) + \chi^{(2)}[E(t)]^2 + \chi^{(3)}[E(t)]^3 + \cdots$.

polarization then starts to have terms which are non-linear in $\mathbf{E}(\mathbf{x}, t)$, with second-($\propto E^2$) and higher-order contributions.

The starting point for a description of the second-order NLO effect in a crystal-line solid is the second-order non-linear susceptibility tensor d_{ijk} defined through

$$P_i(\mathbf{x}, t) = 2d_{ijk}E_j(\mathbf{x}, t)E_k(\mathbf{x}, t) \tag{7.4}$$

where P_i is the non-linear component of the polarization in the ith direction, and the elements of d_{ijk} are related to non-linear elements in χ by $\epsilon_0\chi_{ijk} = 2d_{ijk}$.

The symmetry properties of d_{ijk} must be compatible with those of the crystal, and only crystals without a centre of inversion symmetry (e.g. zincblende lattices) have non-zero second-order susceptibility tensors. If two light beams at frequencies ω_1 and ω_2 propagate through such a medium, the E^2 dependence of the induced polarization component will produce polarization (and hence optical radiation) components at frequencies of $\omega = 2\omega_1, 2\omega_2$, (optical second harmonic generation), $\omega = \omega_1 + \omega_2$ (sum frequency mixing), $\omega = \omega_1 - \omega_2$ (difference frequency mixing) and at $\omega = 0$ (optical rectification) (Guenther, 1990; Boyd, 1992; Bloembergen, 1996).

The efficiency of conversion of energy from the initial beams at ω_1 and ω_2 depends on a number of factors, including the degree to which momentum conservation is satisfied between the phase velocities of the ω_1, ω_2 and ω beams, but is proportional to the intensities at ω_1 and ω_2, and to the strength of the appropriate component in the d_{ijk} tensor. The values of d_{ijk} range over almost three orders of magnitude from e.g. $5170 \times 10^{-24}\,\mathrm{m\,V^{-1}}$ for tellurium down to $3.5 \times 10^{-24}\,\mathrm{m\,V^{-1}}$ for KH_2PO_4. Values for typical semiconductors (e.g. GaAs, for which $d_{14} = 720 \times 10^{-24}\,\mathrm{m\,V^{-1}}$) lie at the upper end of the range (Yariv, 1989). A detailed analysis, though, shows that the majority of this wide spread in the values of d_{ijk} arises from variations in the linear susceptibility χ_{ij}, and that the dimensionless quantity

$$\delta_{ijk} = \frac{d_{ijk}}{\epsilon_0^2\chi_{ii}\chi_{jj}\chi_{kk}} \tag{7.5}$$

varies by less than a factor of 2, being around 2×10^9 for nearly all the materials investigated so far.

This can be understood in terms of a simple classical picture of a solid composed of N anharmonic dipoles per unit volume. The largest degree of anharmonicity and, hence, the highest d_{ijk} value will occur with the most asymmetric electronic orbital configuration for the dipoles in the solid. It can be shown that a dipole composed of charges e and $2e$ a distance $\approx 0.5\,\mathrm{nm}$ apart will give (Yariv, 1989) $\delta_{ijk} \approx 4 \times 10^9$. Electronic bonds in crystals rarely involve the transfer of more than one electron, and the solid interatomic spacing is always of the order of 0.5 nm, so, in broad terms, the scope for developing materials with significantly improved δ_{ijk} values is limited.

At present, the conclusion of the preceding discussion represents a substantial obstacle to the exploitation of non-dissipative NLO effects in practical devices. To achieve usefully large 'non-dissipative' energy conversion efficiencies the optical electric field must be a significant fraction of typical intra-atomic fields ($\approx 10^{11}$ V m^{-1}), and at a wavelength of 1.5 μm an optical field of 10^{9} V m^{-1} corresponds to an intensity of $\approx 2.3 \times 10^{15}$ W m^{-2} in GaAs. These intensities can only be produced using the focussed radiation from sophisticated laboratory picosecond pulsed lasers; they are not yet available from practical optical sources such as laser diodes.

The confinement potentials in typical low-dimensional structures act primarily on the envelope function of the electronic states (see, e.g., Section 1.2.3), and have only a minor perturbative effect on the cell-periodic part of the electronic wavefunctions which are responsible for the d_{ijk} values. LDS technology has thus produced little improvement in the non-dissipative NLO properties of semiconductor materials themselves. Recent years have seen dramatic improvements in solid-state pulsed LDS optical sources and LDS optically confined structures (e.g. semiconductor waveguides), however, and these can substantially improve the efficiency of existing NLO processes by increasing the local optical intensity (Ho *et al.*, 1991). Current research interests tend to concentrate on developing novel device structures to enhance the utility of existing effects as opposed to using LDS techniques to attempt to improve the actual material d_{ijk} values.

7.3   Dissipative NLO Effects

If the optical frequency is tuned close to an electronic resonance (often the semiconductor band gap), then radiation will be absorbed and the electronic structure of the semiconductor becomes perturbed (Fig. 7.3). Under suitable circumstances this perturbation can lead to a substantial modification, $\Delta\alpha(\hbar\omega)$ of the semiconductor absorption coefficient spectrum $\alpha(\hbar\omega)$. This, in turn, will produce a modification of the refractive index spectrum, $\Delta\mathbf{n}(\hbar\omega)$, which can be conveniently calculated through the Kramers–Kronig relation:

$$\Delta\mathbf{n}(\hbar\omega) = \frac{c}{\pi} \int_0^\infty \frac{\Delta\alpha(\hbar\omega)\,\mathrm{d}\omega}{\omega'(\omega - \omega')} \tag{7.6}$$

where c is the speed of light.

Although generically similar (Fig. 7.3), dissipative NLO processes differ widely in the nature of the mechanism relating the perturbation to the change in the optical constants, and the recent advances have stemmed from producing device structures where small amounts of absorbed optical energy produce large $\Delta\alpha(\hbar\omega)$ changes in a narrow spectral region close to the semiconductor band edge. This wavelength region can be tailored by choice of the semiconductor band gap to suit available laser diode sources, or to lie in the 1.33–1.55 μm range of interest for silica fibre optical telecommunications.

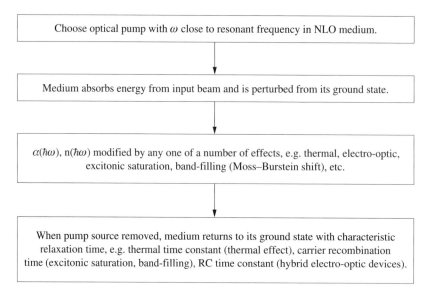

Figure 7.3.   Generic description of a 'dissipative' non-linear optical effect.

The switching dynamics of these processes are determined by a characteristic relaxation time in the semiconductor (Fig. 7.3), and generally the response times are several orders of magnitude slower than the 10^{-14} s inverse of a typical optical bandwidth. The impressive low-power thresholds of dissipative NLO devices are always obtained at the expense of switching speed. In most of the schemes described below, speed and sensitivity are traded off by tailoring the characteristic recombination time to suit a particular device application and to bring the sensitivity of the NLO effect within reach of available optical sources.

Developments in LDS technology in recent decades have produced rapid advances in this field. Band structure and optical engineering techniques have produced a new family of device ideas in which electronic resonances close to the band edge are combined with optical resonances designed into the device, e.g. multilayer Bragg mirrors and Fabry–Perot resonators (Hecht, 1987), to produce devices with useful combinations of switching speed and sensitivity. 'Photonic' all-optical logic elements (Jewell *et al.*, 1985) and prototype photonic computers have been fabricated. Sections 7.5–7.8 concentrate on the LDS physics of a selection of these new device concepts.

7.4   Potential Applications of NLO

7.4.1   Serial Channel Applications

Present electronic computing and communications systems operate with an essentially serial architecture with small numbers of signal channels carrying data in bandwidths which can exceed 10^{10} Hz. For fibre optical telecommunications all optical data switching schemes are only of interest if they offer bandwidths in excess of this, and if they can operate at wavelengths in the 1.3–1.55 μm region, where the absorption and dispersion of silica optical fibres is at a minimum.

Silicon very large-scale integration (VLSI) computers have now reached such a level of complexity that the problem of connecting logic devices within one chip, and from one chip to another, with suitable high-bandwidth, low-crosstalk interconnections is emerging as a major constraint on increasing their complexity and hence computing power. A possible solution is to move towards optical interconnects between functional logic elements. The technical requirements for this application are similar to those applying to long distance optical fibre telecommunications, but with a relaxation on the operational wavelength requirement due to the much shorter beam propagating distances involved.

Both of these applications favour the exploitation of non-dissipative NLO mechanisms. The high optical intensities required are potentially achievable in the form of ultrashort laser pulses in confined waveguide geometries. For small numbers of signal channels the thermal problems associated with these intensities are in principle containable, and a large fraction of the optical bandwidth is available for data transmission.

The dissipative NLO device concepts are also potentially useful for these applications, although as the required bandwidths increase, their sensitivity advantages become less compelling.

7.4.2   Multi-channel Applications: Optical Computing

There is an important class of computing problems in the general area of two-dimensional pattern recognition and image processing (e.g. two-dimensional spatial Fourier transforms, image correlation and image filtering) where the initial data to be processed is in the form of a two-dimensional image of typically 10^6 individual pixel channels in parallel. In conventional electronic computers with their serial architectures, this huge volume of information has to be reduced to a few tens of channels for data processing and storage. The difficulties in constructing any form of electronic serial data processing machine capable of recognizing, handling and reacting to optical image information at anything approaching 'real time' speeds are substantial, and a radically different approach with a highly parallel computing architecture is being considered for these tasks.

As an alternative, various all-optical image processing schemes (Lee, 1981; Feitelson, 1988; Wherrett and Tooley, 1988) have been proposed which exploit the massive parallelism inherent in a two-dimensional pixel array. In the optical domain, switching between an image $I(x, y)$, and its spatial Fourier transform $F(k_x, k_y)$ can be achieved effectively instantaneously simply by passing the image through a lens (Hecht, 1987). Once in the Fourier domain, convolution and deconvolution correlation operations are reduced to simple multiplications at a pixel by pixel level (Fig. 7.4).

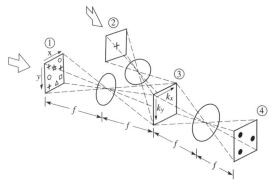

Figure 7.4.   Example of all-optical image correlating scheme. (1) Image containing patterns to be recognized. (2) Pattern template. (3) Non-linear optical image multiplying element, illuminated with intensity distributions $F_1(k_x, k_y)$ and $F_2(k_x, k_y)$ corresponding to the spatial Fourier transforms of $I_1(x, y)$ and $I_2(x, y)$. Transmitted light intensity $\propto F_1 F_2$. (4) Retransformed product of $F_1(k_x, ky)$, $F_2(k_x, k_y)$ corresponding to the image correlation of $I_1(x, y)$ and $I_2(x, y)$.

To implement such concepts requires two-dimensional non-linearly transmitting (or reflecting) screens. A two-dimensional array of bistable optical elements, i.e. an image memory, and some means of efficiently converting electronic data to two-dimensional optical information (a 'spatial light modulator') are among the other basic building blocks of such a system. Operating wavelengths are not critical as long as suitable sources are available, but thermal stability and optical power generation considerations require a sizeable non-linear effect occurring at intensities of less than tens of watts per square centimetre for practical large area arrays. The increase in processing power over conventional electronic image processors comes mainly from the massive parallelism and the inherent suitability of optical methods for this class of problem; even individual pixel switching times as slow as 10^{-3}–10^{-4} s would result in a substantial improvement in computing power over present electronic image processing schemes.

A number of the LDS dissipative NLO device concepts below have already demonstrated performance characteristics which make them worthy of serious consideration for these 'photonic computing' applications, and the maturity of semiconductor device processing technologies coupled with the increasing availability of cheap high-power laser diode sources at suitable wavelengths suggests that practical large-scale photonic computers may become a reality in the near future.

7.5   Excitonic Optical Saturation in MQWs

7.5.1   Excitonic Absorption at Low Intensities

The simplest theory of light absorption in direct-gap semiconductors predicts an absorption curve with an onset at E_g, the semiconductor band gap, and a form

$\alpha(\hbar\omega) \propto (\hbar\omega - E_g)^{1/2}$, corresponding to the form of the joint optical density of states (Section 2.2.1). However, this approach neglects the fact that the electron and hole states involved in the absorption event are electrically charged and can bind together into an entity known as an 'exciton' because of their mutual Coulomb attraction. In the three-dimensional case appropriate to a homogenous bulk semiconductor, an estimate of the excitonic binding energy R_{3D}^* can be obtained from the hydrogenic theory used for impurity states (Section 6.6.1) as:

$$R_{3D}^* = \frac{\mu^*}{m_0 \epsilon_r^2} R \qquad (7.7)$$

where $R = 13.6\,\text{eV}$ is the Rydberg constant, ϵ_r is the static dielectric constant, and μ^* is the reduced effective mass:

$$\frac{1}{\mu} = \frac{1}{m_e^*} + \frac{1}{m_{hh}^*} \qquad (7.8)$$

and m_e^* and m_{hh}^* are the electronic and (heavy) hole effective masses, respectively. Putting in the appropriate values for GaAs gives $R_{3D} \approx 5\,\text{meV}$, with an effective excitonic Bohr radius,

$$a = \frac{m_0 \epsilon_r}{\mu^*} a_0 \approx 10\,\text{nm} \qquad (7.9)$$

In three dimensions this excitonic state is thus weakly bound and spread over a large crystal volume; it is readily field-ionized by the random static electric fields which are present in all but the purest samples due to compensated ionised impurities. At finite temperatures excitons absorb and emit phonons and each phonon-scattering event destroys the phase of the excitonic wavefunction, broadening the energy width of the excitonic absorbtion peak in accordance with the Heisenburg uncertainty principle. Excitonic absorption can be seen in ultrapure bulk GaAs crystals at very low tem-

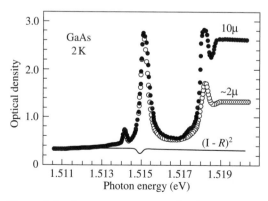

Figure 7.5. Low-temperature excitonic absorption peaks in 2 mm and 10 mm thick ultra-pure bulk GaAs samples (Sell, 1972, with permission).

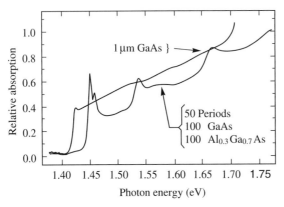

Figure 7.6.   Room temperature absorption coefficient curves measured for thick bulk GaAs film and MQW sample with 10 nm wide GaAs QWs surrounded by 20 nm thick $Al_{0.3}Ga_{0.7}$ As barriers (Schmitt-Rink *et al.*, 1989). (Courtesy D. S. Chemla)

peratures as sharp absorption peaks at $\hbar\omega = E_g - R^*_{3D}$ (Fig. 7.5), but these rapidly broaden and become unresolvable at 300 K (Schmitt-Rink *et al.*, 1989) (Fig. 7.6).

In quantum well (QW) samples, however, the situation is dramatically changed by the heterojunction potentials which confine the electron–hole pair into a layer much thinner than the normal bulk three-dimensional exciton diameter. This forces the electron and hole closer together on average, and increases the effective exciton binding energy by up to a factor of 4 in the limit of perfect two-dimensional confinement. In real QWs this limit cannot be reached because the finite values of the confining potentials allow substantial penetration of the carrier wavefunctions into the barrier layers for very narrow wells. Nevertheless R^*_{2D} values of up to 9–10 meV can readily be achieved at typical QW widths (Leavitt and Little, 1990) of 5–10 nm.

This confinement of the exciton also increases the usable 'oscillator strength' for optical absorption at the excitonic peak. The area under the excitonic absorption peak is constant for a given material, so a large R^*_{2D} value produces narrow absorption peaks in $\alpha(\hbar\omega)$ with high peak absorption values for quasi-monochromatic radiation. This increase in R^*_{2D} in QWs, whilst leaving the phonon-scattering rates substantially unchanged, means that strong excitonic peaks are clearly resolved in 300 K optical absorption spectra, and excitonic effects dominate the room temperature optical properties of typical QW structures.

The confinement potentials in the valence band also act to split the $\mathbf{k} = \mathbf{0}$ degeneracy of the light- and heavy-hole bands, each of which has an associated excitonic peak, leading to a typical MQW absorption spectrum of the form shown in Fig. 7.6.

7.5.2 Saturation of Excitonic Peaks at High Intensities

If a QW sample is illuminated with light of energy greater than or equal to the heavy-hole excitonic absorption peak, a steady-state exciton population is created. The density of the exciton population created this way is proportional to the illuminating intensity and proportional to the lifetime of the excitons, τ_{rec}, against decay by recombination. As the illumination intensity (and hence the mean exciton density) is increased, there comes a point where extra exciton scattering processes set in and broaden the exciton line to a degree where it is no longer resolvable in the $\alpha(\hbar\omega)$ spectrum, and the latter reverts to a form similar to that of a hypothetical QW in which the excitonic interaction is absent (Fig. 7.7).

An approximate estimate of the intensity required to do this can be made by assuming that the exciton absorption peak will be substantially broadened when excitons are present at areal concentrations $n_n \approx (\pi a^{*2})^{-1}$, i.e. when the excitons are beginning to 'rub shoulders'. A typical 10 nm QW has an absorption coefficient $\alpha(\hbar\omega) \approx 10^6\,\text{m}^{-1}$ at the heavy-hole exciton peak, and thus absorbs $\approx 1\%$ of the light incident on it. At 300 K, under an illumination of $P\,\text{W}\,\text{m}^{-2}$, our criterion for the onset of substantial bleaching of the exciton line can thus be written

$$n_{ex} \approx (P/\hbar\omega)\,0.01\,\tau_{rec} \geq (\pi a^{*2})^{-1} \tag{7.10}$$

and inserting typical 300 K GaAs/AlGaAs MQW values of $\hbar\omega \approx 1.5\,\text{eV}$, $\tau_{rec} = 10^{-8}\,\text{s}$, and $a^* = 10\,\text{nm}$, leads to an estimate of $\approx 750\,\text{W}\,\text{cm}^{-2}$ for the onset of excitonic absorption saturation.

A more sophisticated theoretical approach (Schmitt-Rink *et al.*, 1986; Haug and Schmitt-Rink, 1985; Chemla *et al.*, 1984) takes account of the fact that the QW excitonic states are in fact composed of an admixture of Bloch wave states and that excitons are rapidly thermally ionized at 300 K. The broadening of the excitonic absorption peak is due to a complex mixture of state-filling effects and electrostatic screening of the electron–hole attraction by high densities of free

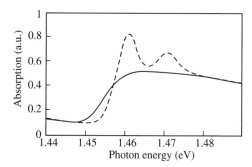

Figure 7.7. Absorption spectra of a 10 nm GaAs QW: broken line – under weak illumination; solid line – under illumination at approximately 800 W cm^{-2} (Chemla and Miller, 1985). (Courtesy D. S. Chemla)

carriers. Nevertheless, these theories give similar estimates for the critical saturation intensity to the simple calculation above.

Typical experimentally measured changes (Miller *et al.*, 1982) for the absorption
strength at the heavy hole exciton peak for a sample in which $\tau_{\text{rec}} \approx 2.1 \times 10^{-8}$ s
were found to follow

$$\alpha(\hbar\omega_{\text{hh-ex}}) = \alpha_0(\hbar\omega_{\text{hh-ex}}) \frac{P}{1 + P/P_s} \tag{7.11}$$

with the saturation intensity parameter $P_s = 580\,\text{W}\,\text{cm}^{-2}$.

The spectrum of the change in absorption $\Delta\alpha(\hbar\omega)$ in Fig. 7.8(a) results,
through equation (7.6), in an associated change in the QW refractive index
$\Delta n(\hbar\omega)$ (Fig. 7.8(b)). This intensity dependent refractive index is sometimes
quoted as an 'effective $\chi^{(3)}$', in analogy with the intensity dependent refractive
index effect arising from the third-order component in the expansion of the susceptibility tensor (Guenther, 1990; Boyd, 1992). MQW effective $\chi^{(3)}$ values
defined this way are typically five orders of magnitude larger than real $\chi^{(3)}$ values

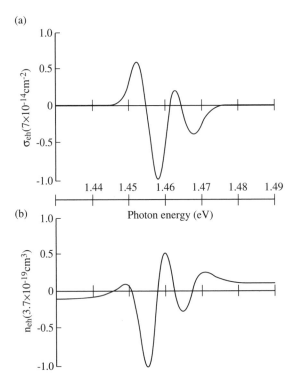

Figure 7.8.   Spectral dependence of (a) the change in the absorption coefficient, $\Delta\alpha(\hbar\omega)$,
and (b) the change in the refractive index, $\Delta n(\hbar\omega)$, due to excitonic saturation in a 10 nm
QW. Data normalized to unity electron–hole pair concentration per unit QW volume
(Chemla and Miller, 1985). (Courtesy D. S. Chemla)

for semiconductors but, when comparing the two processes, it should always be remembered that the QW effect is dissipative and will only work up to a band-width of approximately $(\tau_{rec})^{-1} \approx 10^9$ Hz. In contrast, the genuine $\chi^{(3)}$ third-order non-dissipative process works, in principle, over the majority of the optical band-width of $\approx 3.6 \times 10^{14}$ Hz at QW exciton wavelengths.

The 300 K QW τ_{rec} values can be reduced by more than two orders of magnitude by proton bombarding the QW structures, whilst leaving the strength of the excitonic features in the small signal $\alpha(\hbar\omega)$ spectrum substantially unchanged. This has allowed operation with switching times of ≈ 200 ps, although at the expense of a proportional increase in the P_s parameter. The saturation wave-lengths and intensities required are achievable from focussed diode laser sources and the utility of the effect has been enhanced by inserting the QWs in a Fabry–Perot resonant cavity to produce positive optical feedback and bistable optical operation (Jewell et al., 1985).

This effect has been used to demonstrate a range of prototype all-optical logic elements (Jewell et al., 1985; Silverberg et al., 1985; Venkatesan et al., 1986). Time division demultimplexers (Venkatesan et al., 1984) and reflection-mode devices (Hefferman et al., 1991) have also been fabricated. The latter advance is an important one because typical P_s values result in formidable thermal drift and uniformity problems for the large area arrays of the thin ($\approx 5 \mu m$) MQW semi-conductor layers required for transmissive devices; from an engineering stand-point it is much easier to control the working temperature of reflective devices on top of a thick thermally conducting substrate.

7.6   The Quantum Confined Stark Effect

The electroabsorption properties of GaAs/AlGaAs MQWs have been the subject of much recent research for electro-optic modulator applications, but they have also given rise to an important new class of non-linear and bistable NLO effect, based on the 'self-electro-optic effect device' (SEED) concept (Miller et al., 1984). SEEDs exploit the fact that in a QW an applied electric field perpendicular to the wells, E_z, reduces the energies of the excitonic absorption peaks by an amount much larger than their linewidths. Even at peak shifts up to $\approx 2.5 R^*$ the broad-ening of the exciton peaks is much smaller than in bulk 3D semiconductor sam-ples and they are still clearly resolvable in the $\alpha(\hbar\omega)$ spectrum. This remarkable fact, known as the quantum confined stark effect (QCSE), is due to the presence of the confining heterojunction potentials which limit the separation of the elec-tron and hole even at high E_z, preserving the excitonic electron–hole correlation responsible for the excitonic absorption (Fig. 7.9).

If several such quantum wells are inserted in the undoped 'i' region of a p-i-n diode, which is then reverse biassed through an electrical load (Fig. 7.10), its

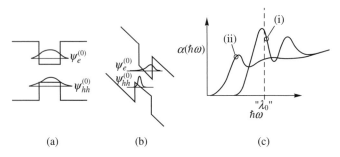

(a)   (b)   (c)

Figure 7.9.   (a) Electronic wavefunctions in a GaAs/AlGaAs QW with an applied perpendicular electric field E_z: (a) $E_z \approx 0$, (b) $E_z \approx 8 \times 10^4\,\mathrm{V\,cm^{-1}}$. (c) Optical absorption coefficient spectrum, $\alpha(\hbar\omega)$, corresponding to situations (a) (curve i) and (b) (curve ii).

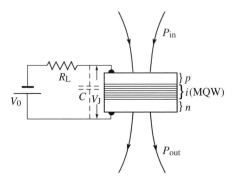

Figure 7.10.   Schematic of a single element transmission-mode 'self-electro-optic effect device'.

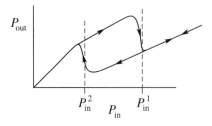

Figure 7.11.   Schematic input output switching characteristic of a SEED device.

optical absorption at a wavelength λ_0 shows a hysteretic (bistable) switching characteristic as a function of the incident light intensity (Fig. 7.11).

When $P_{in} = 0$, the electrical impedance of the reverse biased p-i-n junction is high, only a small current flows in the circuit and a negligible voltage is dropped across R_L. Almost all of the voltage V_0 appears across the p-i-n diode, the QWs are in a region of high E_z and have an $\alpha(\hbar\omega)$ spectrum corresponding to curve (ii) in Fig. 7.9(c). As P_{in} is increased the incident light creates carriers in the MQW which (at 300 K) rapidly escape from the QWs, causing a photocurrent I_L to flow in the external circuit. As P_{in} is increased further, the voltage across the p-i-n junction $V_J = V_0 - I_L R_L$ drops, reducing E_z and shifting the MQW $\alpha(\hbar\omega)$ curve towards curve (i) in Fig. 7.9(c), thus increasing the absorption coefficient at λ_0.

At a critical value of input intensity ($P_{in}^{(1)}$) this positive feedback loop (increasing photocurrent → increasing absorption coefficient → increasing photocurrent) results in a sharp switch from a low optical absorption/low I_L configuration to a highly absorbing one where E_z has adjusted to bring the excitonic peak into close resonance with λ_0. Further increases in P_{in} have only a minor effect on the absorption of the p-i-n structure, but if P_{in} is subsequently decreased the positive feedback means that the SEED will only switch back to its low absorption state when P_{in} has been reduced to a much lower value, $P_{in}^{(2)}$, than the initial switching intensity $P_{in}^{(1)}$ (Fig. 7.11).

The switching time of the device is determined by the electronic time constant formed between R_L and the p-i-n diode capacitance C. Experimental values down to 20 ns (Miller *et al.*, 1985) have been demonstrated in 100 μm diameter devices. Again there is a speed-sensitivity tradeoff, and for typical device parameters (50×9.5 nm QWs in the 'i' region, 100 μm diameter active device area), the product of the DC optical power density required for switching and the switching time gives a nearly constant (Chemla *et al.*, 1983) switching energy density of ≈ 50 pJ cm^{-2}.

Since the development of the basic SEED concept a large number of variations and improvements have been demonstrated. R_L has been replaced by active electronic devices, e.g. phototransistors (Wheatley *et al.*, 1987), heterojunction resonant tunnelling diodes (Chen *et al.*, 1991, Kawamura *et al.*, 1992) and even a second SEED (Miller *et al.*, 1986) to provide increased functionality and the optoelectronic gain essential for cascading multiple devices. The need for substrate removal has been eliminated by using QWs made from narrower gap materials than the GaAs substrate (Chen *et al.*, 1992, Fujiwara *et al.*, 1990) and by using reflection mode structures grown on multilayer mirrors (Sale *et al.*, 1991). The power handling capabilities of the device (and hence its ultimate switching speed) have also been improved by going to designs with extremely shallow MQWs, so that the exciton line is rapidly eliminated by exciton field ionization at modest values of E_z (Morgan *et al.*, 1991; Goosen *et al.*, 1990, 1991).

SEEDs have been fabricated into arrays using conventional photolithographic techniques (Feitelson, 1988), and from a macroscopic systems engineering point of view such an array looks like a sensitive non-linear material which shows intrinsic bistability in either transmission or reflection at rather modest power densities. Switching powers can be engineered to be compatible with currently available diode laser sources, and multiple-pixel array prototype photonic computers have been demonstrated. At present the practicalities of high power optical source generation and power dissipation limit the attainable switching speeds to the MHz range, but this still results in improvements of some orders of magnitude in effective computing power over conventional serial electronic image processing computers.

For all their successes, however, SEED based devices are essentially based on a hybrid optoelectronic technology and they are prone to the problems inherent in conventional VLSI electronics. In particular, making high reliability individual electrical contacts to the large number ($>10^6$) of pixels in an array is a formidable technical challenge, and inevitably a large portion of the wafer (and hence of the input optical signal beam power) is wasted. More serious is the problem of electronic crosstalk between the connecting wires (Govindarajan and Forrest, 1991). This induces transient voltages which randomly shift the optical switching threshold of a particular pixel depending on the configuration of its near neighbours. This produces an unacceptable bit error rate unless all of the system tolerances are broadened to accommodate the random fluctuations, a precaution which increases the optical power densities required to run the system.

All of these problems become progressively more serious as larger and larger arrays are contemplated, and there is still a strong incentive to develop new NLO device concepts which do away with external electronic circuitry completely.

7.7   Doping Superlattices ('*n-i-p-i*' Crystals)

Doping superlattices were first proposed and demonstrated (Dohler *et al.*, 1981) in the early 1980s as a means of producing materials with LDS electronic structure without the lattice matching restrictions which apply to heterojunctions. Epitaxial growth techniques are used to deposit alternately planes of *n*- and *p*-type dopants in an epitaxial layer and, if the dopant sheet concentrations are matched with sufficient accuracy, all the excess electrons from the *n*-type donor planes transfer to the *p*-type acceptor planes to produce alternating sheets of positive and negative charge along the growth axis (Section 3.3.2.1).

The space charge fields associated with this structured charge distribution produce a characteristic 'sawtooth' electrostatic modulation to the conduction and

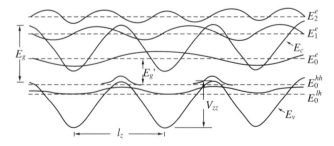

Figure 7.12.   Electronic band structure of a typical '*n-i-p-i*' doping superlattice, showing the envelope wavefunctions of the ground state subbands.

valence band edges (Fig. 7.12) which constrains any free electrons (holes) created by, e.g., photoexcitation to move in two-dimensional subbands in the n- (p-) doped layers. The band structure produced by this potential is necessarily 'Type II' in character, namely, the electron and hole subband wavefunctions are spatially separated. The degree of separation, characterized by the overlap between the normalized electron and hole ground-state wavefunctions averaged along the growth direction, $\langle \psi_e^{(0)}(z)|\psi_{hh}^{(0)}(z)\rangle^2$, can be altered over many decades by choice of the n-/p- sheet concentrations and the superlattice period.

If a n-i-p-i structure is grown in which the spatial extent of the dopant atoms is negligibly small (the so-called 'δ-doped' limit), the electrostatic potential is triangular (Fig. 7.13) and has a depth V_{zz} of

$$V_{zz} = \frac{eN^{2D}L_z}{4\epsilon_r\epsilon_0} \tag{7.12}$$

where N^{2D} is the sheet impurity concentration in the doped layers and L_z is the superlattice period.

For practical purposes it is useful to distinguish between 'Type B' n-i-p-is, where the overlap integral between the electron and hole ground states is less than ≈10% of that in an undoped bulk crystal, and 'Type A' n-i-p-is where it is much greater. This is because, although the n-i-p-i potential creates the possibility of optical absorption at energies E_g' (Fig. 7.13) which are substantially below the band gap E_g of the host semiconductor, the strength of the absorption at these energies is proportional to $\langle \psi_e^{(0)}(z)|\psi_{hh}^{(0)}(z)\rangle^2$ and this quantity must be preserved if useful optical absorption is to be available.

The 'Type B' character of the n-i-p-i potential also results in a substantial increase in the photoexcited carrier recombination time, since electron–hole pairs are rapidly separated by the n-i-p-i potential and scatter into their ground-state subbands before recombining. Under weak excitation the carrier lifetime is enhanced over that in the host semiconductor by a factor proportional to $\langle \psi_e^{(0)}(z)|\psi_{hh}^{(0)}(z)\rangle^{-2}$, which in Type B structures can easily be several orders of magnitude.

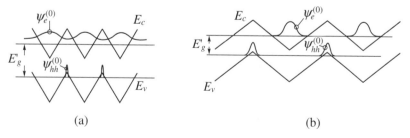

(a)   (b)

Figure 7.13. (a) 'Type A' n-i-p-i with a large overlap integral between the electron and heavy-hole ground-state wavefunctions. (b) 'Type B' n-i-p-i with small overlap integral value, weak optical absorption at E_g' and long recombination time.

In δ-doped *n-i-p-i* structures, the subband confinement energies can be computed analytically for the *i*th subband as

$$E_i = \frac{(i+1)^{2/3}}{2^{7/3}} \left(\frac{e^2 h N^{2D}}{\epsilon_r \epsilon_0 m^{*1/2}} \right)^{2/3}$$

(7.13)

Using a WKB approximation (Section 3.3.1) to calculate the form of carrier wavefunctions in the barrier regions and their overlap integral values yields a critical superlattice period, L_z^{crit}, for the crossover from Type *A* to Type *B* behaviour of the form

$$L_z^{crit} = \tfrac{1}{2} h (2m^* E_0)^{-1/2} + 2 \left[\frac{3h\epsilon_0 \epsilon_r \ln(10)}{4\pi e m^* N^{2D}} \right]^{2/3}$$

(7.14)

Both the terms in the above scale as $m^{*-1/3} \epsilon_r^{1/3}$, implying that for a given two-dimensional sheet impurity concentration, Type *A* behaviour should be easier to achieve simultaneously with large changes in the effective band gap if semiconductors with light carrier effective masses and large dielectric constants are used. In most practical cases the dopants are spread over a non-negligible distance and the subband envelope functions and optical absorption characteristics have to be calculated numerically (Phillips, 1990). For typical achievable MBE doping concentrations of $\approx 2\text{--}4\times10^{18}$ cm^{-3}, Type *A* behaviour can be achieved with reductions in the effective band gap of ≈ 200 meV in a wide range of semiconductor hosts.

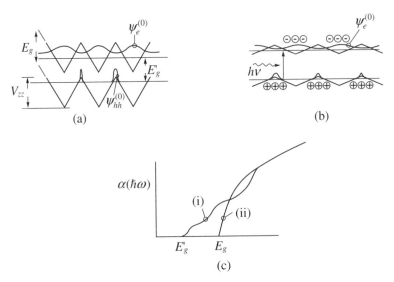

Figure 7.14.   Schematic band structure of *n-i-p-i* samples: (a) unilluminated, (b) under illumination, showing screening effect of photoexcited carriers, (c) absorption coefficient curves for cases (a) (curve i) and (b) (curve ii).

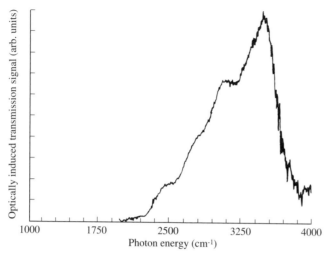

Figure 7.15.   Spectrum of the induced change in absorption of an InAs *n-i-p-i* sample excited with $\approx$600 mW cm^{-2} of $\lambda = 514$ nm radiation (Adapted from Phillips *et al.*, 1993). (Courtesy C. C. Phillips)

Under illumination with photons of energy $\hbar\omega > E_g'$ the photocreated carriers separate into the doped layers and screen the 'grown-in' electrostatic fields (Fig. 7.14(a,b)) produced by the impurity charges. This has the effect of reducing V_{zz} and, in the limit of high excitation density, the *n-i-p-i* potential is completely screened, returning the $\alpha(\hbar\omega)$ spectrum to that of the bulk semiconductor host material and removing the absorption in the spectral region just above E_g' (Fig. 7.14(c)). Since this effect relies on the generation of a non-equilibrium photocarrier concentration, the intensity required is inversely proportional to the photocarrier lifetime and, in Type *B* structures, can be very low.

These non-linear absorption (and the associated refractive index) changes have been demonstrated in *n-i-p-i*s in a range of materials (Fig. 7.15) (Simpson *et al.*, 1986; Hodge *et al.*, 1990; Phillips *et al.*, 1993), and are potentially useful for device applications because they occur in a region below the host semiconductor band gap where the linear absorption can be made very small.

As with all dissipative NLO mechanisms there is a tradeoff between saturation intensity and switching speed. Type *B* structures (large superlattice period, small wavefunction overlap) offer low saturation powers at modest switching speeds and Type *A* structures offer response times close to the carrier recombination times ($\approx 10^{-8}$ s) of the semiconductor host material at the expense of a proportionate increase in saturation power density. Typical values are a saturation intensity of the order of 10^{-1} W cm^{-2} at switching speed of $\approx 10^{-4}$ s (Phillips *et al.*, 1993).

In contrast to the enhancement of the exciton binding energy in a LDS system with 'Type I' offsets (i.e. where the electron and hole states are confined in the same layer), the 'Type II' band structure in *n-i-p-i*s acts to reduce the excitonic binding energies, and excitonic peaks are not seen in 300 K *n-i-p-i* $\alpha(\hbar\omega)$ spectra.

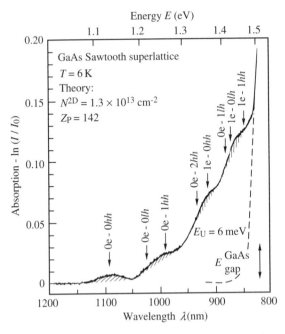

Figure 7.16. Low temperature ($T = 6$ K) absorption coefficient spectrum of a high-quality GaAs δ-doped *n-i-p-i* sample with a period of 14.2 nm and sheet dopant concentrations of 1.3×10^{13} cm^{-2}. Arrows indicate calculated energies of intersubband transitions. (Reprinted from E. F. Schubert, *Surface Science*, **228**, 240, (1990) with kind permission of Elsevier Science-NL, Sara Burgerhartstraat, 25, 1055 KV Amsterdam, The Netherlands)

Moreover, unless the mean spacing between the dopants within the dopant planes is very much less than the spacing $\frac{1}{2}L_z$ between the planes, impurity disorder produces significant fluctuations in V_{zz} at different points in the two-dimensional plane. This blurs the step-like structure expected in the $\alpha(\hbar\omega)$ spectra, and although individual subband transitions can be identified in good samples (Hodge *et al.*, 1990, Schubert, 1990) (Fig. 7.16), the absorption features are generally too broad to be exploited in highly resonant devices. Current research interest centres on linear optoelectronic applications such as radiation detectors and emitters, where the tunability of the band gap and recombination times offer potentially useful device improvements (Phillips, 1990).

7.8   Hetero–*n-i-p-i* Structures

Perhaps the most exciting of the recent developments in LDS non-linear optics have resulted from a combination of the excitonic saturation and QCSE effects demonstrated in QW heterostructures with the tunable carrier lifetimes and optically controlled electric fields available in 'Type B' n-i-p-i structures.

As with the previously discussed effects, the utility of the non-linear properties of these 'hetero–n-i-p-is' can be enhanced by incorporating them in device structures such as Fabry–Perot resonators and multilayer reflector stacks (distributed Bragg reflectors, DBRs) to amplify the effects of small $\Delta\alpha(\hbar\omega)$ and $\Delta n(\hbar\omega)$ changes on macroscopic device properties such as optical reflectivity, absorption and optical phase shift.

The full possibilities of hetero–n-i-p-i structures are only just beginning to receive serious research attention, and what follows should be considered as being a brief description of what amounts to the author's personal choice of some of the more interesting ideas from recent research literature.

7.8.1   Band Filling Effects in Hetero–n-i-p-is

The 'band filling' or Moss–Burstein effect occurs in all three-dimensional semiconductors and is most easily observed in degenerately n^+-doped samples where the Fermi energy lies in the conduction band. It is a consequence of the fact that electrons are fermions and therefore it is impossible (by the Pauli exclusion principle) to optically excite an electron into a **k**-state which is already occupied. In the case of a strongly degenerate n^+-doped sample this has the effect of prohibiting any interband transitions into electron states below the Fermi energy, leading to an upward shift in the effective absorption edge, E'_g, given by

$$E'_g - E_g = (E_f - E_c)(1 + m_e^*/m_h^*) = \frac{\hbar^2(3\pi^2)^{2/3}n^{2/3}(1 + m_e^*/m_h^*)}{2m_e^*} \qquad (7.15)$$

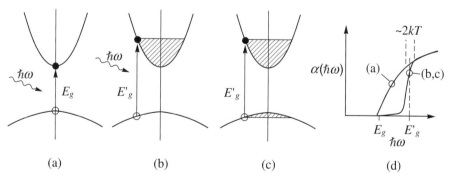

Figure 7.17.   Allowed optical transitions in direct-gap semiconductors: (a) undoped material, absorption threshold of E_g; (b) n^+-doped material, absorption threshold blue shifted to E'_g by the Moss–Burstein effect; (c) optically excited undoped material, absorption threshold blue shifted to E'_g by the 'dynamic Moss–Burstein shift'; (d) schematic absorption coefficient spectra corresponding to cases (a), (b), and (c).

At finite temperatures the broadening of the Fermi function at E_f leads to a corresponding blurring of the absorption edge in n^+-doped samples of $\approx 2kT$. Similar effects occur in p^+ samples, but the $1/m^*$ dependence of the absorption shift makes it much more pronounced at a given carrier concentration in narrow gap n-doped semiconductors with their low electron effective masses.

A similar effect, the 'dynamic Moss–Burstein effect', is seen in lightly-doped bulk semiconductors when a large concentration of excess electron–hole pairs is created by optical excitation (Fig. 7.17(c)). As soon as the excitation is removed, however, the excess carriers recombine, typically with a $\tau_{\rm rec}$ value of $\approx 10^{-8}$–10^{-9} s in high-quality direct-gap materials, and the $\alpha(\hbar\omega)$ curve reverts to its low-intensity shape on this timescale.

In bulk semiconductors this effect produces intensity dependent absorption bleaching and refractive index changes (Vodopyanov $et\ al.$, 1993) which can be used, for example, to switch short pulses of laser radiation (Vodopyanov $et\ al.$, 1991) but the typical required power densities are high. In GaAs ($m_e^* \approx 0.067\,m_0$) at 300 K, for example, to obtain $E_g' - E_g \approx 2kT = 50$ meV requires a mean photon excited carrier density of $n \approx 7 \times 10^{17}\,{\rm cm}^{-3}$. Assuming typical values of $\tau_{\rm rec} = 5 \times 10^{-10}$ s and $\alpha(\hbar\omega) \approx 10^4\,{\rm cm}^{-1}$ for the exciting radiation at $\hbar\omega \approx 1.5$ eV, an optical pump density of

$$P = \frac{n\hbar\omega}{\alpha(\hbar\omega)\tau_{\rm rec}} \approx 3.2 \times 10^4\ {\rm W\ cm}^{-2} \tag{7.16}$$

is required to sustain this excess carrier density. At these intensities spectacular and irrevocable sample damage will occur unless the pump radiation density is present as very short pulses with low energies but high peak intensities.

Hetero–n-i-p-is offer one way to circumvent this problem by increasing the effective carrier recombination time by many orders of magnitude (Ianelli $et\ al.$, 1989; Kobayashi $et\ al.$, 1988) to the point where continuous wave device opera-

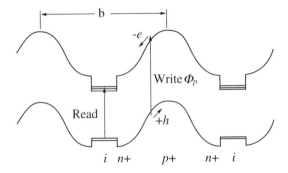

Figure 7.18.   Hetero n-i-p-i structure using the enhancement in recombination time to allow large electron concentrations to build up in the n-type regions under weak illumination, where they produce strong absorption changes due to the dynamic Moss–Burstein effect (Ianelli $et\ al.$, 1989). (Courtesy F. J. Grunthaner)

tion is possible. The structure of Fig. 7.18, for example, shows absorption changes of several per cent at pump densities of only $\approx 50\,\mathrm{mW\,cm^{-2}}$, due to electron state filling effects bleaching the absorption in the region of the quantum well effective band gap. This dramatic reduction in saturation threshold has, of course, been obtained at the expense of a correspondingly large decrease in switching speed, in this case down to $\approx 50\,\mathrm{Hz}$ (Ianelli *et al.*, 1989). The ability to trade speed and sensitivity in these structures, however, does imply a potential for their application in the implementation of photonic computer systems, and their relative insensitivity to operating wavelength (compared with devices operating close to excitonic resonances) may ultimately prove advantageous from the point of view of systems design.

7.8.2   The QCSE in Hetero–*n-i-p-i*s

If a Type *B n-i-p-i* is fabricated with wide undoped '*i*' regions between heavily *n*- and *p*-doped slabs, a structure results which is similar to a number of *p-i-n* junctions back-to-back. The electrostatic screening of the *n-i-p-i* potential under optical excitation has a similar effect on the perpendicular electric fields in the '*i*' regions as does electrically altering the voltage on the single *p-i-n* junctions used in SEED and QCSE electro-optic modulator devices. This similarity has led to the proposal and demonstration of all optical MQW-hetero–*n-i-p-i* structures (Kost *et al.*, 1988) which exploit the changes in $\alpha(\hbar\omega)$ and in $n(\hbar\omega)$ of MQWs in the '*i*' regions resulting from the QCSE driven by the reduction in E_z when the structure is illuminated (Fig. 7.19).

Even if the illuminating beam energy corresponds only to the confined heavy-hole exciton transitions in the MQWs, rapid thermally assisted tunnelling (Fox *et al.*, 1991 and references therein) ensures that the photocarriers escape from the wells with near unity quantum efficiency and are available to screen the built-in perpendicular electric field. Initial devices were operated in the transmission mode (with the substrate removed) and showed saturation switching intensities of $\approx 375\,\mathrm{mW\,cm^{-2}}$ and peak fractional transmission changes of $\Delta T/T \approx 10\%$ (Kost *et al.*, 1988), although no switching speed data were reported.

More recently, devices have been studied with the hetero–*n-i-p-i* structure grown on a passive $\mathrm{Al_{0.2}Ga_{0.8}As/AlAs}$ 'distributed Bragg reflector' multilayer mirror which was designed to be highly reflective in the spectral region of the MQW electron–heavy-hole transition (Fig. 7.20). This allowed reflection-mode operation with the probe beam passing through the hetero–*n-i-p-i* layer twice. The structure showed fractional reflectivity changes $\Delta R/R$ of up to 60% at optical pump densities of only $\approx 20\,\mathrm{mW\,cm^{-2}}$ and response times of $\approx 10^{-4}\,\mathrm{s}$ (Law *et al.*, 1989).

The incorporation of the MQW-hetro–*n-i-p-i* layers described above into an optically resonant structure has two major potential advantages. The first is a simple amplification of the non-linearity at a given pump density; if, for example, the MQW-hetero–*n-i-p-i* were to be placed within a Fabry–Perot resonant cavity

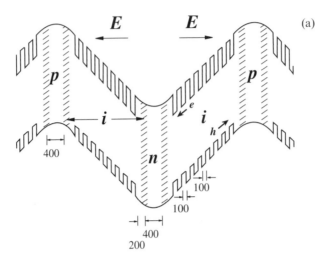

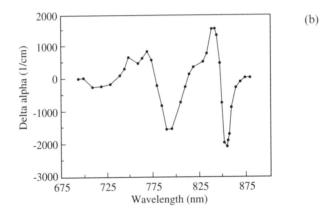

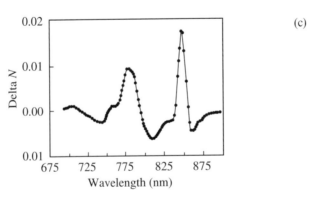

Figure 7.19.   (a) Electronic structure of the MQW-hetero–*n-i-p-i* of Kost *et al.*, (1988) in the absence of exciting illumination. Changes in (b) absorption coefficient and (c) refractive index of the structure under illumination with approximately $800\,\mathrm{mW\,cm^{-2}}$ of illumination at $\lambda = 806\,\mathrm{nm}$. (Courtesy P. D. Dapkus)

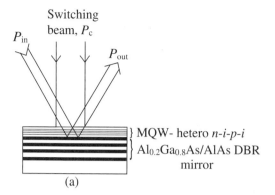

(a)

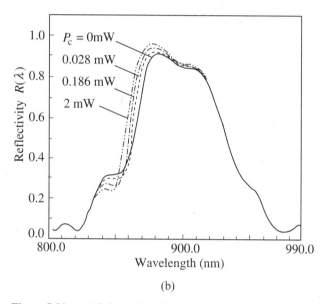

(b)

Figure 7.20. (a) Schematic of the mode of operation of the reflective hetero–*n-i-p-i* optical modulator of Law *et al.* (1989). (b) Experimentally measured reflectivity changes under varying excitation levels from a structure containing four 7.9 nm QWs in each '*i*' region clad with $Al_{0.2}Ga_{0.8}As$. In the absence of illumination the electric field, E_z, in the '*i*' regions was $\approx 1.5 \times 10^5\,\mathrm{V\,cm^{-1}}$. (Courtesy A. C. Gossard)

with a finesse factor F, then small changes in the absorption of the MQW result in fractional changes in the optical transmission of the overall Fabry–Perot resonant cavity which are of the order F times larger (Hecht, 1987, Section 8.2). This increased sensitivity is achieved at the expense of narrowing the spectral range of operation, but most practical implementations of dissipative NLO effects use narrow linewidth laser sources and in principle this restriction is not a major one.

The second advantage of optically resonant structures is that, if the relationship between the photon energy of the optical resonance and that of the electron–

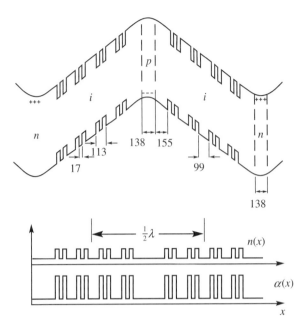

Figure 7.21. Schematic of the 'Bragg hetero–*n-i-p-i*-doubly-resonant optical modulator' (BH-DROM) concept. The period of the hetero–*n-i-p-i* is chosen to produce a modulation in the refractive index and absorption coefficient which repeats in depth with a period corresponding to half the wavelength of the heavy-hole exciton transition. This produces a saturable DBR mirror where the wavelengths corresponding to the DBR and heavy-hole excitonic resonances can be chosen to provide built-in optical feedback.

heavy-hole absorption resonance is chosen judiciously, then a form of positive feedback can be designed into a structure. In this case intrinsic hysteresis (bistability) in the transmission (or reflection) versus incident intensity of the structure can result. This approach has been used to achieve bistable operation due to the excitonic saturation non-linearity (e.g. Schmitt-Rink *et al.*, 1986, Haug and Schmitt-Rink, 1985) and can in principle yield extremely low-power optical bistable devices (i.e. prototype optical memories) if used with MQW-hetero–*n-i-p-i*s.

The 'BH-DROM' is an example of such an idea (Poole *et al.*, 1992). In this case the layered nature of the MQW-hetero–*n-i-p-i* itself is used to introduce the additional optical resonance. This is done by choosing the hetero–*n-i-p-i* period such that a light beam of wavelength λ, close to the electron–heavy-hole resonance in the MQWs, sees a series of layers of different refractive index/absorption.

These repeat (when allowance is made of the reduction of λ in the material by the large linear refractive index), with a period which corresponds to $\frac{1}{2}\lambda$, (Fig. 7.21) to form a multilayer 'distributed Bragg reflector' at the heavy hole exciton wavelength with a highly resonant ($F \approx 40$) small signal reflectivity peak (Fig. 7.22). The initial unoptimized design of Poole *et al.*, (1992) showed frac-

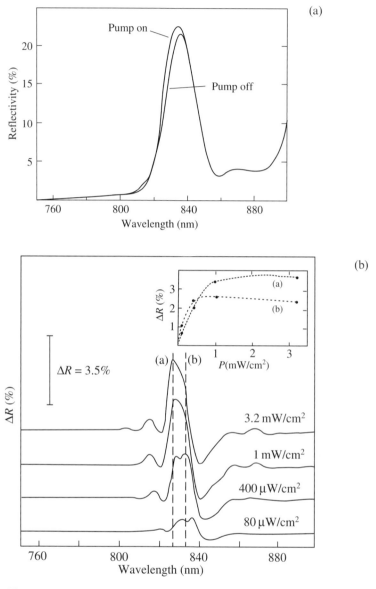

Figure 7.22.   (a) Small signal reflectivity spectrum of the sample in Fig. 7.21, showing the sharply resonant ($F \approx 40$) Bragg peak in reflectivity. (b) Reflectivity change spectra from the same sample under varying levels of optical excitation. (After Poole *et al.*, 1992.) (Courtesy C. C. Phillips)

tional reflectivity changes of $\Delta R/R \approx 31\%$ using non-resonant pump densities of $\approx 0.8\,\mathrm{mW\,cm^{-2}}$, but second-generation designs (Poole *et al.*, 1994) have since demonstrated $\Delta R/R$ figures of $\approx 300\%$ with similar pump densities and switching times of $\approx 10^{-4}$ s. In common with the non-resonant designs described above, the

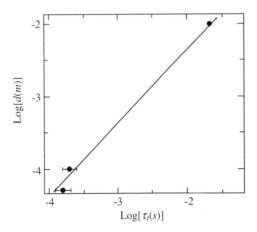

Figure 7.23. Switching speed versus side length of square pixels photolithographically fabricated from the BH-DROM epilayer structure of Fig. 7.21. The switching speed is determined by the carrier recombination time, determined in this case by the time taken for the carriers to diffuse to the pixel edges and recombine non-radiatively at edge defects.

switching speed is controlled by extrinsic carrier recombination, and can be adjusted by reducing the active area using photolithographic techniques (Fig. 7.23).

In functional terms the BH-DROM behaves very much like a SEED device, but with the feedback arising from the internal electric field screening mechanism as opposed to the changing voltages applied via an external electronic circuit. This is potentially a major technical advantage for the fabrication of large low crosstalk pixel arrays, and the speed-sensitivity product already makes it worthy of serious consideration as a candidate for all-optical image processing applications.

7.9   Concluding Remarks

An attempt has been made here to provide a brief introduction to a wide variety of non-linear optical mechanisms in layered semiconductor structures, and briefly to discuss the types of emerging technologies in which they may find application in the future.

At the time of writing the science of producing LDS structures has reached a degree of maturity, but the science of LDS device design can only be considered to be in its infancy. Many of the ideas and structures described here exist at present only in one or two research laboratories around the world and it may well be that many of them will never make the difficult transition to the commercial world of practical and cost-effective devices.

Nevertheless the pace of advance and the burgeoning research literature in this area imply that important advances will continue to be forthcoming. Now, perhaps more than ever before, there appear to be serious grounds for optimism that in the near future some of these exciting new ideas will find their way into practical devices, and that the much vaunted 'photonic computer' will become a reality.

EXERCISES

1. Explain why semiconductor microstructures in which the number of dimensions in which the charge carriers can move freely has been reduced (e.g. quantum wells) offer enhanced potential for exploiting excitonic effects in non-linear optical devices compared with conventional (three-dimensional) bulk samples. Illustrate your discussion with a description of at least two device concepts which use excitonic effects to produce non-linear optical properties.

2. Figure 7.24 shows a hetero–*n-i-p-i* wafer fabricated from GaAs wells ($E_g = 1.42$ eV) and AlGaAs barriers ($E_g \approx 1.8$ eV). The figure below is a spectral plot of the *change*, $\Delta\alpha$, in its absorption coefficient when it is illuminated with $\sim 1\,\mathrm{W\,cm}^{-2}$ of laser light with a photon energy somewhere between 1.42 and 1.8 eV. In qualitative terms, but with as much detail as possible, describe the physical processes giving rise to the features 'a' and 'b'.

3. A 0.1 µm thick MBE layer of single crystal InAs grown on a transparent GaAs substrate is to be used as a 300 K all-optical shutter, exploiting the 'dynamic Moss–Burstein effect' to switch one infrared laser beam with another. Both beams have wavelength $\lambda = 2.94$ µm. When the wafer is hit with an intense negligibly short 'switching' pulse, enough carrier pairs are instantaneously created to fill the conduction and valence bands and to increase the threshold for optical absorption of the InAs to an energy E'_g equal to the photon energy and rather larger than E_g, the normal InAs band gap (Fig. 7.25).

   (a) Using the data below, calculate the wavevector k_{opt}, of the electrons and holes involved in optical transitions in InAs $\lambda = 2.94$ µm.

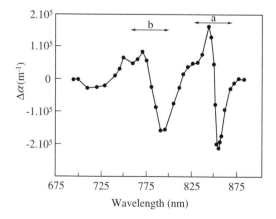

Figure 7.24.   Figure for Exercise 2.

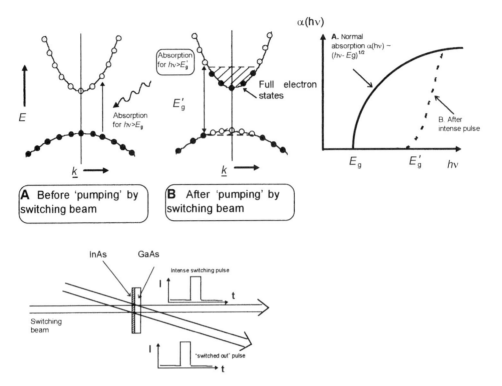

Figure 7.25.   Figure for Exercise 3.

(b) What excess carrier density needs to be created to fill the conduction band
up to this value of wavevector? Assume that the Fermi occupation curve
has negligible width here, i.e. that the Fermi sphere for electrons has a
sharp boundary, and that the electrons and holes are uniformly distributed
throughout the InAs.

(c) Given that all these electrons were initially created by absorption of
$\lambda = 2.94\,\mu$m photons, calculate the energy density (in $\mathrm{J\,m^{-2}}$) of optical
energy absorbed by the InAs layer from the pump beam to instantaneously
create these carriers.

(d) If the excess electron–hole pairs have a mean recombination time of 1 ns,
calculate the steady-state intensity that would be needed to keep the shutter
transparent (i.e. 'open') by sustaining the carrier density you calculated in (b).

(e) Given that the specific heat capacity of (bulk) InAs is $1.739 \times 10^6\,\mathrm{J\,K^{-1}\,m^{-3}}$,
how long can the shutter be held open for optically, before its temperature
exceeds 600 K and it becomes irreversibly damaged?

Data for InAs at 300 K: $E_g = 0.35\,\mathrm{eV}$, $m_e^* = 0.022\,m_0$, $m_h^* = 0.4\,m_0$.

4. For a particular application it is proposed to make a *n-i-p-i* detector sensitive to $\lambda = 1.3\,\mu$m light using GaAs (the required materials parameters of which are below). Because it is crucially important that the *n-* and *p*-sheet dopant densities are well matched, so that the residual carrier concentration in the *n-i-p-i* is negligible, the dopant concentrations used must be low enough to be reproducible from one growth run to the next.

   (a) Because of this the maximum usable doping density corresponds to a bulk value of $2 \times 10^{19}\,\mathrm{cm}^{-3}$. Calculate the mean impurity separation at this density and use this value to calculate the maximum sheet impurity density, N_{2D}, which can be incorporated into a δ-doped plane.

   (b) Assuming this value for N_{2D}, and neglecting confinement energy effects for the moment, calculate the minimum *n-i-p-i* period, L_z, needed to result in a modulation of the band-edge energies, V_z, large enough to reduce the GaAs band gap to the energy of a $\lambda = 1.3\,\mu$m photon.

   (c) Use your answer to (b) with the simple analytical method, applicable in the 'δ-doped limit,' to estimate the electron and heavy hole confinement energies.

   (d) Use your answer to (c) to recalculate an increased L_z value to estimate the period required to produce a detector sensitive out to $\lambda = 1.3\,\mu$m when the quantum effects are accounted for. (Of course this problem can be solved more accurately with an iterative numerical method, but it is simple enough to be programmable into a basic computer or calculator with minimal effort.)

   (e) To have a useful sensitivity we would like the absorption strength at the new *n-i-p-i* band gap to be not too much less than that of bulk GaAs. Test this factor by calculating whether or not your answer to (d) is above or below the period, L_z^{crit}, defining the cross-over between 'Type *A*' and 'Type *B*' *n-i-p-i* behaviour.

References

N. Bloembergen, *Nonlinear Optics* (World Scientific Press, 1996).

R. Boyd, *Non-linear Optics* (Academic Press, New York, 1992).

D. S. Chemla, T. C. Damen, D. A. B. Miller, A. C. Gossard and W. Wiegmann, *Appl. Phys. Lett.* **42**, 864 (1983).

D. S. Chemla and D. A. B. Miller, *J. Opt. Soc. Am. B* **2**, 1155 (1985).

D. S. Chemla, D. A. B. Miller, P. W. Smith, A. C. Gossard and W. Wiegmann, *IEEE J. Quant. Electron.* **QE-20**, 265 (1984).

L. Chen, R. M. Kapre, K. Hu and A. Madhukar, *Appl. Phys. Lett.* **59**, 1523 (1991).

L. Chen, K. Hu, R. M. Kapre and A. Madhukar, *Appl. Phys. Lett.* **60**, 422 (1992).

G. H. Dohler, H. Kunzel, D. Olego, K. Ploog, P. Ruden, H. J. Stolz and G. Abstreiter, *Phys. Rev. Lett.* **47**, 864 (1981).

D. G. Feitelson, *Optical Computing: A Survey for Computer Scientists* (MIT Press, Cambridge, MA, 1988).

A. M. Fox, D. A. B. Miller, G. Livescu, J. E. Cunningham and W. Y. Yan, *IEEE J. Quant. Electron.* **QE–27**, 2281 (1991).

K. Fujiwara, K. Kawashima, K. Kobayashi and N. Sano, *Appl. Phys. Lett.* **57**, 2234 (1990).

K. W. Goosen, J. E. Cunningham and W. Y. Jan, *Appl. Phys. Lett.* **57**, 2582 (1990).

K. W. Goosen, L. M. F. Chirovsky, R. A. Morgan, J. E. Cunningham and W. Y. Yan, *IEEE Phot. Tech. Lett.* **3**, 448 (1991).

M. Govindarajan and S. R. Forrest, *Appl. Optics* **30**, 1335 (1991).

R. Guenther, *Modern Optics* (Wiley, New York, 1990).

H. Haug and S. Schmitt-Rink, *J. Opt. Soc. Am. B* **2**, 1135 (1985).

E. Hecht, *Optics*, 2nd edn (Addison-Wesley, Reading, MA, 1987).

J. F. Hefferman, M. H. Moloney, J. Hegarty, J. S. Roberts and M. Whitehead, *Appl. Phys. Lett.* **58**, 2877 (1991).

S. T. Ho, C. E. Soccolich, M. N. Islam, W. S. Hobson, A. F. J. Levi and R. E. Slusher, *Appl. Phys. Lett.* **59**, 2558 (1991).

C. C. Hodge, C. C. Phillips, R. H. Thomas, S. D. Parker, R. L. Williams and R. Droopad, *Semicond. Sci. Technol.* **5**, S319 (1990).

J. M. Ianelli, J. Maserjian, B. R. Hancock, P. O. Andersson and F. J. Grunthaner, *Appl. Phys. Lett.* **54**, 301 (1989).

J. L. Jewell, Y. H. Lee, M. Warren, H. M. Gibbs, N. Peyghambarian, A. C. Gossard and W. Wiegmann, *Appl. Phys. Lett.* **46**, 919 (1985).

Y. Kawamura, H. Asai, S. Matsuo and C. Amano, *IEEE J. Quant. Electron.* **QE-28**, 308 (1992).

H. Kobayashi, Y. Yamauchi and H. Ando, *Appl. Phys. Lett.* **52**, 359 (1988).

A. Kost, E. Garmire, A. Danner and P. D. Dapkus, *Appl. Phys. Lett.* **52**, 637 (1988).

K.–K. Law, J. Maserjian, R. J. Simes, L. A. Coldren, A. C. Gossard and J. L. Merz, *Opt. Lett.* **14**, 230 (1989).

R. D. Leavitt and H. W. Little, *Phys. Rev. B* **42**, 11774 (1990).

S. H. Lee, ed., *Optical Information Processing Fundamentals, Topics in Applied Physics* **48** (Springer-Verlag, Berlin 1981).

D. A. B. Miller, D. S. Chemla, D. J. Eilenberger, P. W. Smith, A. C. Gossard and W. Wiegmann, *Appl. Phys. Lett.* **41**, 679 (1982).

D. A. B. Miller, D. S. Chemla, T. C. Damen, A. C. Gossard, W. Wiegmann, T. H. Wood and C. A. Burrus, *Phys. Phys. Lett.* **53**, 2173 (1984).

D. A. B. Miller, D. S. Chemla, T. C. Damen, T. H. Wood, C. A. Burrus, A. C. Gossard and W. Wiegmann, *IEEE J. Quant. Electron.* **QE-21**, 1462 (1985).

D. A. B. Miller, J. E. Henry, A. C. Gossard and J. H. English, *Appl. Phys. Lett.* **49**, 821 (1986).

R. A. Morgan, M. T. Asom, L. M. F. Chirovsky, M. W. Focht, K. G. Glogovsky, G. D. Guth, G. J. Przybylek, L. E. Smith and K. W. Goosen, *Appl. Phys. Lett.* **59**, 1049 (1991).

C. C. Phillips, *Appl. Phys. Lett.* **56**, 151 (1990).

C. C. Phillips, E. A. Johnson, R. H. Thomas, H. L. Vaghjiani, I. T. Ferguson and A. G. Norman, *Semicond. Sci. Technol.* **8**, S373 (1993).

P. J. Poole, C. C. Phillips and O. H. Hughes, *Appl. Phys. Lett.* **60**, 1126 (1992).

P. J. Poole, C. C. Phillips, C. Roberts and M. Paxman, *IEEE J. Quant. Electron.* **30**, 1027 (1994).

T. E. Sale, J. Woodhead, A. S. Pabla, R. Grey, P. A. Claxton, P. N. Robson, M. H. Moloney and J. Hegarty, *Appl. Phys. Lett.* **59**, 1670 (1991).

S. Schmitt-Rink, C. Ell and H. Haug, *Phys. Rev. B* **33**, 1183 (1986).

S. Schmitt-Rink, D. S. Chemla and D. A. B. Miller, *Adv. Phys.* **38**, 89 (1989).

E. F. Schubert, *Surf. Sci.* **228**, 240 (1990).

D. D. Sell, *Phys. Rev. B* **6**, 3750 (1972).

Y. Silverberg, P. W. Smith, D. A. B. Miller, B. Tell, A. C. Gossard and W. Wiegmann, *Appl. Phys. Lett.* **46**, 701 (1985).

T. B. Simpson, C. A. Pennise, B. E. Gordon, J. E. Anthony and T. R. AuCoin, *Appl. Phys. Lett.* **49**, 590 (1986).

T. Venkatesan, B. Wilkens, Y. H. Lee, M. Warren, G. Olbright, H. M. Gibbs, N. Peygham-barian, J. S. Smith and A. Yariv, *Appl. Phys. Lett.* **48**, 145 (1986).

T. Venkatesan, Wiegmann, J. L. Jewell, H. M. Gibbs and S. Tarng, *Opt. Lett.* **9**, 297 (1984).

K. L. Vodopyanov, H. Graener, C. C. Phillips and T. J. Tate, *J. Phys. D* **73**, 1 (1993).

K. L. Vodopyanov, A. V. Lukashev, C. C. Phillips and I. T. Ferguson, *Appl. Phys. Lett.* **59**, 1658 (1991).

P. Wheatley, G. Parry, J. E. Midwinter, G. Hill, P. Mistry, M. A. Pate and J. S. Roberts, *Electron. Lett.* **23**, 1249 (1987).

B. S. Wherrett and F. A. P. Tooley, *Optical Computing* (SUSSP Publications, University of Edinburgh, 1988).

A. Yariv, *Quantum Electronics*, 3rd edn (Wiley, New York, 1989).

8 Semiconductor Lasers

A. Khan, P. N. Stavrinou and G. Parry

8.1 Introduction

Previous chapters in this book have described in detail how low-dimensional structures affect the optical properties of semiconductor materials. It should therefore be no surprise to readers to find that the main applications of low-dimensional materials have been in optical devices which emit light – particularly the semiconductor laser. The semiconductor laser, even without the use of low-dimensional structures, has become the most common form of laser and new operating wavelengths; new characteristics and new applications appear at an amazing rate. This chapter could not hope to provide a comprehensive review of all these developments. Specialist texts (Agrawal, 1986; Zory, 1993; Coldren and Corzine, 1995) will do that far more effectively. Instead we hope to introduce some of the key concepts, presented in the context of developments in semiconductor physics, which will lead the reader to the more advanced texts. Consequently, many of the sources quoted are review papers and not the original texts.

Many of the advances in low-dimensional semiconductors have been motivated by the fascinating new range of physical phenomena which arise when electrons and holes are confined in very small dimensional structures (Bastard, 1988; Schmitt-Rink *et al.*, 1989; Weisbuch and Vinter, 1991). Advances in semiconductor lasers have driven this fascination, but there has also been a clear focus for the development. The most important of these, to date, has been the need to develop very high-performance semiconductor lasers for optical-fibre-based communication systems (Koch and Koren, 1990). The fact that extremely good lasers are now readily available is one of the major contributing factors to the success of optical communications and, consequently, to the development of communications networks for computing and global data transfer (Midwinter, 1988; Dagenais *et al.*, 1990). Software engineers often forget (even if they ever knew!) the crucial impact of a single development like the semiconductor laser!

It is worthwhile pursuing the optical communication story because it illustrates the essential interdependence between physics, engineering and material science which must occur when developing a component to be used in a real system. The first optical fibres were designed to operate in the wavelength range of around

800–900 nm. GaAs and AlGaAs structures were the appropriate materials for the development of lasers and light-emitting diodes (LEDs) for this application, and by 1973 simple lasers had been demonstrated, albeit with rather short lifetimes (around 1000 h). Extending the lifetimes (and other characteristics) required more sophisticated designs and improved materials. Whilst this was happening, systems engineers decided that the minimum loss windows in the attenuation spectrum of glass fibres was at 1300 nm. Moreover, the intramodal dispersion was minimized at this wavelength, allowing higher data rate transmission. So the need for new lasers arose. This was not just an extension of the GaAs technology, but required the development of phosphorus-containing quaternary materials, such as GaInAsP, which had to be grown epitaxially on InP substrates. Within a short time, system engineers decided that they really wanted to work at 1550 nm, where there was an even lower loss! Fortunately, sources at this wavelength could be produced by modification of the composition of the material developed for 1300 nm. Today both the 1300 nm and 1550 nm wavelengths are used, and both low loss and low dispersion can be achieved by dispersion-shifted and dispersion-flattened optical fibres. The key parameters which are now the focus for improvements are the laser threshold, optical linewidth and the maximum modulation rate.

That was just one example. We could have chosen another based on the need for lasers in optical storage systems. Indeed, this is now proving to be a much bigger market for semiconductor lasers than even the communications market and is the focus for much of the interest world-wide in II–VI and nitride-based semiconductor materials (Nurmikko and Gunshor, 1996; Prior, 1996). Table 8.1 lists some of the materials that are now used for semiconductor lasers and LEDs. It is not a complete list of all lasing materials but it does illustrate the current dominance of the elements Ga, In, As, and P in various combinations. We shall see later that the list of material combinations is not constrained to lattice-matched systems but that strain induced by the use of lattice-mismatched materials does indeed help improve the performance of the lasers.

Table 8.1 Material systems and wavelength ranges.

Active layer	Confining layer	Wavelength	Substrate
GaAs	(AlGa)As	800–900 nm	GaAs
GaInP	(AlGa)InP	630–650 nm	GaAs
InGaAs	GaAs	900–1000 nm	GaAs
InGaAs	InGaAsP	1550 nm	InP
InGaAsP	InGaAsP	1300–1550 nm	InP
InAsP	InGaAsP	1060–1400 nm	InP
InGaAs	InP	1550 nm	InP

8.2   Basic Laser Theory

In this section we will discuss some of the basic characteristics of semiconductor lasers (see, e.g. Sze, 1981). The discussion applies to semiconductor lasers in general and not specifically to quantum well lasers, but in a later section the key differences from the quantum well will be introduced. The essential components of the semiconductor laser are an active region where electrons and holes may recombine to emit photons, a p–n junction which, in foward bias, injects electrons and holes into the active region, and an optical cavity formed in the crystal structure to provide optical feedback into the active region. Laser emission was first demonstrated using structures containing just these features, but the absence of any physical constraint on the location of the electrical carriers after injection, together with the absence of any mechanism for control of the transverse optical mode in the laser, meant that high current densities were required and sources usually operated in a pulsed mode to minimize electrical heating effects. The double heterostructure such as that shown in Fig. 8.1 for the GaAs/(AlGa)As solved many of these problems and is an essential design feature in all lasers today.

The structure shown in Fig. 8.1, without any dimensions indicated, clearly resembles a quantum well. The layer thicknesses are, however, considerably greater than those found in quantum wells and the role played by the heterostructure is rather different from its role in a quantum well. Typically, the GaAs active layer is 0.5–1 µm in thickness and the AlGaAs layers are about 1 µm thick. The two important characteristics here are the fact that the wider band-gap layers AlGaAs confine electrons and holes, so recombination is constrained to occur in the active region and the lower refractive index of the AlGaAs located on either side of the GaAs provides some optical waveguiding, ensuring good overlap of the optical mode and the active region. In a quantum-well laser the quantum well is located in the gain region which is itself still surrounded by the double heterostructure. The use of low-dimensional structures must therefore be viewed as a

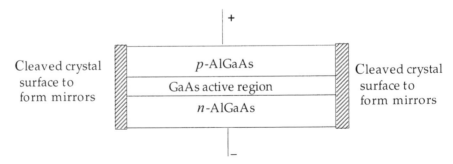

Figure 8.1.   Schematic diagram showing the essential components of a double heterostructure semiconductor laser.

perturbation on what is already a complex optoelectronic structure. Materials which can be used to produce the heterostructure are idential to those listed in Table 8.1, with the exception of those materials which are not lattice matched. Coherent strain is perfectly acceptable for quantum wells located in the active region of a laser, but the use of lattice mismatched systems for the thick layers needed to form double heterostructure lasers would simply lead to the production of a large number of dislocations and, consequently, to the formation of a number of undesirable non-radiative recombination centres.

The laser geometry described here is that of the in-plane laser (IPL). Gain, and hence amplification, is achieved when the electron and hole carrier injection levels are such that the quasi-Fermi levels in the conduction and valence bands, denoted $E_{F,c}$ and $E_{F,v}$, respectively, are separated by the band-gap energy of the semiconductor, denoted E_g:

$$E_{F,c} - E_{F,v} > h\nu > E_g \tag{8.1}$$

where $h\nu$ is the photon energy. This is effectively the condition referred to in other lasers as population inversion and is derived in detail in Section 8.3. As mentioned above, optical feedback and, hence, lasing action is achieved through the provision of an optical cavity. The cavity in a simple IPL is formed by cleaving the laser material, epi-layers plus substrate, into sections typically 300 μm in length. The reflectivities $R_{F/B}$ of the front and back mirrors formed in this way are given by the Fresnel reflection coefficient evaluated for the air-semiconductor interface and are of order 30% for most semiconductors where the refractive index, n_s, is of order 3.5. The exact values are given by

$$R_{F/B} = \left| \frac{n_0 - n_s}{n_0 + n_s} \right|^2 \tag{8.2}$$

The cavity formed between the two cleaved facets of an IPL is, of course, a Fabry–Perot (FP) cavity. Consequently the optical field within the cavity is a standing wave. Figure 8.2 shows a schematic example of the field along the length of an IPL device. Two possible field distributions, or modes, are shown.

In general, the cavity will sustain a number of modes and we can associate a slightly different wavelength with each mode. The emission spectrum of this

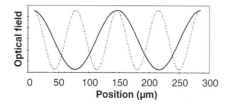

Figure 8.2.   Schematic field profiles within an FP cavity, e.g. GaAs bounded by air.

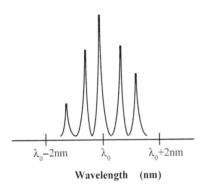

Figure 8.3.   Schematic emission spectrum for an FP IPL.

laser will therefore contain a number of wavelengths, as shown schematically in Fig. 8.3. The envelope of the modes and, hence, the number of modes present is determined by the spectral shape and extent of the gain spectrum. The spacing of these modes, $\Delta\lambda$, is given by

$$\Delta\lambda = \frac{\lambda^2}{2Ln_{\text{cav}}} \tag{8.3}$$

where n_{cav} is an appropriate average refractive index. For a typical FP cavity IPL the mode spacing is of order $10\,\text{Å}$. Obviously, such multimode behaviour is a problem when wavelength stability is required, for example, in communication systems where high data rates are necessary. In such cases, these higher-order modes may be eliminated by incorporating wavelength selective loss into the laser. For example, a grating may be integrated with the laser. If etched directly into the epi-layers this grating provides feedback only to the desired lasing wavelength (Hansmann, 1992; Coldren and Corzine, 1995).

Of course, the optical mode is confined in only one dimension here and additional steps have to be taken to provide optical confinement in the perpendicular direction. This can involve selectively contacting the device so that current flows only in a resticted region of the structure. The contact is called a stripe contact and could consist of a metal stripe a few microns wide running along the length of the device. The optical mode is then determined by the localization of the gain region and the devices are usually referred to as gain-guided lasers. An alternative and often preferred method of optical confinement involves index-guiding, where a channel or ridge structure imposes a constraint on the transverse modes which can propagate. This is shown in Fig. 8.4. The large number of variations in laser design which have been proposed or demonstrated to achieve good optical mode control will not be discussed further here but the reader should be aware that these design features are essential to producing high-quality lasers.

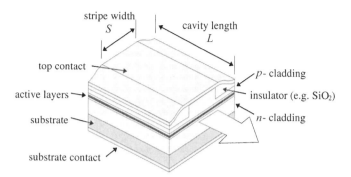

Figure 8.4.   Schematic of stripe contact double heterostructure in-plane laser. The active layers in this case may be quantum wells.

8.2.1   Laser Threshold

The threshold gain of a laser is defined as the gain required to balance the optical losses within the structure. More generally, it may be defined as the gain required in a structure to allow light of a given intensity, I, to traverse the optical cavity, distance $2 \times L$, without attenuation. The optical losses include scattering, diffraction and absorption terms, amongst others.

The reader will recall that light of intensity I_0 passing through a slab of medium with gain g and loss α will experience an attenuation or amplification of $e^{(g-\alpha)L}$ with distance L. If the medium is bounded by reflectors of reflectivity R then on one round trip the total intensity will be given by

$$I = I_0 R^2 \, e^{2(g-\alpha)L} \tag{8.4}$$

By equating the final intensity I to the initial intensity I_0 and re-arranging this expression we arrive at

$$g = \alpha + \frac{1}{L} \ln(1/R) \tag{8.5}$$

This equation, which defines the threshold gain of a laser, basically states that the laser gain must equate to the losses of the structure plus the output losses due to the imperfect reflectors. This second loss term is associated with the (useful) light exiting the output facet of the laser.

Equation (8.5) is, in practice, an over-simplification that neglects many important parameters relating to the laser structure. The threshold condition for a typical laser is, more generally, given by an expression of the form

$$(g_t - \alpha_{\mathrm{fc}})\Gamma_a = \alpha_n \Gamma_n + \alpha_p \Gamma_p + \frac{1}{L}\ln(1/R) + \alpha_s \tag{8.6}$$

where g_t is the threshold gain, α_{fc} is the loss within the active layer due to free carriers and unpumped gain material, $\alpha_{n/p}$ accounts for losses within the n- and

p-type cladding layers, α_s is a summation of all the scattering losses caused by surface and interfacial roughness and R is the average facet reflectivity, $R = \sqrt{R_{\mathrm{F}} R_{\mathrm{B}}}$. It must be remembered that many of the terms are dynamic quantities dependent upon the injection level (Agrawal, 1986; Zory, 1993; Coldren and Corzine, 1995).

Equation (8.6) can be seen to include terms representing various sections of the laser. In particular, losses are attributed to individual layers. These losses are weighted by factors, $\Gamma_{a/n/p}$, which account for the relative magnitude of the optical field within that layer of the structure. These factors are called confinement factors and, in this case, they are for the active-, n-type and p-type regions of the laser, respectively. Confinement factors represent the fraction of the total field within a structure confined to a given layer. The sum of the individual layer confinement factors, Γ_x, is therefore numerically equal to unity. The confinement factor for a given layer is given by

$$\Gamma_x = \frac{\int_{-x}^{x} |E|^2 \, \mathrm{d}s}{\int_{-\infty}^{\infty} |E|^2 \, \mathrm{d}s} \tag{8.7}$$

where from this equation we readily see that sum of Γ_x over all layers is equal to unity.

Before continuing the discussion of laser threshold we will briefly discuss the importance of the optical confinement factor. It is evident that in an IPL the active region is confined to a narrow layer with a typical length of 300 μm and a transverse dimension of typically 50 μm. The active layer itself may be microns thick in a simple heterojunction laser or just a few nanometres in a quantum-well laser. In either case only the light confined to the active layer will be amplified. It is therefore beneficial to maximize the light in this layer, i.e. to maximize Γ_a. This is achieved through the use of optical confinement layers which combine to produce an optical waveguide, as discussed previously and shown in more detail in Fig. 8.5.

A waveguide is produced by bounding the active layer with cladding layers of different, generally lower, refractive index. In its simplest form the light is channelled along the waveguide through internal reflections experienced at each active

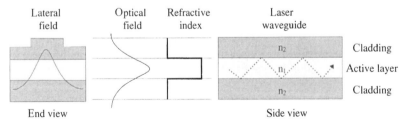

Figure 8.5.   Simple IPL waveguide. Indices are chosen to produce a guiding structure. The stripe isolation laterally confines both the optical mode and the injection.

layer and cladding boundary, as shown schematically in Fig. 8.5. In the lateral direction the optical field may also be confined through the use of an etched stripe. This produces a lateral refractive index variation that acts to confine the optical field. In this way the lateral confinement factor may be optimized. By performing both these optimizations the volume confinement factor, the product of the two in-plane confinement factors and the vertical confinement factor (Γ_a), for the active region may be optimized. For a typical IPL, Γ_a is of the order of a small percentage.

Optimization of the active layer confinement factor ensures that the percentage of photons within the laser that are confined to the active layer is maximized. The modal gain, defined as the product of the actual gain and the confinement factor, is then maximized and the laser threshold is reduced. This can be seen clearly from equation (8.6) above. The reader will note that all gains and losses within the laser are multiplied by the corresponding confinement factors. In this way the interaction of the photons and the loss and gain sections can be precisely accounted for (Agrawal, 1986; Coldren and Corzine, 1995; Prior, 1996).

It can be seen from equation (8.6) that a number of parameters can be varied to lower the threshold of a simple IPL. First, a simple increase in the cavity length may be used to increase the available gain. Second, the facet reflectivities may be changed through the use of high reflectivity facet coatings. Finally, structural and material changes may be made to the laser to increase its active layer confinement factors and minimize these factors for the loss-inducing cladding layers.

8.2.2   Threshold Current Density

Given the laser threshold gain it would, at first glance, appear a simple task to calculate the actual threshold current of a laser. Calculation of the power output might then seem a simple step. Unfortunately, within real laser structures there is no simple relationship between the threshold gain and the threshold current. In fact, a large number of physical mechanisms come into play to make the task a highly complex one (Agrawal, 1986; Zory, 1993; Coldren and Corzine, 1995).

Figure 8.6 shows the gain spectrum and spontaneous emission spectrum of an InGaAs quantum well (QW). The calculation of the gain in quantum wells is discussed later – here we need to use a typical curve to identify the generic features. The gain can be seen to increase rapidly with increasing carrier injection and can also be seen to saturate due to band filling at higher carrier levels. Figure 8.7 shows this in a clearer form by plotting the gain against carrier density. It can be seen that below a minimum carrier injection level the gain drops below zero, indicating that the quantum well is absorbing. There is therefore a critical carrier injection level, known as the transparency injection level, at which the active medium is lossless and provides no material gain.

For a QW structure the gain-current relationship can be approximated by

$$g(\mathrm{QW}) = b_0 J_0 \ln(J_{\mathrm{nom}}/J_0) \tag{8.8}$$

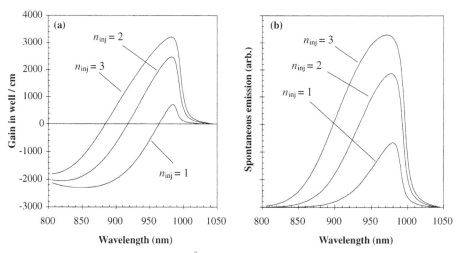

Figure 8.6.   (a) Gain spectra for 80 Å In$_{0.2}$Ga$_{0.8}$As QW with GaAs barriers with carrier levels indicated. (b) Spontaneous emission spectra for the QW. Carrier injection levels are $\times 10^{18}$ cm^{-3}.

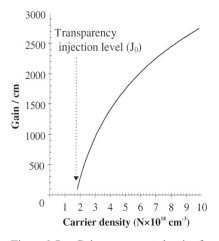

Figure 8.7.   Gain vs. current density for an 80 Å In$_{0.2}$Ga$_{0.8}$As QW with GaAs barriers.

where J_0 represents the transparency current density, J_{nom} represents the nominal current density (at the QW) and b_0 is the gain-current coefficient – this is essentially a fitting parameter, although its value can be derived from fundamental band-structure and gain calculations. By substituting this expression into the expression for the threshold equation for a laser in (8.5) (with the inclusion of a confinement factor), we obtain the following expression for the threshold current density

$$J_{\text{th}}(\text{QW}) = \frac{J_0}{\eta_i} \exp\left[\frac{\alpha + (1/L)\ln(1/R)}{\Gamma b_0 J_0}\right] \qquad (8.9)$$

where the internal quantum efficiency of the active material is given by η_i (in high-quality material it is typically in excess of 90%). This equation shows again how the confinement factor can affect the lasing threshold. It also shows how the internal losses of the structure dictate the threshold. By improvements in growth both the internal losses and the internal quantum efficiency of the laser can be improved. Typical losses are now less than $5\,\mathrm{cm}^{-1}$ in the commercially-important InP and GaAs based lasers.

Figure 8.8 shows the calculated threshold current density for an IPL. Equation (8.9) has been used with $b_0 = 0.5\,\mathrm{cm/A}$ and $J_0 = 450\,\mathrm{A/cm}^2$. These are typical values for a simple double heterostructure (DH) FP in-plane laser. The figure shows the effect of changing both the confinement factor of the active region and the losses within the cavity region. In all cases, as expected, the threshold current density reduces as the laser cavity and, hence, the active region length are increased.

As mentioned earlier the real threshold current density calculation is far more complex. The above, highly simplified, expression neglects a number of important mechanisms that act to deplete carriers in a non-radiative, and hence useless, manner. These parasitic mechanisms invariably act to increase the lasing threshold and to produce device heating.

Carriers injected into a laser initially have to overcome various losses at the device contacts. They must then travel to the active layers with a mobility and probability of arrival dictated by the quality of the material layers. The carriers are thus injected into the active layers with some injection efficiency, η_{inj}, that falls

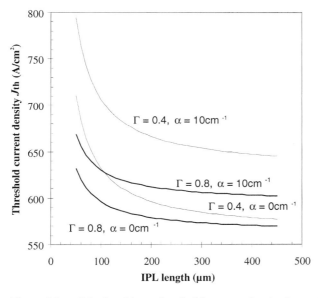

Figure 8.8.   Calculated laser threshold current density for various IPLs. Both the confinement factor and loss are varied.

below 100% and is typically of order 80%. Within the active layers a fraction of the injected current will contribute to spontaneous emission, a fraction will typically recombine non-radiatively in the active layer barriers and cladding, some fraction will be lost to Auger recombination and some will leak from the QWs and recombine by various other mechanisms. The total injected current may therefore be expressed as a summation of useful terms, including the current contributing to population inversion and material transparency, and the loss terms,

$$J(n) = \frac{1}{\eta_{\text{inj}}} \left[J_{\text{spon}}(n) + J_{\text{barr}}(n) + J_{\text{Auger}} + J_{\text{leak}} + \cdots \right] \qquad (8.10)$$

with most terms being functions of the number of injected carriers, n.

A number of methods exist to help alleviate the problems of carrier leakage (Agrawal, 1986; Coldren and Corzine, 1995). These include simply increasing the number of QWs in a laser structure along with an optimization of the QW barrier compositions and the QW width. More generally, confinement layers are used to clad the active material. These layers have higher band-gaps than the active QW layers and act to confine injected carriers to the region near the QWs. This reduces the number of carriers leaking to the barriers. Phonon-assisted thermalization into the QWs then occurs. A range of confinement structures exist, starting with the simple use of single high band-gap cladding layers, producing a structure called a separate confinement heterostructure (SCH), to the use of graded composition layers. The latter, though difficult to grow, has advantages in material quality and in optical confinement. It is regularly referred to as a graded index separate confinement heterostructure (GRINSCH). Figure 8.9 shows a selection of typical confinement structures.

8.2.3 Power Output

Once the threshold current of a laser is known, the output power P for a given drive current may be calculated in terms of the injected current, I:

$$P = \frac{hc}{e\lambda} \, \eta_{\text{inj}} \, \eta_{\text{opt}} (I - I_{\text{th}}) \qquad (8.11)$$

Note that an injection efficiency appears in this expression. This is often used as a fitting parameter. An optical efficiency, η_{opt}, also appears in the expression. This

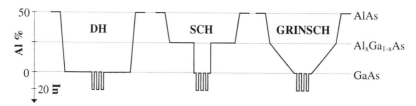

Figure 8.9.   Schematics of possible cavity configurations within a laser. DH represents the simple double heterostructure arrangement.

relates the amount of useful light exiting a structure to the amount generated within. Evidently, the light generated within the laser is either emitted or lost to absorption and scattering. If the loss within a structure due to these effects is α and the structure length is L, then the total incremental loss, L_{loss} is $2\alpha L$. The useful incremental loss exiting the structure, or transmission loss T_{loss}, is simply a function of the average mirror reflection coefficients and is

$$T_{loss} = \ln\left(\frac{1}{R_1 R_2}\right) \tag{8.12}$$

The optical efficiency is then defined as the ratio of the transmission loss to the total loss:

$$\eta_{opt} = \frac{T_{loss}}{L_{loss} + T_{loss}} \tag{8.13}$$

It can be shown that this expression is equivalent to dividing the threshold gain of a laser with losses set to zero by the threshold with losses set to α.

Equation (8.11) is clearly a linear function of current with an intercept dictated by the threshold current. In practical situations the power, P, is not a linear function of current. Lasers experience a number of current-dependent effects that act, in general, to lower the power output. These effects include carrier leakage and temperature-related effects. In fact, as the laser heats up it often becomes less efficient and hence heats up further. In extreme cases, this process can prevent laser operation. More generally, the power curve, versus current, flattens and turns over at higher injection levels. It is noted that the threshold current of the laser also changes with temperature, T. In many practical cases the relationship is a simple exponential one, given in terms of a characteristic temperature T_0:

$$I_{th} \sim \exp(T/T_0) \tag{8.14}$$

Owing to various parasitic effects, as the laser heats up, its material refractive indices and active-region band gap will change. These mechanisms are not discussed in detail here but it is noted that for a simple FP cavity laser the lasing wavelength shifts in correspondence to the peak of the gain spectrum shifting with temperature. Typically this occurs at a rate of $\sim 3\,\text{Å}/°C$. This rate of shift can be dramatically reduced by including wavelength selective grating structures, as in the case of a distributed feedback or DFB laser, in the laser cavity (Koch and Koren, 1990; Hansmann, 1992; Coldren and Corzine, 1995). The shift with temperature is then due primarily to changes in the refractive indices of the grating layers which occur at a rate of $\sim 1\,\text{Å}/°C$.

Figure 8.10 shows the calculated power output of a simple IPL using equation (8.11). The effect of changing the optical efficiency is shown. This change may be brought about through a change in the losses within the structure or, more commonly, through a change of one or both of the facet reflectivities from their original

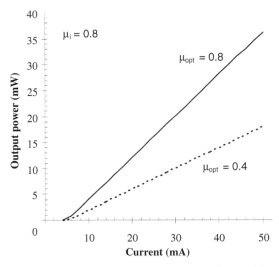

Figure 8.10.   Calculated power output of IPL with two different optical efficiencies.

cleaved, 30%, values. The use of high-reflectivity optical coatings is a common way of achieving this. Note the dramatic increase in output power for the more efficient device. The use of optical coatings also reduces the lasing threshold current.

8.3   Fundamental Gain Calculations

In the next two sections we will look more closely at the factors which contribute to optical gain in semiconductors, and how the nature of the active region, i.e. whether bulk or (un)strained quantum well (QW), affects the spectral gain. At the end we hope to have justified two main points regarding the active region of lasers:

(1) how QWs improve certain laser characteristics over their traditional bulk counterparts;
(2) how the use of biaxial strain in the QW active region can further enhance laser characteristics.

8.3.1   Electronic Band Structure and Densities of States

Optical gain in semiconductors is caused by photon-induced transitions of electrons from the conduction band to the valence band in a direct gap semiconductor. We define $E(\mathbf{k})$ as the energy of an electron with a wavevector $\mathbf{k}$. For $E(\mathbf{k})$ dispersions close to these band edges (near $\mathbf{k} = \mathbf{0}$) we define, as usual, an effective mass for each band as

$$(m_j^*)^{-1} = \frac{1}{\hbar^2}\frac{\partial^2 E_j(\mathbf{k})}{\partial \mathbf{k}^2} \tag{8.15}$$

where j labels either the conduction or valence band. The usual interpretation (Chapter 6; Bastard, 1988; Burt, 1992) of (8.15) is to refer to an electron effective mass for the conduction band and a hole effective mass for the valence band.

For bulk-like active regions, which have fairly large-scale dimensions ($>0.1\,\mu m$), an electron in a particular band is free to move and acquire momentum in all directions, such that we can derive $E(\mathbf{k})$ where $\mathbf{k} = k_x, k_y, k_z$. Now consider when the active region is some low-dimensional structure such as a QW, where dimensions are typically $\sim 100\,\text{Å}$. Within an effective mass description, the problem of finding the electron energy levels of the quantum well can be solved with a Schrödinger-type equation and the 'particle-in-a-box' approach (Chapters 2 and 6). Each quantized level, E_n, possesses its own in-plane energy band or subband as indicated in Fig. 8.11. An electron in one of these subbands is trapped or confined along the growth direction but allowed to move freely within the plane of the well and acquire in-plane momentum along the other two directions. This *two*-dimensional motion of an electron may still be described by a dispersion, $E(\mathbf{k})$, but now $\mathbf{k} = (k_x, k_y)$ with z, say, as the growth direction. An analogous situation occurs in the valence band, although here the degeneracy that occurs in bulk semiconductors gives rise to both heavy-hole and light-hole quantized states.

Given the dispersion relation, $E(\mathbf{k})$, for some active region, our next concern is to find the density of states (DOS) per unit energy. This is discussed at length in Chapter 2, but here we quote again the results which have particular relevance for quantum well-lasers. It is useful to consider an idealized case where the $E(\mathbf{k})$ dispersions have a parabolic nature, i.e. $E \propto k^2$ (in fact, this description is not too bad for the relatively isolated conduction band, but is quite poor for the valence band where the close proximity of other bands results in a highly non-parabolic dispersion). The DOS function for a bulk-like active region starts at the band-edge energy and increases smoothly with increasing energy (Fig. 8.11). In contrast, for a QW active region, a *constant* density of states function may be defined for *each* (parabolic) subband of a quantized energy level, E_n. The DOS

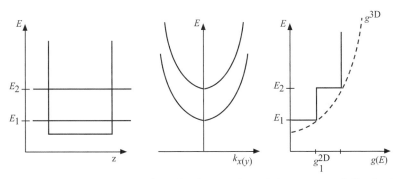

Figure 8.11.   Illustration of confined states in a QW structure (left), the in-plane $E(\mathbf{k})$ dispersions for the two states (middle) and the resulting density of states (right).

functions for quantum-well and bulk materials, within a parabolic description, can be written as (cf. equations (2.9) and (2.14)),

$$g_{3D}(E) = \frac{1}{2\pi^2} \left(\frac{2m^*}{\hbar^2} \right)^{3/2} \sqrt{E} \quad \text{(bulk)} \tag{8.16a}$$

$$g_{2D}(E) = \frac{m^*}{\pi\hbar^2 L_w} \quad \text{(quantum well)} \tag{8.16b}$$

where m^* is the effective mass of the particular band or subband, E is the energy *above* the band edge and L_w is the quantum-well width. Note that, for infinite well energies E_n, $g_{2D}(E_n)L_w = g_{3D}(E_n)$, as in Fig. 6.11. It is straightforward to see an overall step-like DOS for the QW (Fig. 8.11) results from adding together $g_{2D}(E)$ for each subband.

As we have already remarked, optical gain involves transitions between states in the conduction and valence bands. Expressions for gain are often written in terms of a *joint* (or reduced) density of states, (JDOS), which takes account of the local density of states in *both* participating bands (or subbands) and is defined as

$$g_j^{-1} = g_c^{-1} + g_v^{-1} \tag{8.17}$$

In keeping with a parabolic description, the functional form of the JDOS, for bulk or QW active region, turns out to be identical to the expressions in (8.16), only now the effective mass will refer to the reduced mass of the participating bands, i.e. $m^{*-1} = m_c^{*-1} + m_v^{*-1}$. The function JDOS tells us how many states (per unit energy per unit volume) are available for the gain process and is therefore of fundamental importance in any calculation of gain. However, to describe gain completely we shall also require some measure of the population inversion *and* some idea of the strength of interaction between the participating bands. We leave, for the moment, a discussion on the latter until we are ready to compare gain for different active regions.

8.3.2 Carrier Density and Inversion

The question of population inversion is directly related to the carrier density in the active region. In a typical operation, electrons and holes are injected into the active region using a forward-biassed p–i–n diode, i.e. holes are injected from the *p* side and electrons from the *n* side towards the intrinsic (under equilibrium conditions, i.e. with no injection) region comprising the active material. The carrier density in a given band may be found for a given quasi-Fermi level by integrating the product of the DOS function and the occupation probability over the entire band. Explicitly, the electron density in the conduction band can be written

$$n_c = \int_{E_c}^{\infty} g(E - E_c) f_c(E, E_{F,c}) \, dE \tag{8.18}$$

where E_c is the conduction band edge in the active region and $g(E - E_c)$ is the DOS function for the band (in the case of a QW, the *total* 'step-like' DOS function should be used). The occupation probability, or Fermi function for electrons in the conduction band, is defined as

$$f_c(E, E_{F,c}) = \frac{1}{1 + \exp[(E - E_{F,c})/k_B T]} \tag{8.19}$$

where $E_{F,c}$ refers to the quasi-Fermi level in the conduction band, E is the energy of the electron, k_B is the Boltzmann constant and T is the temperature of the region. Typical calculations start with a given density of carriers injected into the active region, a value of $E_{F,c}$ is found (self-consistently) which satisfies equation (8.18), whereupon the occupation probability (8.19) can be specified for a given energy. By requiring charge neutrality in the active region, i.e. $n_c = n_v$ where n_v is the density of holes in the valence band, the occupation probability of electrons in the valence band, $f_v(E, E_{F,v})$, may be found in a similar manner. Therefore the difference between the two functions, $f_c(E, E_{F,c}) - f_v(E, E_{F,v})$, gives a measure of population inversion.

Before we go on to look at the expression for gain, it is worth mentioning some further points regarding the subjects discussed so far. Firstly, laser characteristics are typically referred to *sheet* carrier density (per unit area). In a basic picture, this is the carrier density (per unit volume) specified over the length of the active region. For example, this might be the intrinsic region width in a bulk laser or the well width in a QW structure. For a greater degree of accuracy, particularly in the case of QW structures, carriers residing in the surrounding regions (e.g. the barriers) are also included. This requires equation (8.18) to be evaluated for carriers distributed with a 2D DOS (in the well) *and* a 3D DOS (in the active region). The second point relates to the limitation of the charge neutrality condition, taken here as $n_c = n_v$. This is strictly applicable to intrinsic active regions and is not suitable in cases where doping in the active region is present. For this latter case, the ionized dopants in the close vicinity of the active region must be accounted for in the charge neutrality condition; it is also noted that the effects of the dopants should also be included in the electronic description and, hence, in the DOS functions. We make the point here that our subsequent description of laser characteristics assumes that the active regions are situated in an intrinsic environment under equilibrium.

Finally, since we started with the energy band dispersions $E(\mathbf{k})$, it seems appropriate to end with some comments about this quantity. It should be apparent by now that both the DOS functions and the degree of population inversion depend heavily on the form of the $E(\mathbf{k})$ dispersion. Therefore, an accurate calculation of laser characteristics (in fact, any device characteristic!) rests largely on starting with an accurate description of the band structure in the active region. This is particularly important for the valence band, where the degeneracy of the heavy-hole and light-hole bands (and, in some materials, the close proximity of the spin-orbit split-off band) result in QW valence subbands that show $E(k_x, k_y)$ dispersions that are highly non-parabolic (see later in Fig. 8.16). In this case, it is often easier to calculate exclusively in momentum-space (**k**-space) with a density of **k**-vector states given as $g(\mathbf{k})$. The more useful energy-space interpretation may then be found from

$$g(E) = \frac{g(k)}{\mathrm{d}E/\mathrm{d}k} \tag{8.20}$$

8.3.3   Gain Expression

An expression for the optical gain in a semiconductor is commonly derived through the application of Fermi's Golden Rule (equation (6.38) and see, e.g., Merzbacher, 1970). In this approach, the time-dependent Schrödinger equation for the semiconductor in the presence of the optical field is written with the Hamiltonian represented as the sum of the photon field, which is classically represented by a time-dependent vector potential, and the original Hamiltonian (which describes the electron in the solid). The interaction between the photons and the electrons is then treated as a time-dependent perturbation which induces transitions between the conduction and valence bands, i.e. transitions to and from the conduction band, such that absorption (which in lasers counts as loss) may also be described in the same manner (Agrawal, 1986; Zory, 1993; Coldren and Corzine, 1995). The rate of these transitions is calculated using first-order time-dependent perturbation theory. The optical gain *g* experienced by a photon with energy *E* is then obtained as

$$g(E) = \frac{\pi h e^2}{\varepsilon_0 m_0 n_r^2 E} |M_T|^2 g_j(E) \times \frac{1}{v_g} [f_c(E, E_{F,c}) - f_v(E, E_{F,v})] \tag{8.21}$$

The three main groupings on the right-hand side of this equation are as follows. The first term is essentially the expression obtained from Fermi's Golden Rule and gives the total number of vertical transitions occurring per second inside the active region. The emphasis in this expression is on *vertical* transitions, which refers to the derivation being carried out by using what is known as the **k**-selection rule, i.e. only transitions between bands (or subbands) such that $\mathbf{k}_e = \mathbf{k}_h$ are considered in order to conserve momentum. Notice that, since typical photon wavelengths are much longer than electron wavelengths, the photon momentum is assumed negligible and is not included. Apart from the physical constants, the other terms refer to the refractive index of the active region, n_r, and the joint DOS function, $g_j(E)$. The interband matrix element, M_T, governs the strength of transitions and depends on several factors, including the overlap of the electron wavefunctions that correspond to the initial and final band states as well as the polarization state of the optical field. This quantity will be examined in more detail in the later sections. The second term contains the electronic group velocity, v_g, which is the velocity at which energy travels inside the active region. Its appearance in (8.21) essentially converts the transition rate per second (from the first term) to transitions per unit length. As we mentioned earlier, the third term is a measure of the population inversion that has been achieved for a given electron density in the active region.

Some final points to note are that the expression for gain in equation (8.21) describes single band-to-band transitions, e.g. in bulk we have the conduction band to valence band. However, in QWs there are typically several subband-to-subband transitions to consider, e.g. $e1$–$hh1$, $e2$–$hh2$, $e1$–$lh1$, etc. Therefore, calculating gain in QW active regions involves evaluating (8.21) for each subband-to-subband transition. Additionally, because of phonon interactions, the lifetime of a given state is finite and so realistically (8.21) should account for the spectral broadening of each transition, through the use of some line shape function, e.g. a Lorentzian distribution.

8.3.4   Optical Gain in 2D and 3D Active Regions

Gain is achieved in a semiconductor when we have a population inversion, i.e. $f_c - f_v > 0$. Then, from (8.21), we have $g(E) > 0$, so an incoming photon of energy E will be amplified by the material (conversely, without inversion $g(E) < 0$ and the material will be absorbing at that energy). This inversion condition for gain can be written (cf. (8.1)) in terms of the quasi-Fermi levels, $E_{F,c}$ and $E_{F,v}$:

$$E_{F,c} - E_{F,v} > E > E_g' \tag{8.22}$$

In other words, to achieve optical gain, the quasi-Fermi level energy separation must be greater than the energy separation of the participating bands, E_g', i.e. the band gap in bulk material or typically the $e1$–$hh1$ transition energy in a QW. This condition is known as the Bernard–Duraffourg condition and is satisfied above a certain carrier density referred to as the transparency carrier density, N_{tr}. Optical gain in the material is then generated as the injection current density increases beyond N_{tr}.

In any laser device, a particular value of optical gain, known as the threshold gain (g_{th}), is required to start the device lasing. In 3D structures, the build-up of inversion to reach g_{th} requires the lower-lying energy states near the band edge to be filled first. As more carriers are injected the peak of the gain spectra increases to g_{th} *but* shifts away from the bottom of the band, E_g' (Fig. 8.12(a)). Thus, all carriers at energies below E_{th} are essentially redundant. Consider now the situation with a 2D active region (Fig. 8.12(b)). Injecting carriers increases the spectral peak gain to g_{th}. However, the main difference is that added carriers contribute to the gain at its peak, which is fixed to the bottom of the band, i.e. $E_{th} = E_g'$. In this sense the injected carriers are used more efficiently than in the 3D case, so that overall it may be stated that, for a given value of gain, the carrier density required

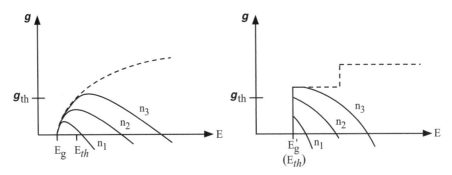

Figure 8.12.   Illustration of gain formation in bulk (left) and QW (right) structures for three different carrier densities, n_i. The same gain threshold value, g_{th} (required for lasing), is indicated in each figure. Broken lines depict full inversion which trace out the JDOS function for each structure.

in 2D structures is typically lower than that needed in 3D structures, a direct result of the sharp step-like edge of the 2D DOS.

Another difference between 2D and 3D lasers appears as an increasing number of carriers are injected into the respective active regions. The peak gain in the 2D band *temporarily* saturates, whereas in 3D structures it continues to increase (linearly, to a good approximation). We have referred to 'temporary saturation' since, for some value of injected carrier density, population inversion can be achieved for a second set of subbands (e.g. *e2–hh2*). At this instant, the peak gain will shift from the band edge of the first quantized level to the second level. Note that the second level produces more gain because it contains con- tributions from *both* quantized levels. Ultimately, for both 2D and 3D active regions, the gain spectra will reach *complete inversion*. We can picture this by saying that all of the carriers which were once in the valence band and created the absorption spectra through transitions to the conduction band are now all in the conduction band creating the identical gain spectra through downward transitions. In terms of the gain expression in (8.21), we have complete absorp- tion, $f_c - f_v \approx -1$, going to complete inversion, $f_c - f_v \approx 1$, for increasing carrier density injection.

Of course, in practice, the interest is simply to attain some threshold value of gain, which is largely determined by the device design, as discussed in Section 8.2. However, a potentially realistic problem is when the saturated gain of the first quantized level is not enough! To overcome this, several QWs are used in the active region, with the gain characteristics related to that of a single QW by a simple scaling transformation.

We have discussed the considerable advantage available by using QW active regions in place of bulk-like material, i.e. a lower carrier density is required to attain the same value of gain. It should be remembered we have illustrated these advantages assuming a nice step-like JDOS, although, as we shall later see, using an accurate description of the valence band, the reality is somewhat different. Nevertheless the improvements noted are by and large observed in practice, although as we shall shortly see, further improvements (quite dramatic ones!) are available when the use of biaxial strain is employed in the active region.

To illustrate the motivation for pursuing greater improvements via strained layers, it is useful to point out a common problem that both bulk and unstrained QW active regions exhibit. The origin lies in the fact that the effective mass in the conduction band (or subband) is much smaller than that seen in the valence band, and referring to equation (8.16), a similar consideration also applies for the DOS in the bands. For an increasing carrier injection, the difference in the DOS results in the quasi-Fermi level for electrons in the conduction band ($E_{F,c}$) moving at a faster rate than ($E_{F,v}$) in the valence band (recall that, for a given injection density, their absolute positions reflect the charge neutrality between the bands, cf. (8.18) and (8.19)). Therefore, when $E_{F,c} - E_{F,v} = E_g'$, i.e. the transparency condition, we find $E_{F,c}$ is well into the conduction band before $E_{F,v}$ has reached the valence band edge (Fig. 8.13). The large spread of electron energies in the conduction band (which increases further with increasing inversion) is rather useless, since only particular electrons will have the correct energy and momentum to contribute to the laser output. In fact, the excess of electrons often degrades the performance, for example, by contributing to non-radiative losses.

Consider now the situation depicted in the lower plot of Fig. 8.13(b), where the effective masses (and hence DOS) are assumed equal. At the transparency condition, the quasi-Fermi levels simultaneously reach the band edges, therefore greatly reducing the spread of electron energies. The result, from the point of view of device properties, is a reduction in the current density required for transparency and, importantly, an increase in the *differential* gain, dg/dN, which describes the

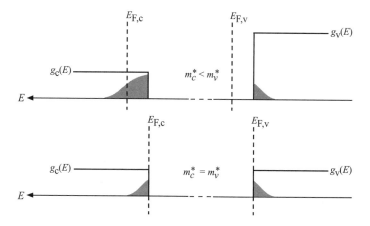

Figure 8.13.   Illustrated transparency condition for two types of QW active region. The Fermi levels are shown as broken lines with the carrier filling in each band depicted by shading.

rate with which the gain increases as more carriers are injected. For symmetrical bands ($m_c^* \approx m_v^*$), the differential gain is directly related to how quickly the band-edge carrier density can be increased and, therefore, benefits from the quasi-Fermi level being at or near the band edge at transparency.

With a clearer idea of some of the benefits that can be obtained by reducing the valence band DOS (i.e. effective mass), we now outline the approach commonly used which relies on incorporating strain into the active region (O'Reilly, 1989; O'Reilly and Adams, 1994).

8.4   Strained Layers

When a thin layer of a semiconductor has a different bulk lattice constant in comparison with the rest of the structure, which includes the substrate, then it is possible that during growth this layer may elastically deform to take on the dominant lattice constant of the structure, i.e. typically that of the substrate. The resulting strained structure has the same in-plane lattice constant (i.e. parallel to the growth direction) throughout. Under these circumstances the strain has been coherently accommodated and the deformed layers may be described as being under either compressive or tensile strain. Strained-layer systems are discussed in Chapters 1, 6 and 10.

It is important to realize that the coherent accommodation of strain in the layer is only possible because of the layer thicknesses typically used. In other words, there is a certain critical thickness above which it becomes energetically more favourable to relieve the strain energy arising from the deformation through the generation of dislocations. To a good approximation, the critical thickness of a layer can be expressed in terms of a percentage strain thickness product, e.g. typical values lie between 100 and 200 Å%, so taking, say, 150 Å%, then for a 2% strain the layer will be coherent up to a thickness of 75 Å. To highlight how the effects of strain can ultimately improve laser characteristics we restrict the following discussions to structures grown on (100) substrates and where only the well material is strained, since this configuration is the most commonly used in practice.

We first clarify the two types of biaxial strain mentioned above. If a layer (well) has a larger bulk lattice constant than the dominant one then, in the coherent structure, the layer will be under compressive strain. Similarly, a layer with a smaller bulk lattice constant is said to be under tensile strain. In either case, the total strain in the layer may be resolved into a hydrostatic component and an axial component. Improvements to laser characteristics can be traced to the changes these components of strain impart on the electronic band structure, $E(\mathbf{k})$. Before considering the effect on QW structures, it is useful to look first at the effects of strain on the bulk band structure.

The hydrostatic component of strain acts on the band edges, thereby changing the band gaps. However, it is the axial strain component, acting on degenerate bands such as in the valence bands, which leads to many of the improvements in strained QW lasers. In the valence band, the axial component lifts the degeneracy that exists for the heavy-hole and light-hole bands at the band edge (Γ-point). The nature of this splitting depends on whether it is brought on by either tensile or compressive strain. For example, under compressive strain, the highest valence band is the heavy-hole band, whereas it is the light-hole band under tensile strain (Fig. 8.14). As illustrated in Fig. 8.14, the axial component of strain renders the bulk valence band anisotropic, i.e. heavy along one direction and light along the other. In Fig. 8.14 the z-axis is along the growth direction, which is the confinement direction in QWs.

In a strained QW, the lifting of the degeneracy has a direct bearing on the confined states (Fig. 8.15). Consider an unstrained QW (lattice matched). The bulk band edges of the well material ($E_{\Gamma,hh(lh)}^{w}$) are degenerate. However, owing to

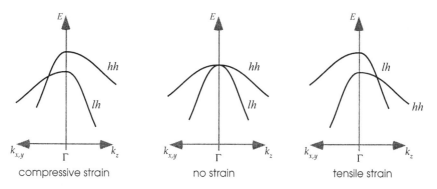

Figure 8.14.   Schematic representation of the effects of strain on the valence band structure near the Γ-point. Note that it is common to refer to the k_z-direction as the direction of growth (and hence confinement).

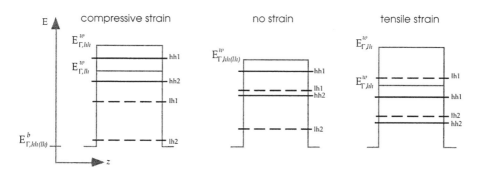

Figure 8.15.   Illustration of the positions of confined states in strained and unstrained QW structures. Note that in the left and right figures, only the well material is assumed to be strained, which can be seen by considering the bulk band edge labels to the left of each figure.

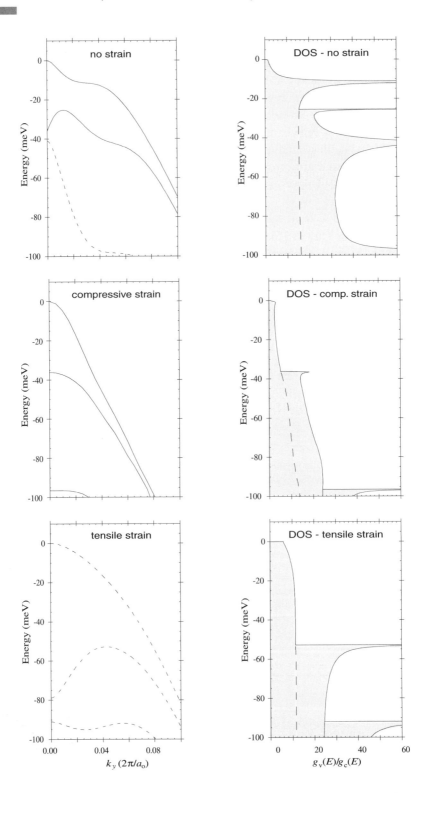

the difference in effective masses, quantum confinement gives rise to a distinct set of heavy- and light-hole states with the heavy-hole state ($hh1$) being the highest quantized state, as shown in the middle plot of Fig. 8.15. Now, when the well material is strained, we still, of course, have the confinement effects, but in addition we have the consequences of the non-degenerate bulk band edges, $E_{\Gamma,hh}^{w}$ and $E_{\Gamma,lh}^{w}$.

For a layer under compressive strain, the heavy-hole band edge ($E_{\Gamma,hh}^{w}$) is higher in energy. Thus, even with the quantized states occurring at the same energies from their respective band edges, it is clear that the separation between $hh1$ and $lh1$ will increase, while still leaving $hh1$ as the highest valence state. In a sense, compressive strain acts in the same manner as the effect of confinement. Similar reasoning can also describe a well layer under tensile strain, although here the band edge shifts work against the direction of confinement and so, for increasing tensile strain, initially the $hh1$–$lh1$ separation would decrease to zero but then increase again as $lh1$ becomes the highest quantized state.

The effect of strain on the highest valence band states has profound consequences for the in-plane $E(\mathbf{k})$ dispersions and, consequently, for the DOS. In Fig. 8.16, calculations of the in-plane $E(\mathbf{k})$ dispersions for three types of QW structure are presented, i.e. unstrained, compressive and tensile strained QWs. A realistic treatment for the valence band, via a multi-band effective mass approach, has been used. Alongside each dispersion, the resulting valence band DOS, $g_{v}(E)$, is plotted, which is expressed as a ratio to the DOS of the conduction band state, $g_{c}(E)$.

If we first consider the highest quantized state ($hh1$) in the unstrained case, the calculated dispersion is seen to be non-parabolic, which is a consequence of the close proximity of the other states (particularly $lh1$). Strictly speaking the band is only purely heavy-hole at the band edge ($k_{y} = 0$), for $k_{y} \neq 0$ the contributions from the other bands result in a mixed states (i.e. partly heavy- and light-hole). It is this in-plane mixing which leads to the non-parabolic behaviour. The effect of in-plane mixing on the DOS is quite dramatic: near the band edge $g_{v}(E) = 3 \times g_{c}(E)$. However, as the in-plane mixing becomes significant the total DOS ratio is seen to increase rapidly. Now the problems associated with a large $g_{v}(E)$ have already been noted (Fig. 8.13). Then to minimize $g_{v}(E)$, it appears we need to minimize the in-plane mixing, i.e., increase the separation

---

Figure 8.16.   (Left) Examples of in-plane valence subband structure from three types of In$_x$Ga$_{1-x}$As/InP quantum wells taken along k_y. In each dispersion, the solid lines represent heavy-hole subbands and the broken lines are light-hole subbands. From top to bottom: unstrained (80 Å/$x = 0.53$), 1.2% compressive strain (80 Å/$x = 0.28$) and 1.6% tensile strain (120 Å/$x = 0.68$). (Right figures) The resulting density of states is shown (with respect to the first conduction state); the dotted line indicates the DOS from the highest subband only. Note that zero slopes in the dispersion result in the poles seen in the DOS (cf. equation (8.20)). To a large extent these would be softened if broadening were included.

between the highest quantized states in the valence band. As we have seen, this is exactly what happens in a strained QW (Fig. 8.15). The other plots in Fig. 8.16 explicitly show this for both compressive and tensile strain, where the in-plane mixing on the highest state has been suppressed, resulting in a parabolic subband dispersion and, consequently, reducing the DOS. In fact, the total DOS begins to look more like the ideal step-like function we expect from 2D structures (cf. Fig. 8.11).

It is worth noting that, in compressive strain, a reduction in DOS will always occur, regardless of well width, since confinement and the strain-induced band edge shifts act in the same way. In tensile strained QWs, the DOS is very sensitive to the choice of well width and the amount of strain. The lighter in-plane mass and, consequently, smaller $g_v(E)$ found in compressive strain simply reflects the anisotropy in the bulk structure, where a heavy hole in the confinement direction has a light-hole type dispersion in the plane of the layers where the DOS is calculated.

Before we go on to consider our final subject, namely the effects of interband matrix element, it is useful to clarify how the reduction in valence band effective mass via strain governs the optical gain properties. In a parabolic subband picture, the maximum optical gain obtained from a QW layer can be expressed as

$$g_{\max} \propto \frac{m_c^* m_v^*}{m_c^* + m_v^*} \tag{8.23}$$

with m_c^* and m_v^* referring to the effective masses near the band edge (see the equation (8.15)). For a symmetric band structure the transparency carrier density may be expressed as $n_{\mathrm{tr}} \propto \sqrt{m_c^* m_v^*}$ and, consequently, the differential gain can be written as

$$\frac{dg}{dN} \propto \frac{\sqrt{m_c^* m_v^*}}{m_c^* + m_v^*} \tag{8.24}$$

The transparency and threshold currents are, therefore, expected to reduce, while the differential gain should increase as the valence band effective mass is reduced. We mentioned earlier that for equal effective masses, both transparency and differential gain will be optimal, although in practical structures, $m_c^* = m_v^*$ is generally not achieved (as our results above show $g_v(E)/g_c(E) \neq 1$, though it comes close in the case of tensile strain). Nevertheless, measurements on numerous strained QW laser structures have indeed verified the general trends suggested by these simple expressions (Thijs et al., 1994).

8.4.1   Optical Interband Matrix Element

The interband matrix element M_T which appeared in (8.21) and governs the strength of the optical transitions is discussed more fully in Section 6.8. For our purposes we will consider $|M_T|^2$ as some multiplying constant in the gain expression (which for transitions close to the band edge is not a bad approxima-tion). Its magnitude is largely determined by a fundamental constant of the material ($\mathbf{P}^2$), but it also depends on the overlap of the electron wavefunctions between the transition states and the polarization of the optical field.

For the two types of active regions we have looked at, the following expressions for $|M_T|^2$ can be written in a way to highlight these various factors:

$$|M_T|^2 = \tfrac{2}{3}\mathbf{P}^2 \quad \text{(bulk)}$$

$$|M_T|^2 = \begin{cases} \tfrac{4}{3}\mathbf{P}^2|M_{e-lh,n}|^2 & \text{(QW; TM)} \\[2mm] \mathbf{P}^2(|M_{e-hh,n}|^2 + \tfrac{1}{3}|M_{e-lh,n}|^2) & \text{(QW; TE)} \end{cases} \tag{8.25}$$

where

$$M_{e-lh,n} = \int F_{lh,n}^* F_{e,n}\, \mathrm{d}z \tag{8.26}$$

and

$$M_{e-hh,n} = \int F_{hh,n}^* F_{e,n}\, \mathrm{d}z \tag{8.27}$$

As we have remarked, $\mathbf{P}^2$ can be regarded as a fundamental constant of the bulk material (which is most useful in calculating the band structure of solids). It is used to quantify interactions between states having s-type symmetry (like the conduction band) and p-type symmetry (like the states making up the valence band). Having said this, we note that its value is often determined experimentally. The other terms that appear in the QW expressions are the electron wavefunc-tions, $F_{i,n}$ (or actually the envelope function since all the expressions are derived within an effective mass approach, as discussed in Section 6.2). The label i denotes the states the envelopes refer to, i.e. light hole (lh), heavy hole (hh) and conduction band (e), whereas n refers to the subband index. In fact, the integral part of the expressions highlight selection rules governing an allowed transition, i.e., an opti-cal transition will only take place when the integral is non-zero. For a symmetric potential (as we have considered the QW to be) the allowed transitions have $\Delta n = 0$. Hence, transitions $e1$–$hh1$, $e1$–$lh1$, $e2$–$hh2$, and so on, are expected while $e1$–$hh2$, $e2$–$lh1$, etc. are forbidden transitions.

In what follows, we will just concentrate on the lowest energy transition invol-ving carriers close to the band edge. Recall the situation in unstrained and com-pressive strained QW regions. The highest valence band is the heavy-hole state ($hh1$). Therefore $F_{lh,n} = 0$ here, so M_T is only non-zero for photons polarized perpendicular to the confinement direction (TE). That is, above transparency, carriers injected into the active region only contribute to the TE gain (ignoring spontaneous emission). The situation in tensile strain QW with a light hole ($lh1$) as the highest valence band state is that carriers contribute to both TE and TM gain but the strength of TM gain is considerable larger.

The approximations we assumed above are really not that bad for strained QW structures. In both cases the highest valence band is typically well separated from the other bands and the carriers are generally confined to small **k** values near the band edge (since the effective mass is reduced, cf. (8.15)). However, in unstrained QWs, the light-hole state ($lh1$) may not be too far (in energy) from $hh1$ at the top of the valence band. Therefore, the TM gain can be comparable to TE gain, possibly reducing the TE to TM mode discrimination in a laser. Therefore, we find another benefit of strained QW active regions, which is the suppression of either one of these polarization states.

We have now finished describing the factors which govern the optical gain in semiconductors. As a very brief summary we can roughly say that the carrier density required for a given value of gain tends to be lower in QW active regions than in bulk regions, and the introduction of strain helps improve carrier threshold density, differential gain and polarization properties. By illustrating how these improvements came about we hope to have justified the statements made at the beginning of Section 8.3.

8.5   Some other Laser Geometries

To date, the mainstream of laser research and development has centred around the production of in-plane lasers, that is, semiconductor lasers with an optical cavity parallel to the substrate with light emission occurring in the plane of the laser heterojunctions. Recently, however, a demand has arisen for lasers offering light emission orthogonal to the substrate. This change in geometry offers advantages in the optical interconnection of systems requiring a high degree of parallel information throughput (Craft and Felbblum, 1992; Banwell et al., 1993). A number of devices, fulfilling this purpose to various degrees, have emerged and we will briefly introduce these and discuss the features common to them and their important differences.

A range of different methods can be employed to produce lasers that emit in the vertical direction. Vertically emitting lasers are generically referred to as surface-emitting lasers or SELs. Figure 8.17 schematically shows some of the available devices. Most SELs employ a simple IPL structure. The laser output, however, is turned through 90° through the use of a reflector. In the simplest case this might be a simple 45° mirror formed by etching an angled facet on to the laser or into the surrounding material (see, e.g., Kim et al., 1990).

Figure 8.17(a) shows the case where a reflector is etched at a precise 45° to the laser in-plane cavity. In this case the light output from the laser is highly asymmetric. It is well known that the output from a standard IPL is highly divergent, because the light exiting the laser is apertured by the guiding layers such that when

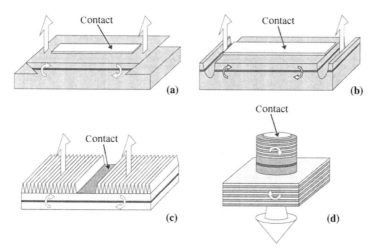

Figure 8.17.   Schematic diagram of a range of surface-emitting lasers.

exiting the laser it experiences diffraction effects. For a typical laser the effective aperture is approximately the cross-section of the active layer. Typically this is less than 1 μm in the vertical direction and 50 μm in the lateral direction. When this divergent output is incident upon a 45° reflector it simply continues to diverge. However, the reflector, with a reflectivity of order 30% (air/GaAs interface), will contain scattering imperfections due to the quality of the etch. This will diverge (scatter) the beam further.

Figure 8.17(b) shows an improved situation where, again, a reflector is etched to steer the beam into the vertical direction. In this case, however, the reflector has a parabolic form and is not simply a flat facet. The advantage in this case is that the reflector can compensate for some of the divergence introduced by diffraction at the IPL aperture. Again, this laser suffers from beam problems associated with the quality of the etched reflector – the reflector in this case is highly complex in both design and fabrication. It is noted that, because they can be fabricated to arbitrary length, these lasers can produce very high output powers – often a magnitude of many watts. They are therefore finding applications in optical pumping and cutting. In this case many devices may be run together in arrays.

Optical gratings may also be etched into the layers of an IPL to diffract the in-plane beam into the vertical direction (Fallahi *et al.*, 1992). This is shown schematically in Fig. 8.17(c). The etching of simple gratings into IPLs is a standard technique developed originally to control the in-plane modes of the laser. In such a scheme the grating is chosen to feedback only the wavelength desired for lasing. In more complex forms the grating is used to couple first-order light back into the laser, to lower threshold, and second-order light up into the vertical direction. Further complexity is introduced by angling the etched features to produce blazed structures. The vertical coupling efficiency of these gratings can be very high. As with the etched reflector structures the simple grating steered lasers suffer from asymmetric output beams. However, circular gratings may also be used. This allows for a circularly symmetric output beam, although in the near field the field distribution may contain minima in the lateral direction due to the top contact, and bondwires, effectively windowing certain regions of the output. In this case the quality of the grating structure determines the output beam quality. Grating lasers, in particular circular grating structures, can suffer problems associated with packing density and power output. In the first case many lasers cannot be placed in close proximity because of the large extent of the grating, and coupling between lasers must also be avoided. In the case of power the circular grating lasers are limited by the contact area that may be effectively accessed – the gain area is limited by the grating. The uniformity of the injected current may also be a problem.

An advantage common to most of the above geometries, and to IPLs in general, is that the light output from one facet of the laser may be used to monitor the output power of the laser. The other facet may then be used to output in the vertical direction. Integrated detectors may be fabricated from the same material as the laser, operated in a reverse-biassed scheme, which can help to control, via feedback, the stability of the source.

A fourth class of SEL involves actually growing the laser such that its primary optical cavity is vertical and not defined in the lateral direction. This relatively new geometry, known generically as the vertical cavity SEL (or VCSEL), offers many advantages over the other SEL types (Iga *et al.*, 1988; Jewell *et al.*, 1991). VCSELs utilize planar mirrors grown, or deposited, during laser production. The optical cavity and active layers are also defined during the growth phase, unlike IPLs where the cavity and gain length are defined through sample cleaving. By avoiding the need for cleaving of mirror facets, VCSELs gain an advantage in fabrication – the yield of IPLs depends critically upon the quality of the cleave.

Figure 8.18 schematically shows the structure of a VCSEL. The laser comprises a lower mirror, typically a distributed Bragg reflector (DBR) (discussed later), a cavity and active region and a top mirror. For a VCSEL in the InGaAs/GaAs,

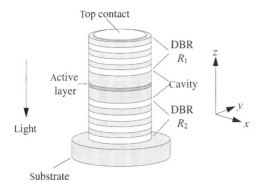

Figure 8.18.   Generic (bottom emission) index-guided VCSEL.

900–1000 nm, system the laser can be made to emit either vertically away from the GaAs substrate or down through the effectively transparent substrate. It is evident from this diagram that the active region of a VCSEL is very thin compared with the total structure. As with IPLs, QWs are used as the active region. However, unlike IPLs, light within a VCSEL is propagating along a normal to the plane of the QWs. Thus, only a narrow gain region is available, a feature which must be accounted for in the design of VCSELs (Corzine *et al.*, 1989).

For growth reasons the active region of a VCSEL cannot be made very thick. This is to be contrasted with IPLs, which may be cleaved to arbitrary length. The mirrors of the laser must therefore be made very highly reflective such that the small amount of gain may be compensated for. Following the IPL example in the previous sections, we may write down an expression for the threshold gain, g_{VC}, of a VCSEL with effective cavity length L_{VC}, average mirror reflectivity R_{VC} and active region volume confinement factor Γ_{VC} as

$$\Gamma_{VC}g_{VC} = \alpha_{VC} + \frac{1}{L_{VC}}\ln(1/R_{VC}) \tag{8.28}$$

It must be noted that the ratio of the VCSEL active region thickness to the cavity length is of the order of 1/100. For an IPL this ratio approaches unity. However, for an index-guided structure (a semiconductor pillar bounded by air) like the one in the figure, the lateral confinement factor approaches unity. The volume confinement factor is therefore $\sim 1\%$. This value is similar to that found within IPLs. If we now equate the threshold gain for an IPL and a VCSEL and assume that the optical losses within the two structures are similar (this is not a strictly valid assumption) we arrive at

$$\frac{1}{L_{VC}}\ln(1/R_{VC}) = \frac{1}{L_{IP}}\ln(1/R_{IP}) \tag{8.29}$$

The reader will recall that a typical IPL is of order 300 μm long with a facet reflectivity of 30%. By substituting these numbers into the above expression the average reflectivity required from the mirrors of a VCSEL may be derived. The value is in excess of 98%.

In order to achieve such a high reflectivity, special mirrors must be used. Typically these are DBRs which are grown into the laser structure. Post-growth deposition techniques may also be used. A DBR comprises a series of material layers of alternating high and low refractive index. The materials must be of extremely low loss for a reflectivity of 99% to be achieved. For an InGaAs-based device the DBR typically comprises alternating layers of GaAs (high index, ~ 3.5) and AlAs (low index, ~ 2.9) with each layer optically being a quarter of a wavelength thick. The combination of a single GaAs and AlAs layer is referred to as a period. Reflections at subsequent interfaces of the composite structure add in phase. By growing a number of periods, a layered structure with a high field reflectivity may be obtained. The magnitude of the reflectivity

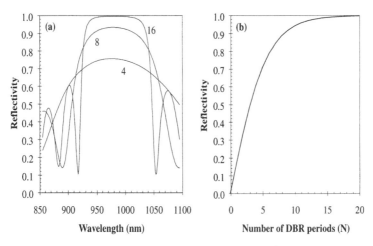

Figure 8.19.   Theoretical graphs of reflectivity for $\frac{1}{4}\lambda$ GaAs/AlAs (830 Å/696 Å) DBR. (a) and (b) show reflection spectrum and peak reflectivity vs. number of periods.

is dictated by the number of periods used and the ratio of the two refractive indices. The bandwidth of the DBR is also dictated by the index ratio. Figure 8.19 shows the reflectivity spectrum for a DBR centred at 980 nm. Note the flat top of the reflectivity spectrum and the 100 nm-wide stop-band (bandwidth) when the number of periods is 16. The figure also shows the effect of changing the number of DBR periods. All spectra are calculated using transfer matrix solutions to Maxwell's equations (see, e.g., Macleod, 1986).

DBRs typically have a thickness of ∼3 μm. The total VCSEL, with a cavity length of only a few wavelengths, is therefore of order 5–6 μm. With such a short optical cavity, VCSELs have a number of inherent features. Equation (8.3) gives the mode spacing of a VCSEL with a cavity of only one wavelength as ∼100 nm. If the primary mode is centred at 980 nm then, using the DBR of Fig. 8.18, the higher-order modes of the laser will experience a very low reflection at the DBRs. These higher-order modes will therefore not lase and so it can be seen that a VCSEL is inherently a single-longitudinal-mode laser. This is important for communications applications. It is noted that IPLs employ similar techniques through the use of in-plane gratings to eliminate higher-order modes. Because higher-order modes cannot lase, a VCSEL produces a very stable wavelength output over a wide temperature range. The laser will not hop between modes as the gain spectrum shifts with temperature. Figure 8.20 shows the calculated output from a VCSEL. The single mode is evident. Note how the DBR suppresses spontaneous emission around the main mode.

It is clear from the above discussions that VCSELs offer a number of advantages over IPLs. Firstly, because of the vertical cavity the laser may be tested before cleaving. Next the designer has a simple means of controlling the mirror reflectivities for the laser. This means that the laser may be optimized for power output or a low lasing threshold. Another distinct advantage of moving to a vertical cavity geometry is that the lateral dimensions of the laser can be controlled. This offers advantages in fibre coupling in which case the laser's dimensions can be tailored to match the fibre core. Alternatively, the use of small mesas allows the packing density of vertical cavity lasers to be made very high. This allows the fabrication of high-density two-dimensional arrays; in-plane lasers cannot be easily fabricated into such arrays (Fig. 8.21).

Because the effective output aperture of a VC laser is the mesa cross-section, the output of a VCSEL can be made circularly symmetric. This offers advantages for both fibre coupling and free space communications. The laser's divergence (generally 10°) can also be controlled by integrating lenses onto the output surface. This can result in output divergence angles of order 2°. It is noted, however, that for VCSELs with an in-plane dimension exceeding $\sim$10 μm a number of in-plane

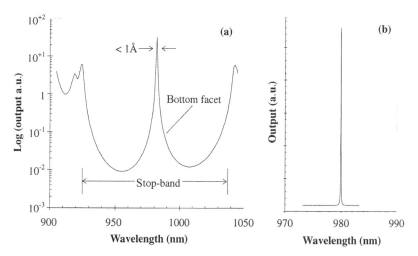

Figure 8.20.   (a) Emission spectra for 980 nm VCSEL at threshold. (b) Emission spectra on linear scale.

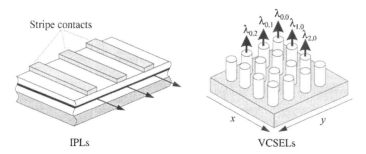

Figure 8.21.   Schematics of in-plane and vertical cavity laser arrays. IPLs cannot easily form 2D arrays.

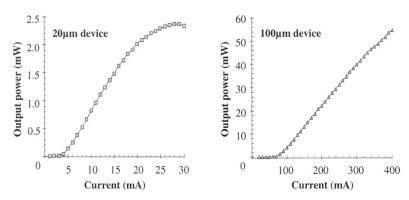

Figure 8.22.   Pulsed light output vs. current input curves for simple VCSEL.

or transverse modes can occur. This results in the appearance, in the emission spectrum, of a number of extra wavelengths. In general only higher-power VCSELs suffer from this effect, which can cause problems in fibre coupling. Figure 8.22 shows the power output characteristics for two VCSELs, one of 20 μm diameter and one of 100 μm diameter. The difference in output power is dramatic. Essentially this is an area scaling argument, although sidewall recombination and other mechanisms mean that this is not exactly the case. The higher-power laser in this figure is operating in a multi-transverse mode manner.

Having shown some of the advantages of VCSELs we must also mention some of the disadvantages. VCSELs, because of their complex multilayer structure, suffer a number of thermal problems (Scott *et al.*, 1993). These are primarily due to the high series resistance of the DBRs through which current is injected. Figure 8.22 shows that at high injection levels the power output of the VCSEL is saturating and turning over. This is because of the series resistive heating effect occurring within the DBRs. The heating de-tunes, in wavelength, the peak of the gain spectrum and the primary FP mode of the cavity. Heating also increases the leakage currents within the active region.

The VCSEL is probably the most sophisticated of semiconductor lasers today. In concluding, it is worth comparing its sophisticated structure exploiting quantum wells, strain and carrier confinement together with integrated Bragg structures, with the first lasers developed only 20 years earlier. The advances are, by any standards, remarkable.

EXERCISES

1.   The emission wavelength of a quantum-well laser is governed largely by both the width and the bulk band gap of the well material.

(a) Assuming that the quantum well has infinitely high barriers, derive an expression to find the required well width to obtain lasing at a given energy.

(b) A well material has a band gap of 0.75 eV at room temperature and effective masses of 0.05 m_0 and 0.5 m_0 for the conduction and valence bands, respectively. Using the result in (a), find the well widths required to obtain lasing at 1.55 µm and at 1.3 µm. (These wavelengths are commonly used for long-haul optical fibre communication systems.)

2. Suppose that the quantum well in Exercise 1 now has finite potential barriers of 220 meV and 330 meV for the conduction and valence bands, respectively.

(a) Qualitatively discuss how the introduction of finite barriers will change the results found in Exercise 1(b). In particular, comment on which energy state, i.e. the electron or hole state, is more likely to be strongly affected by finite barriers.

(b) Assuming the barrier material has the same effective masses as the well material, compute the well width required to obtain lasing at 1.55 µm.

3. Given the following parameters of a 1550 nm InGaAsP laser, $g_{th} = 35$ cm^{-1}, $\alpha = 10$ cm^{-1}, and $n_{av} = 3.3$, calculate the spacing between laser modes if the laser facets are uncoated. If the facets are subsequently coated to give effective reflectivities of 90% and 70%, what is the change in the threshold gain?

4. Starting with equation (8.11), derive an expression for the output power in terms of the cavity length L, the facet reflectivity R (assuming equal reflectivities), and the loss α. Hence, show that the external quantum efficiency, η_{ext}, of a laser can be expressed in the following form:

$$\eta_{ext} = \eta_{inj} \left[\frac{\alpha L}{\ln(1/R)} + 1 \right]$$

5. The threshold current density of a semiconductor laser is found to obey the formula (cf. (8.14))

$$J_{th}(T) = J_{th}(T_0) \exp(T/T_0)$$

(a) Discuss whether you would seek to design a laser with a high or low value of T_0.

(b) Find out whether T_0 increases or decreases with band-gap and comment on the observed behaviour.

6. Consider equation (8.24) for the differential gain, dg/dN. Sketch the band structure (valence and conduction bands) when maximum differential gain is achieved in an ideal semiconductor.

7. The maximum reflectivity R_{max} of a Bragg mirror at the centre wavelength is dependent on the ratio $r \sim n_A/n_B$ and the number of layer pairs, N, according to

$$R_{max} = \left[\frac{(n_m/n_s) - r^{2N}}{(n_m/n_s) + r^{2N}} \right]^2$$

where n_m is the refractive index from which light is incident and n_s is the index of the exit medium. Use this expression to design two mirrors for a 980 nm VCSEL using GaAs and AlAs containing (a) one InGaAs/GaAs quantum well of width 8 nm, and (b) three InGaAs/GaAs wells. Sketch the basic structure of the VCSEL, neglecting any absorption and scattering loss. Assume a value of $3000\,\text{cm}^{-1}$ for the gain.

References

G. P. Agrawal, *Long Wavelength Semiconductor Lasers* (Van Nostrand Reinhold, New York, 1986).

T. C. Banwell, A. C. V. Lehman and R. R. Cordell, *IEEE J. Quantum Elect.* **29**, 635 (1993).

G. Bastard, *Wave Mechanics Applied to Semiconductor Heterostructures* (Les Editions de Physique, Paris, 1988).

M. G. Burt, *J. Phys. Condens. Matter* **4**, 6651 (1992).

L. A. Coldren and S. W. Corzine, *Diode Lasers and Photonic Integrated Circuits* (Wiley, New York, 1995).

S. W. Corzine, R. S. Geels, J. W. Scott, R. H. Yan and L. A. Coldren, *IEEE J. Quantum Elect.* **25**, 1513 (1989).

N. C. Craft and A. Y. Felbblum, *Appl. Optics* **31**, 1735 (1992).

M. Dagenais, R. F. Leheny, H. Tempkin and P. Battacharya, *J. Lightwave Technol.* **8**, 846 (1990).

M. Fallahi, F. Chatenoud, I. M. Templeton, M. Dion and M. Wu, *IEEE Phot. Tech. Lett.* **4**, 1087 (1992).

S. Hansmann, *IEEE J. Quantum Elect.* **28**, 2589 (1992).

K. Iga, F. Koyama and S. Kinoshita, *IEEE J. Quantum Elect.* **24**, 1845 (1988).

J. L. Jewell, J. P. Harbison, A. Scherer, Y. H. Lee and L. T. Florez, *IEEE J. Quantum Elect.* **27**, 1332 (1991).

J. H. Kim, R. J. Lang, A. Larsson, L. P. Lee and A. A. Narayanan, *Appl. Phys. Lett.* **57**, 2048 (1990).

T. L. Koch and U. Koren, *J. Lightwave Technol.* **8**, 274 (1990).

H. A. Macleod, *Thin Film Optical Filters*, 2nd edn (Adam Hilger, London, 1986).

E. Merzbacher, *Quantum Mechanics* (Wiley, New York, 1970).

J. E. Midwinter, *Physics Technology* **19**, 101 (1988).

A. Nurmikko and R. L. Gunshor, *Physica Scripta* **T68**, 72 (1996).

E. P. O'Reilly, *Semicon. Sci. Technol.* **4**, 121 (1989).

E. P. O'Reilly and A.R. Adams, *IEEE J. Quantum Elect.* **30**, 366 (1994).

K. Prior, *Comtem. Phys.* **5**, 345 (1996).

S. Schmitt-Rink, D. S. Chemla and D. A. B. Miller, *Adv. Phys.* **38**, 89 (1989).

J. W. Scott, R. S. Geels, S. W. Corzine and L. A. Coldren, *IEEE J. Quantum Elect.* **29**, 1295 (1993).

S. M. Sze *Physics of Semiconductor Devices*, 2nd edn (Wiley, New York, 1981).

P. J. A. Thijs, L. F. Tiemeijer, J. J. M. Binsma and T. van Dongen, *IEEE J. Quantum Elect.* **30**, 477 (1994).

C. Weisbuch and B. Vinter, *Quantum Semiconductor Structures: Fundamentals and Applications* (Academic Press, New York, 1991).

P. S. Zory, Jr, *Quantum Well Lasers* (Academic Press, New York, 1993).

Mesoscopic Devices

T. J. Thornton

9.1   Introduction

The semiconductor industry is a massive business and has a seemingly inexhaustible appetite for new devices, new materials and new applications. A single type of device can open up markets worth hundreds of millions of dollars if it can fill a suitable gap or demonstrate superior performance to those currently available. The rewards are even greater if the same device can then be integrated with many others. The high-electron mobility transistor, or HEMT (Chapter 10), is a good example of such a device. Conceived in the late seventies (Dingle *et al.*, 1978), it was the focus of substantial research and development during the early 1980s (Mimura *et al.*, 1980) and is now arguably the most important element in high-speed, low-noise communications systems such as those used in direct broadcast satellite television.

With such a powerful driving force it is not surprising that so much effort is devoted to researching new semiconductor device technologies. Low-dimensional systems have received particular attention during the last ten or fifteen years and as a result some elegant physics has emerged. It is therefore natural for physicists and engineers to explore ways in which the unique properties of low-dimensional structures might be exploited in devices of the future. Because these structures are usually in the size regime which lies somewhere between the microscopic world of atoms and the macroscopic world we live in, they are often lumped together under the single title of *mesoscopic devices*.

Given the scope of this chapter it is impossible to describe all the mesoscopic devices that have been proposed. Instead we will consider a few examples which represent particular classes of device. These examples will follow something of a historical perspective, starting with quantum interference transistors, which are intended to exploit the wavelike properties of electrons and were some of the first mesoscopic devices to be considered because of their obvious analogies with optical devices. As material quality improved it became clear that the electron mean-free path could easily exceed the minimum feature size of the device and various ballistic electron devices were proposed. Electron tunnelling has promised exciting device applications ever since the development of the Esaki tunnel diode, but the majority of these structures were 2-terminal devices. Transistors based on negative differential resistance would require 3-terminal devices and resonant tunnelling through quantum dots of variable diameter might be one way to achieve this. The last and perhaps most promising class of mesoscopic device is based on the so-called Coulomb blockade. The Coulomb blockade is used to control the flow of single electrons in circuits which are known collectively as single electronics.

Before discussing the proposed operation of each device we shall look at the relevant physics involved. It turns out that for each class of device mentioned above the physics is dominated by one important length scale or energy scale. Other physical phenomena can occur but may be neglected on the first treatment as they only contribute second-order effects to the device action. For the ballistic electron devices the relevant length scale is the mean free path of the electron, while for the resonant tunnelling and Coulomb blockade devices the pertinent energies are the quantum confinement energy and the charging energy, respectively. But we shall begin our look into mesoscopic devices by starting with quantum interference transistors where the important length scale is the phase coherence length.

9.2   Quantum Interference Transistors

A convenient picture of electron transport in a weakly disordered degenerate metal is that of a non-interacting, point-like electron which propagates at the Fermi velocity between random scattering events. As we shall see in Section 9.3 this picture works well for the case of ballistic electron transport but fails to take into account the wavelike nature of the electron. At low temperatures, electron interference can take place and we have to make corrections to the Drude conductivity. The first correction to consider is often called weak localization (Chapter 5) and the associated negative magnetoresistance provides a useful way of determining the phase coherence length, which is the relevant length scale for quantum interference devices. A useful way to tune the quantum interference is by means of the Aharonov–Bohm effect and we shall look at both the magnetic and electrostatic variants. Another phenomena of relevance to quantum interference devices is the fluctuation in the conductance of nominally identical structures due to the random impurity distribution. The fluctuations in conductance have a root-mean-square value of the order of e^2/h, regardless of the degree of disorder in the sample and are known as universal conductance fluctuations. We shall consider each of these in turn before looking in detail at quantum interference devices.

9.2.1   Quantum Interference and Negative Magnetoresistance

To understand quantum interference it is important to appreciate the difference between elastic and inelastic scattering events. An elastic collision, for example, with a charged impurity, will scatter the electron into a new momentum state, i.e. it will move in a different direction but the magnitude of the momentum and the energy will be the same before and after the collision. On the other hand, when an electron suffers an inelastic collision with, for example, a phonon, its energy will change in a discontinuous and random fashion. The distinction between the two types of collisions has important consequences for the phase information carried by the electron. The phase of an electron wave will develop according to its phase velocity, $v_\varphi = \omega/k$, which can be rewritten as $v_\varphi = E/p$, i.e. the ratio of the energy to the magnitude of the momentum. After an elastic collision v_φ is unchanged and the phase memory is preserved. However, after an inelastic collision both E and p change in an essentially random fashion and the phase information is lost. The two scattering mechanisms are often characterized by different scattering rates given by τ_e^{-1} and τ_φ^{-1} where τ_e is the elastic mean-free time, i.e. the average time between elastic collisions, and τ_φ is the phase coherence time, i.e. the average time between phase breaking (inelastic) collisions. We are interested in the regime where $\tau_\varphi > \tau_e$, and invariably this means low temperatures, where the probability of an electron scattering with a phonon is less than that with a charged or neutral impurity. It is often convenient to think in terms of the corresponding length scales, i.e. the mean-free path $\ell_e = v_F \tau_e$ and the phase coherence length $\ell_\varphi = (D\tau_\varphi)^{1/2}$, where v_F is the Fermi velocity and D is the diffusion coefficient given by $D = \frac{1}{2} v_F^2 \tau_e$ in two dimensions (Exercise 1).

Some of the first observations of quantum interference were made using thin metal films (Bergmann, 1979) and silicon inversion layers (Kawaguchi *et al.*, 1978). In these systems the quantum interference is said to be two-dimensional (2D) because the phase coherence length is larger than the thickness of the conducting layer. In what follows we shall mostly be concerned with 2D systems although the same ideas can be applied to three-dimensional (3D) and one-dimensional (1D) systems. At liquid helium temperatures the conductance, g, of thin films and inversion layers decreases as $\ln T$ and shows a pronounced negative magnetoresistance at low fields. The decrease in conductance was at first thought to be a precursor of strong localization (Abrahams *et al.*, 1979) for which a 2D system with finite disorder would become an insulator at zero temperature (Chapter 5). As a result both the $g \sim \ln T$ behaviour and the negative magnetoresistance became known as weak localization and this name is still used even though we now know that both effects are a result of quantum interference. The formal theory of weak localization is well developed but Bergmann (1983) gave a physical description which brings out clearly the unique features of the quantum interference and it is this approach we outline below.

Consider an electron wave propagating from a point A to B as shown in Fig. 9.1(a). On the way it will suffer elastic collisions but we will assume that its phase is not destroyed by an inelastic event and that its last phase-breaking event occurred momentarily before it arrived at A. There are an enormous number of possible trajectories for the electron to take and we should consider averaging

(a)

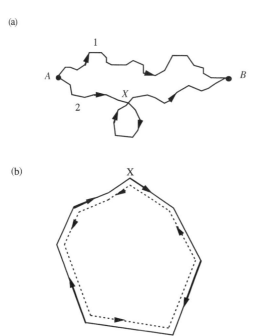

(b)

Figure 9.1.   (a) Electron propagation in a 2D system. Electrons which reach B by a self-intersecting path (2) will contribute to the weak localization but those which do not intersect (1) only contribute to the Drude conductance. (b) The self-intersecting paths can follow time-reversed trajectories and will suffer constructive interference at the origin X.

over all possible paths. The path lengths will vary and electrons which reach B by different paths will have no fixed phase relation – for every electron pair that is in phase at B there will be another pair π out of phase and we should sum them incoherently, i.e. add their intensities. However, there is a special set of trajectories for which the phase relationship is well-defined and electron interference can take place. The special trajectories are those which follow time-reversed paths which self-intersect at the point X. A pair of time-reversed paths are shown in Fig. 9.1(b). The electrons in each path follow the same trajectory but in opposite directions and suffer the same elastic scattering events but in the reverse order. As a result, when the electrons return to point X they have exactly the same phase and the two wavepackets add coherently. The coherent superposition results in constructive interference and the probability of finding the time-reversed electrons at the point X is twice as likely as compared with other electrons which return to X via non-time-reversed paths. Because the probability of finding the electron somewhere in between A and B has been enhanced, its chances of actually reaching point B have been reduced. The quantum interference has, therefore, slightly increased the resistance of the sample.

At first sight it might seem that the probability of two electrons following time-reversed paths is so small that the effect they have on the conductance would be negligible. But we must remember that the number of conduction electrons is enormous and that they are sampling all available trajectories. In addition, the uncertainty principle slightly relaxes the constraints on exact time-reversal and overall the quantum interference correction to the conductivity can contribute as much as 10% to the total resistance at low temperatures. The quantum interference corrections have been derived rigorously by a number of authors (Anderson *et al.*, 1979; Abrahams and Ramakrishnan, 1980; Gorkov *et al.*, 1979) but for the sake of some physical insight we shall instead give a heuristic argument based on Bergmann's picture of weak localization (Bergmann, 1983).

Consider an electron in a two-dimensional electron gas in the state $+k$ marked in Fig. 9.2. It is free to scatter around the Fermi circle which is broadened by an amount $\Delta k_F = \pi/\ell_e$ due to elastic scattering. Therefore the available area for backscattering in **k**-space is $2\pi k_F \Delta k_F = 2\pi^2 k_F/\ell_e$. During a time t greater than several τ_e the electron will have diffused a distance $L = (Dt)^{1/2}$ and, provided $L < \ell_\varphi$, all the states within $q = 1/(Dt)^{1/2}$ of the state at $-k$ will interfere constructively. The proportion of those electrons which suffer coherent backscattering, I_{coh}, is given by

$$I_{coh} = \frac{\pi q^2}{2\pi^2 k_F/\ell_e} = \frac{\ell_e}{2\pi k_F Dt} \tag{9.1}$$

In the Drude approximation we switch on an electric field at time $t = 0$ and consider how the Fermi circle is displaced with time. After a time Δt the current will have increased to ΔJ where

$$\Delta J = ne\Delta v = ne(\Delta p/m) = ne^2 E \Delta t/m \tag{9.2}$$

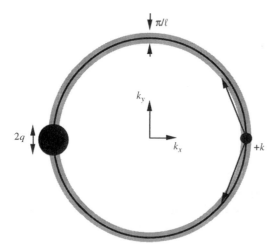

Figure 9.2.   Electron scattering around the Fermi circle in a 2DEG. The spot representing coherent back-scattering has been enlarged for clarity.

where Δp is the change in momentum due to the impulse $eE\Delta t$. Usually we would neglect quantum interference and integrate from $t = 0$ to $t = \tau_e$ to obtain the Drude conductivity $\sigma = ne^2\tau_e/m$. However, for times $\tau_e < t < \tau_\varphi$, the effect of the quantum interference is to reduce the total momentum that contributes to the integral by an amount $1 - I_{coh}$. The expression for the conductivity now becomes

$$\sigma = \frac{ne^2}{m}\int_0^{\tau_e}\mathrm{d}t - \frac{ne^2 v_F \tau_e}{2\pi m k_F D}\int_{\tau_e}^{\tau_\varphi}\frac{\mathrm{d}t}{t}$$

$$= \frac{ne^2\tau_e}{m} - \frac{e^2}{2\pi^2\hbar}\ln(\tau_\varphi/\tau_e) \tag{9.3}$$

where we have made use of the fact that in two dimensions the Fermi wavevector is given by $k_F = (2\pi n)^{1/2}$. As a rule, the phase coherence time follows a power law dependence of the form τ_φ^{-p} and the conductivity decreases as $\ln T$.

In zero magnetic field the quantum interference reduces the conductivity by an amount, $\delta\sigma = -(e^2/2\pi^2\hbar)\ln(\tau_\varphi/\tau_e)$. However, this weak localization term is easily suppressed by a magnetic field because, as we shall see below, the Aharonov–Bohm effect changes the phase of two counter-propagating electrons. For the time-reversed electrons in Fig. 9.1(b) the change in phase is given by $\delta\varphi = 2(e/\hbar)BA$ where A is the area of the loop. For the special case of $B = 0$ all time-reversed electron pairs will return to X with zero phase difference and will interfere constructively. However, for $B > 0$ the phase difference associated with each pair of time-reversed electrons is no longer zero and will vary depending on the area enclosed. As a result, different pairs of electrons will experience different degrees of constructive or destructive interference and for large enough magnetic fields the quantum intereference averages to zero.

The suppression of the weak localization by a magnetic field can be observed by measuring the low field magnetoresistance. As the field is increased, the conductivity correction $\delta\sigma(B)$ gets smaller and the resistance decreases until it approaches the Drude value. The field required to quench the weak localization is given by $B_c \sim \hbar/4eD\tau_\varphi$. Hikami et al., (1980) have calculated the variation of $\delta\sigma(B)$ for a two-dimensional system, with the result

$$\delta\sigma = \frac{e^2}{2\pi^2\hbar}\left[\Psi\left(0.5 + \frac{\hbar}{4eB\ell_\varphi^2}\right) + \ln\left(\frac{4eB\ell_\varphi^2}{\hbar}\right)\right] \tag{9.4}$$

where $\Psi(x)$ is the digamma function. The corresponding results have been derived for 1D and 3D systems. Equation (9.4) is important because we can use it to extract the phase coherence length, ℓ_φ, by fitting the equation to the experimental results. In Fig. 9.3(a) we show the magnetoresistance of the 2DEG in a modulation-doped Si:SiGe quantum well (Prasad et al., 1995). A pronounced negative magnetoresistance can be seen at low fields before the development of Shubnikov–de Haas oscillations above 1 Tesla. Fig. 9.3(b) shows the low-field

a)

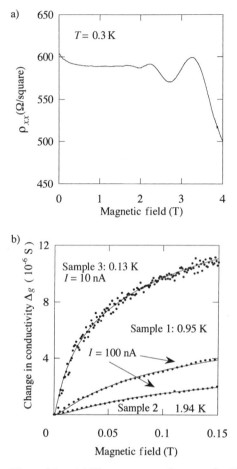

b)

Figure 9.3. (a) The magnetoresistance of a Si:SiGe quantum well measured at a temperature of 0.3 K. (b) The fit to the low-field magnetoconductance data at three different
temperatures (Prasad *et al.*, 1995). (Courtesy T. J. Thornton)

conductance and the fit to equation (9.4). The fit is excellent and we can extract a
phase coherence length of 0.45 μm at a temperature of 0.13 K.

The size of ℓ_φ governs the operation of quantum interference transistors and it
is important that this quantity be as large as possible. Ikoma *et al.* (1992) have
measured the phase coherence time, $\tau_\varphi = \ell_\varphi^2/D$, for a 1D quantum wire as a
function of temperature by a number of techniques and a typical set of results
for a GaAs sample is reproduced in Fig. 9.4. The data in the figure can be
explained by considering the various mechanisms which are responsible for inelastic electron scattering and combining them in the spirit of Matthiessen's rule to
obtain the total scattering rate. At high temperatures electron–phonon scattering
dominates and is very efficient at destroying the phase memory. However, at
lower temperatures ($T < 100$ K) electron–electron scattering around the Fermi

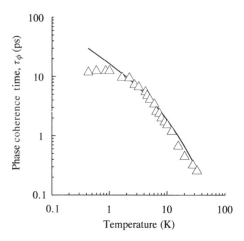

Figure 9.4.   The phase coherence time as a function of temperature as measured in a GaAs quantum wire (Ikoma *et al.*, 1992). (Courtesy K. Hirakawa)

surface takes over and leads to the well known Landau–Baber scattering rate, $1/\tau_{ee} \propto (kT/E_F)^2$. The Landau–Baber scattering (Ashcroft and Mermin, 1976) involves a pair of electrons colliding with a relatively large transfer of energy between them ($\sim kT$). At still lower temperatures, a new mechanism becomes favourable, one in which only a small transfer of energy is involved but after a few collisions the phase is randomized. This mechanism has been called Nyquist scattering because it can be likened to a single electron scattering from the random, noise-like fluctuations in the background charge due to all the other electrons. For a quantum wire in the Nyquist regime the scattering rate is given by $1/\tau_{ee} \sim T^{2/3}$ (Altshuler *et al.*, 1982). Combining the Landau–Baber and Nyquist scattering rates gives the total phase-breaking scattering rate as

$$1/\tau_\varphi = AT^2 + BT^{2/3} \tag{9.5}$$

The data in Fig. 9.4 agree well with equation (9.5) for temperatures above 2 K but τ_φ appears to saturate at lower temperatures for reasons which are still not well understood. The saturation value of τ_φ is about 10 ps which corresponds to a phase coherence length of 0.7 μm. Any device which relies on phase coherence will therefore require active dimensions much less than about 1 μm especially if they are to operate at temperatures higher than 1 K.

9.2.2   The Aharonov–Bohm Effect

The Aharonov–Bohm (AB) effect (Aharonov and Bohm, 1959) provides us with a mechanism for tuning the quantum interference between electron waves by means of an electric or magnetic field. The magnetic AB effect has been demonstrated in metal (Webb *et al.*, 1985; Washburn and Webb, 1986) and semiconductor rings

(Timp *et al.*, 1987; Ford *et al.*, 1989) and the simplest geometry to consider is that of a ring as shown in Fig. 9.5(a). The circumference of the ring is assumed to be much less than ℓ_φ to ensure that no inelastic scattering events randomize the phase. An electron wave which enters the ring from the left will split into two partial waves which propagate along the lower and upper arms of the ring. In the ideal case, the amplitudes of the two partial waves and the distance they travel are the same and, in the absence of a magnetic field, they exit the ring with identical phases and therefore interfere constructively. The constructive interference localizes the electron at the output and the probability of transmission through the ring has been enhanced. The quantum interference has therefore reduced the resistance of the ring. We now switch on a magnetic field applied perpendicular to the ring. In 1959 Aharonov and Bohm (Aharonov and Bohm, 1959) showed that the magnetic vector potential, **A**, changes the phase of the electron by an amount

$$\Delta\varphi = -(e/\hbar) \int \mathbf{A} \cdot \mathbf{dl} \tag{9.6}$$

where **dl** is an element of the path length. From the identity

$$\int \mathbf{A} \cdot \mathbf{dl} = \int \nabla \times \mathbf{A} \cdot \mathbf{ds} = \int \mathbf{B} \cdot \mathbf{ds} \tag{9.7}$$

(a)

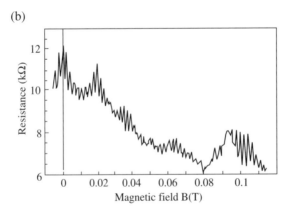

(b)

Figure 9.5.   (a) A typical ring geometry used for investigation of the AB effect and (b) magnetoresistance oscillations in a split-gate defined AB ring for a range of confining bias from -0.7 to -1.01 volts (Adapted from Ford *et al.*, 1989). (Courtesy T. J. Thornton) we can write the phase difference between the waves when they exit the ring as $\Delta\varphi = -(e/\hbar)B\pi r_i^2$ where r_i is the internal radius of the ring. By increasing the magnetic field we can continuously vary the phase difference and each time $\Delta\varphi$ changes by 2π the electron interference moves through a complete cycle from constructive to destructive and back again. The variation in the electron interference means that the resistance of the ring is now an oscillating function of magnetic field and the period is $\Delta B = h/e\pi r_i^2$ which corresponds to increasing the flux through the ring by an amount h/e.

Ford *et al.*, (1989) have observed very clear AB oscillations in a GaAs split-gate defined ring and the results are reproduced in Fig. 9.5(b). The period of the oscillations is 0.21 mT corresponding almost exactly to the measured internal diameter. It is also clear from the figure that the AB oscillations are superimposed on an aperiodic background. The background fluctuation is reproducible and caused by the magnetic field penetrating the arms of the ring as described below in the section dealing with universal conductance fluctuations.

The electrostatic AB effect relies on an electric potential, V, to modulate the phase and is more attractive for device applications because it is easier to apply an electric rather than a magnetic field. The electron phase is now given by $\Delta\varphi = (e/\hbar)\int^\tau V\,dt$. The upper bound to the integral, τ, is the time for which the electron experiences the electric potential or the phase coherence time, τ_φ, whichever is the shorter. Equating $\Delta\varphi$ to 2π gives the voltage period of the electrostatic AB oscillations as $\Delta V = h/e\tau$.

In an attempt to observe the electrostatic AB effect Washburn *et al.*, (1987) placed a sub-micron antimony metal ring between the electrodes of a parallel plate capacitor as shown schematically in Fig. 9.6. With no bias applied to the capacitor electrodes, the ring displayed very clear magnetic AB oscillations (Fig. 9.6(b)). When an electric field was established between the plates of the capacitor the

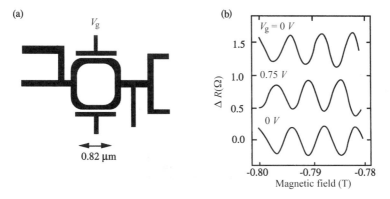

Figure 9.6.   (a) A schematic diagram of the square ring used by Washburn *et al.* (1987) to explore the electrostatic AB effect. (b) When an electric field is established between the capacitor electrodes the phase of the magnetoresistance oscillations could be changed by 180°. V_g is the gate voltage. (Courtesy S. Washburn)

phase of the oscillations could be reproducibly shifted by π radians. From measurements of the phase coherence time Washburn *et al.*, (1987) estimated that a potential difference of approximately $40\,\mu V$ would be sufficient to change the phase by this amount. However, the measured value was many orders of magnitude larger perhaps because the electric field was efficiently screened by the metal.

9.2.3   Universal Conductance Fluctuations

During the 1980s the search for the AB effect in the solid state proved to be a fascinating theoretical and technological challenge and eventually led to a greater understanding of transport in disordered systems. Of particular relevance was the development of the theory of what are now called universal conductance fluctuations which were first observed in very narrow metallic rings and wires but have now been confirmed in a wide range of material systems. Umbach *et al.*, (1984) published data from a $0.3\,\mu m$ wide wire ring made from gold. The data showed a rich structure in the magnetoresistance but despite tantalizing glimpses of periodic structure the fluctuations were mostly aperiodic and it was impossible to identify a single peak in the Fourier transform which could be said to originate from the h/e flux period. In many ways the fluctuations resembled noise in the measurement except they were remarkably reproducible provided the sample was not temperature cycled. We shall see that the aperiodic fluctuations are caused by the magnetic field penetrating the metal wires that make up the ring. The first rings had a fairly small aspect ratio, i.e. the diameter of the ring was only a few times the wire width and the flux penetrating the wires was comparable to that threading through the ring itself. The importance of large aspect ratio rings was demonstrated by Webb *et al.*, (1985), who used ring diameters almost 20 times larger than the wire width. The h/e oscillations in these samples were very pronounced but again an aperiodic fluctuating background could be observed.

The aperiodic conductance fluctuations are another manifestation of quantum interference in disordered metallic conductors which are small enough for phase coherence to be maintained over a significant portion of the structure. However, unlike the previous examples, where the interference was governed by a specific set of time-reversed trajectories or by the ring geometry of the sample, it is now the random impurity distribution that determines the conductance of the sample. The importance of the impurity distribution was made clear in a body of theoretical work starting with the numerical simulations of Stone (1985). In the simulations the samples were modelled on a mesh of 100×10 sites and the disorder could be introduced very conveniently by changing the local potential at each mesh point. The total disorder was chosen to give a mean sample conductance of $\sim 1.5e^2/h$ while the fluctuation in the conductance between samples with different impurity configurations was also of order e^2/h. In addition, changing the magnetic field or the Fermi energy in a single sample also produced fluctuations in the conductance of order e^2/h. Qualitatively, the amplitude and period of the fluctuations with B

or E_F resembled those observed in metallic wires and silicon MOSFETs despite almost three orders of magnitude difference in the mean conductance.

The numerical models showed in a very clear way that any measurement that involved the conductance of nominally identical samples would differ by $\sim e^2/h$ because of the microscopic variations in the impurity configuration. To see why the impurity configuration has such a profound effect consider the diagram in Fig. 9.7(a), which represents a quantum wire with a width and impurity separation which are both comparable to the mean free path. The electron waves will be scattered by the impurities and by the walls of the quantum wire and some of the electron trajectories will form well-defined loops. If the circumference of the loop is smaller than the phase coherence length the electron waves will interfere at the points where the loops intersect and the conductance of nominally identical samples will fluctuate from device to device.

(a)

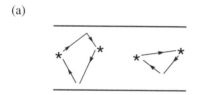

(b)

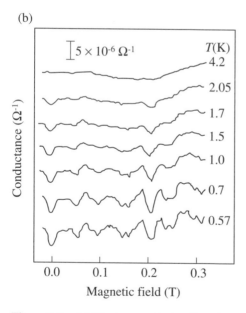

Figure 9.7.   (a) Electron scattering from impurities can result in well-defined loop-shaped trajectories. (b) Universal conductance fluctuations in the magnetoconductance of a GaAs quantum wire at different temperatures (Thornton *et al.*, 1987). (Courtesy T. J . Thornton)

Let us imagine that we now apply a magnetic field which penetrates the quantum wire. Each of the impurity-defined loops will behave as a microscopic Aharonov–Bohm ring. However the flux penetrating each loop will be different because of their different areas. As a result each microscopic loop will contribute a different Fourier component to the magnetoconductance and the result will appear as an aperiodic but reproducible magnetoconductance. The magnetoconductance for a GaAs quantum wire is shown in Fig. 9.7(b). A similar effect will occur if we vary the electron density in the wire. The Fermi wavelength will increase as the electron density decreases and this will alter the interference condition for each self-intersecting loop. As a result the conductance as a function of Fermi energy will appear as a fluctuating signal on a smoothly varying background as has been observed in quantum wire MOSFETs (Skocpol et al., 1986).

The theory of universal conductance fluctuations is now well developed and the root-mean-square fluctuation correlation function can be calculated for any geometry and at finite temperatures. At zero temperature the root-mean-square amplitude of the fluctuations is given by

$$(\langle\delta g^2\rangle - \langle\delta g\rangle^2)^{1/2} = \alpha e^2/h \tag{9.8}$$

where the angle brackets denote ensemble averaging and

$$\delta g = g(E_F, B) - \langle g(E_F, B)\rangle \tag{9.9}$$

The constant α depends upon the shape of the sample and is of order unity (e.g. $\alpha = 0.729$ for the quantum wire of Fig. 9.7(b)). At finite temperatures the root-mean-square amplitude will depend on the dimensionality of the sample. For a long quantum wire where $W < \ell_\varphi < \ell_z$ the result is

$$(\langle\delta g^2\rangle - \langle\delta g\rangle^2)^{1/2} = \alpha(e^2/h)\frac{\ell_T}{\ell_z}\left(\frac{\ell_\varphi}{\ell_z}\right)^{1/2} \tag{9.10}$$

where $\ell_T = (hD/kT)^{1/2}$ is the thermal diffusion length.

The aperiodic nature of the conductance fluctuations allows us to define a correlation function

$$F(B, \Delta B; E_F, \Delta E_F) = \langle g(B, E_F)g(B + \Delta B, E_F + \Delta E_F)\rangle - \langle g(B, E_F)\rangle^2 \tag{9.11}$$

which has been calculated by Lee, Stone and Fukuyama (1987) and Altshuler and Khmelnitskii (1986). For $\Delta B = \Delta E_F = 0$, the correlation function reduces to the variance as defined in equation (9.8). The energy and magnetic correlation lengths, ΔE_c and ΔB_c respectively are obtained from the half-widths of the correlation function $F(\Delta B = 0, \Delta E_F)$ and $F(\Delta B, \Delta E_F = 0)$. In the case of the 1D quantum wire discussed above, the results of the calculation give

$$E_c = h\pi^2 D/\ell_z^2 \tag{9.12a}$$

$$B_c = 1.2\phi_0/W\ell_\varphi \tag{9.12b}$$

where ϕ_0 is the flux quantum h/e. The correlation lengths tell us by how much we can change the magnetic field and Fermi energy before there is a significant change in the conductance. It is useful to think of them as the average 'period' of the aperiodic fluctuations.

The universal conductance fluctuations place very severe constraints on any device based on quantum interference as we shall see in the last part of this section, but it is not just mesoscopic devices where the random impurity config-uration will be a profound nuisance. The latest field effect transistors (FETs) are heavily doped and have gate lengths which are expected to drop below 0.1 μm in the near future. Once again the effect of random positioning of the dopants is expected to be important as shown by the numerical simulations of Jacobini and Ferry (1995). Electrons propagating between source and drain will not experience a smooth drop in potential. Instead the potential landscape will be extremely irregular resulting in threshold voltages and $I–V$ curves which will vary somewhat between different devices.

9.2.4   Quantum Interference Transistors

Quantum interference transistors were some of the first mesoscopic devices to be considered (Fowler, 1984) and the basic principle can be explained by means of Fig. 9.8. Imagine an electron wave entering the ring from the left. It can propagate from A to B by splitting into two partial waves which travel in different arms of the ring. If the electron waves travel the same distance they will have the same phase when they reach B and will interfere constructively. The constructive inter-ference at the output of the device reduces the resistance of the ring, as described in Section 9.2.1. Suppose we now change the phase of the wave in one of the arms so that the two have a phase difference of π when they recombine at B. The interference is now destructive, the electron is less likely to be localized at B and the resistance of the ring is increased. Now all that is needed to make a quantum interference transistor is some mechanism to control the phase differ-ence between different pairs of electron waves. Various approaches have been suggested, most of which have some analogy with optical devices. In this section

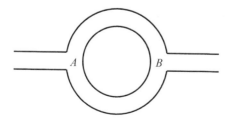

Figure 9.8.   A simple quantum wire ring in which electrons can propagate from A to B via different arms of the ring. The interference of the electron waves at B control the conduc-tance of the device.

we shall consider two types of quantum interference transistor: one based on an electron interferometer similar to the Mach–Zender laser interferometer and another which resembles a microwave stub tuner. In both cases the starting material is considered to be a modulation-doped heterojunction (Chapter 3). This results in a high mobility 2DEG situated $\sim 500\,\text{Å}$ below the surface of the semiconductor which can then be further confined, for example, by surface gates, or by physical etching to form quantum wire and ring structures.

9.2.4.1 The Gated Ring Inteferometer

The gated ring interferometer (de Vegvar *et al.*, 1989) is an extension of the basic ring structure discussed above but now a short gate is placed across one of the arms (Fig. 9.9). The gate forms a Schottky barrier to the heterojunction and by applying a reverse bias the electron density in the underlying electron gas can be reduced. In a 1DEG with n electrons per metre, the electron wavelength at the Fermi energy, λ_F, depends on the concentration as $\lambda_F = 2/n$ and can therefore be controlled by the gate voltage. Suppose the gate in Fig. 9.9 is 2000 Å long and the Fermi wavelength in the ungated material is 500 Å. When the gate is grounded we can assume that the wavelength of electrons beneath it will be the same as in the rest of the ring and two partial waves propagating around different arms will interfere constructively at B, corresponding to a minimum in the resistivity. We can now reverse bias the gate; the Fermi wavelength will increase and at some point only 3.5 electron wavelengths, rather than $4\lambda_F$, will be accommodated under the gate. There is now an effective path difference of $\frac{1}{2}\lambda_F$ between the two partial waves emerging at B. The corresponding phase difference is π, resulting in destructive interference and an increase in the resistance of the ring. In principle, the resistance of a single moded ring (i.e. one with only a single subband occupied) can be modulated by 100% in this fashion with only a relatively small change in the gate voltage, resulting in a high transconductance, $g_m = \partial I_d / \partial V_g$. However, most quantum wires that can be made and measured today have several 1D

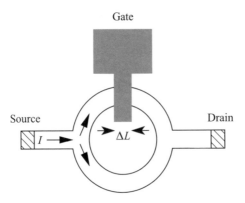

Figure 9.9.   A gated ring electron wave interferometer.

subbands occupied, each of which has its own wavelength given by the de Broglie relationship $\lambda_n = 2\pi\hbar/[2m(E_F - E_n)]^{1/2}$, where E_n is the lowest energy in the nth subband. Because of this there is usually no unique point at which all the modes will suffer destructive interference and the largest possible modulation of the resistance will be significantly less than 100%. Unfortunately, incomplete modulation is not the only problem associated with this and other quantum interference transistors but, before looking at the difficulties, we should first consider a related device, the electron wave stub tuner.

9.2.4.2 The Stub Tuner

The so-called electron wave stub tuner (Sols *et al.*, 1989; Datta, 1989) gets its name from a technique commonly used in microwave engineering to improve the matching between different waveguides. A schematic picture of the device is given in Fig. 9.10. Electron waves travelling between the source and drain can propagate between them directly or via the longer route around the stub. If the length of the stub can be changed then the phase difference of the two waves when they reach the drain can be varied and the conductance of the device modulated. The conducting length of the stub could be varied by using the fringing field from a surface gate electrode, i.e. when the electron gas under the gate is completely depleted the extent of the lateral depletion can be increased by further increasing the reverse bias.

As with the gated ring structure, 100% modulation of the conductance can be achieved with a single mode device but will be much less when multiple modes are occupied. Some calculated results (Datta, 1989) are shown in Fig. 9.11.

9.2.4.3 Problems with Quantum Interference Transistors

At present the problems with quantum interference transistors seem to be so overwhelming that it is hard to imagine any useful devices emerging. We have already seen that at room temperature the phase of an electron will be randomized predominantly by collisions with phonons and that this mechanism is so efficient that the phase coherence length is too small to measure by a transport experiment. As the temperature is reduced, the importance of phonon scattering diminishes

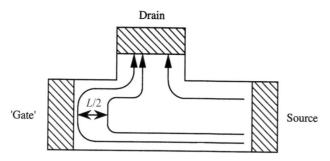

Figure 9.10.   The electron wave stub tuner.

(a)

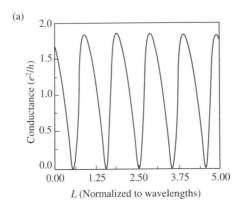

(b)

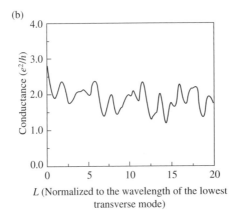

Figure 9.11.   (a) The conductance of a single-mode stub tuner as a function of the non-conducting length of the gated stub. (b) The same as (a) but for a quantum wire with eight modes occupied showing the reduction in modulation efficiency. (Datta 1989, with permission)

and the time between phase breaking events increases, as shown in Fig. 9.4. However, if quantum interference is to be used in a transistor with active dimensions of less than a micron the transistor must be kept at temperatures significantly below 10 K. This is a serious constraint and any device would have to display remarkable properties if it were to justify the extra cost and effort of maintaining it at these low temperatures.

As well as the need for low-temperature operation, quantum interference transistors will suffer from the universal conductance fluctuations discussed in Section 9.2.3. All the structures we have considered so far have been ideal in the sense that the confining potentials were assumed to be very abrupt and to vary smoothly along the edges of the device and there were assumed to be no defects located within the current carrying path. It is well known from experiments that quantum wires have rough boundaries (Thornton et al., 1989) and that impurities can enter the wire so that a more realistic geometry might look like that shown in Fig. 9.12. Here the ideal ring structure now has rough edges which may result from the fabrication but may also be influenced by the close proximity of charged impurities. The electron trajectories are no longer defined by just the two arms of the ring and an electron which initially enters one arm may be scattered at the boundary into the other. The arms themselves might be split into ring-like geometries by defects within the wire. Electrons which reach the output of the ring may have arrived there by a number of different paths which will vary randomly from device to device. As a result the conductance as a function of gate voltage is very difficult, if not impossible, to predict without a detailed microscopic picture of each device.

It might seem that we could improve the technology and reduce the number of impurities until they reach an insignificant level. Unfortunately, the impurities are the charged donor ions which are responsible for producing the electron gas in the first place and a conservative estimate would suggest that in any device with an active area of $1 \, \mu m^2$ there will be at least ten impurity atoms close enough to affect the behaviour of the device. Various groups are now investigating mechanisms for introducing donor impurity atoms in a controlled periodic fashion either by direct growth or by atomic manipulation using a scanning tunnelling microscope but these developments are perhaps many years away.

To make matters worse there are two important effects which occur in a quantum wire which mean that quantum interference transistors can only carry very small currents. The first effect is the conductance quantization (Wharam *et al.*, 1988; van Wees *et al.*, 1988) which, as discussed in the next section, ensures that the resistance of a short quantum wire with only one occupied 1D subband is close to $h/2e^2 \sim 13 \, k\Omega$. In principle the resistance can be reduced by increasing the number of occupied channels, but we have already seen that quantum interference transistors work best if only one subband is occupied.

Since the resistance of a quantum interference transistor is quite high it might seem that we can increase the current flowing through it by increasing the voltage

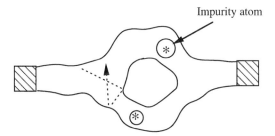

Impurity atom

Figure 9.12.   A more realistic ring geometry. Impurities marked as * create extra loops in each of the arms. The rough edges can also influence the trajectories as shown by the dashed arrow for an electron that initially enters the lower arm but is eventually scattered into the upper.

dropped across the source and drain. Unfortunately, voltage drops of more than a few millivolts will lead to significant electron heating and the phase coherence length will begin to fall. In a ballistic device in which an electron will suffer no scattering events, we are limited to, at the very most, a voltage drop of 36 mV between the source and drain, which corresponds to the energy of the optic phonon in GaAs. (Ferry, 1991). If the electric field across the quantum interference transistor is large enough for an electron to acquire 36 meV before it reaches the drain it will have a high probability of emitting an optic phonon. As a result the electron will scatter almost immediately with a subsequent loss of all the phase information. The power supply rails of any quantum interference based circuits would therefore be limited to 36 mV, with individual transistors drawing less than $36\,mV/13\,k\Omega$ in current (i.e. $< 3\mu A$). The total current could in principle be increased by stacking many wires one on top of the other, but this would only add to the fabrication difficulties. Circuits with low power consumption are of course very attractive, but the individual devices must have very reproducible characteristics. Unfortunately, the presence of even a single impurity in a ring or stub tuner device can drastically affect its characteristics.

In principle it might be possible to overcome the individual difficulties mentioned above by, for instance, developing new fabrication technologies, using new materials or designing new circuits. But taken together the problems associated with quantum interference transistors seem to more than outweigh the one advantage of a high transconductance. The obvious conclusion would seem to be that although quantum interference devices have demonstrated a great deal of interesting physics the possibility of any practical applications seems quite remote.

9.3   Ballistic Electron Devices

The physics of ballistic electron structures is governed by the elastic mean-free path, which can approach $100\,\mu m$ in modulation-doped GaAs/AlGaAs heterojunctions cooled to low temperatures (Chapter 10). This length scale is much larger than the minimum feature sizes which can be patterned by present-day lithography and we can conceive of a new class of ballistic electron device in which an electron experiences no scattering events except perhaps at the boundaries of the structure. In addition, the elastic mean free path is usually less temperature dependent than the phase coherence length and therefore devices which are based on ballistic electron transport could in principle operate at higher temperatures than quantum interference transistors.

A current-carrying electron in the 2DEG of a GaAs/AlGaAs heterojunction will travel at the Fermi velocity, $v_F = \hbar k_F/m$. For a typical electron concentration of $5\times10^{15}\,m^{-2}$, we would expect velocities of order 3×10^5 m/s. Although this is very fast, even higher velocities are obtained in short-gate GaAs FETs and the ballistic electron devices discussed below are therefore not really of interest for their high-speed properties. However, ballistic transport does lead to results which are markedly different from our usual expectations and it is these differences that might be exploited in future devices.

9.3.1   Electron Transmission and the Landauer–Büttiker Formula

When modelling conventional electronic devices, an equivalent circuit approach is often used. The equivalent circuit consists of a number of idealized components which are considered to be either intrinsic or extrinsic to the device. Taking the MOSFET as an example, one important intrinsic component is a voltage controlled current source while an extrinsic component is the parasitic resistance in the source contact. For a mesoscopic device the notion of intrinsic and extrinsic components of the device becomes less useful. This is because an electron wavefunction will sample a large proportion of the entire device so that the active region and connecting leads have to be considered on the same footing. In 1957 Landauer (1957) proposed that the transmission and reflection of electrons by an object would govern its conductance. Büttiker developed this idea for mesoscopic systems (Büttiker, 1986) and the resulting Büttiker–Landauer formalism is outlined below.

Consider the 4-terminal structure shown schematically in Fig. 9.13. Current can be fed into the active region through leads connected to electron reservoirs i (in this case, $i = 1, 2, 3, 4$) each at some chemical potential μ_i, which is measured relative to some reference value such as the lowest potential on any of the reservoirs. The current injected by reservoir i is

$$ev_i g_i(E)\mu_i \tag{9.13}$$

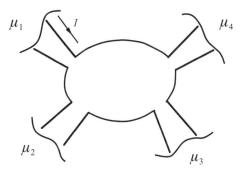

Figure 9.13.   A model system for discussing the transmission of electrons through a multiterminal device using the Landauer–Büttiker formalism. The contacts are held at fixed chemical potentials, μ_i, and are considered to be highly disordered. An electron entering a contact will be absorbed and quickly lose all its phase information by inelastic scattering. This is in contrast to the mesoscopic conductor which is assumed to be fully coherent, i.e. an electron in the conductor only suffers elastic scattering.

where v_i is the electron velocity. For a one-dimensional wire with just one occupied subband, the density of states, $g_i(E)$, equals $2/hv_i$ and the current injected into lead i is $(e/h)\mu_i$. Consider the total current in lead 1. The injected current is reduced by that reflected back into reservoir 1, $(e/h)R_{11}\mu_1$ and that transmitted into reservoir 1 from all the other reservoirs, i.e. $(e/h)(T_{12}\mu_2 + T_{13}\mu_3 + T_{14}\mu_4)$. Similar expressions can be written for each of the remaining leads so that in general we obtain the result

$$I_i = (e/h)\left[(1 - R_{ii})\mu_i - \sum_{j \neq i} T_{ij}\mu_j\right] \tag{9.14}$$

If we now consider the situation where the i leads are multi-moded (i.e. N_i subbands are occupied in the ith lead), we obtain

$$I_i = (e/h)\left[(N_i - R_{ii})\mu_i - \sum_{j \neq i} T_{ij}\mu_j\right] \tag{9.15}$$

The R_{ii} now include all the electrons in lead i, mode m, which are reflected back to reservoir i in mode n, i.e. $R_{ii} = \sum R_{ii,mn}$. Similarly, $T_{ij} = \sum T_{ij,mn}$. In each case the summation is made over all indices m and n.

Equation (9.15) can be used to calculate, in qualitative and quantitative detail, the conductance in a variety of 2-terminal and 4-terminal devices.

9.3.2   Quantized Conductance in Ballistic Point Contacts

The Büttiker–Landauer formula can be used to explain why the conductance of a short and narrow wire is quantized (Wharam *et al.*, 1988; van Wees *et al.*, 1988). These wires are short enough for electrons to pass through them ballistically, i.e. without suffering collisions of any kind; hence, they are commonly known as ballistic, or quantum point, contacts. Quantum point contacts can be formed in a 2DEG by, for example, electrostatic depletion from a pair of split surface gates (Thornton *et al.*, 1986). A suitable geometry is shown in Fig. 9.14(a) (see also the discussion in Section 3.4). A reverse bias applied to the gate electrodes will first deplete the underlying electrons but, at a certain threshold voltage, current can only flow from source to drain via a narrow constriction between the gate electrodes. The fringing field from the edge of the gate electrodes can be used to reduce the width of the constriction simply by increasing the reverse bias to the split gates. In a typical experiment the width of the constriction can be varied in the range 1.0 to 0.1 μm. The number of 1D subbands in the constriction is given approximately by the integer value of Wk_F/π and will, therefore, decrease as the width, W, of the constriction is reduced. If the resistance of the constriction is measured as a function of gate bias it will increase in a step-wise fashion, as shown in Fig. 9.14(b). Note that this is consistent with the conductance variation in Fig. 3.16). The plateaux are quantized to within a small percentage of the value given by $h/2Ne^2$.

(a)

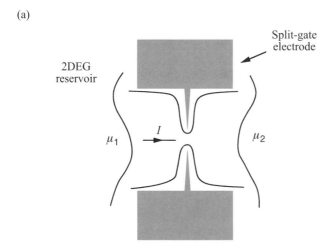

(b)

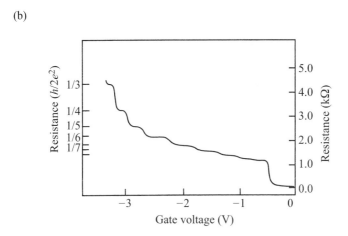

Figure 9.14. (a) The split-gate geometry used to confine a narrow constriction in an underlying 2DEG. (b) As the gate voltage is made progressively more reverse biassed the width of the constriction and therefore the number, n, of occupied 1D subbands decreases and the resistance increases as $h/2ne^2$ (Wharam et al., 1988). (Courtesy T. J. Thornton)

To understand the origin of the quantized conductance we rewrite equation (9.15) in a form suitable for the 2-terminal geometry of the quantum point contact. An alternative derivation is given in Chapter 3. Current conservation demands that all of the electron modes incident in wire 1 are either reflected back into wire 1 or transmitted into wire 2, i.e. $N_1 = R_{11} + T_{21}$. For the simple 2-terminal geometry, $T_{21} = T_{12} = T$, at least for zero applied magnetic field. Substituting this into equation (9.15) gives the conductance, G, of a 2-terminal device as

$$G = \frac{I}{\Delta V} = \frac{eI}{\mu_1 - \mu_2} = (e^2/h)T \qquad (9.16)$$

For a long quantum wire there is nothing particularly remarkable in this result. The value of T will lie between 0 and 1, depending on the degree of backscattering in the channel. However, for the quantum point contact, no scattering takes place as the electron propagates through the constriction. Any electron injected into the constriction is guaranteed to pass through because there is no backscattering to reverse the direction. The total transmission can now be written as $T_{12} = \sum T_{12,mn} = \sum \delta_{mn} = N$, where N is the number of propagating modes (sub-bands) in the constriction. Substituting this value for T in equation (9.16) and multiplying by 2 to account for the spin degeneracy of the electrons gives the conductance of a 2-terminal ballistic wire as $G = 2Ne^2/h$. Although there are some subtle details concerning the mode coupling between the wide and narrow regions that we have omitted (Szafer and Stone, 1989), this simple application of the Büttiker–Landauer formula describes the behaviour of quantum point contacts quite accurately.

9.3.3 Multi-terminal Devices

Low-dimensional structures with multiple probes have demonstrated a wealth of interesting physics. Usually the experiment is set up such that a current is forced to flow between one pair of terminals, m and n, while a voltage drop is measured across another pair, k and l. The voltage probes are assumed to draw no current, and equation (9.15) leads to four simultaneous equations such that $I_m = -I_n$ and $I_k = I_l = 0$. By solving these equations Büttiker was able to derive the following formula for the generalized resistance of a 4-terminal mesoscopic conductor:

$$R_{mn,kl} = (h/e^2)(T_{km}T_{ln} - T_{kn}T_{lm})/D \qquad (9.17)$$

The positive, dimensionless number D is a sub-determinant of the matrix equation (9.15) and can be calculated for specific cases. $R_{mn,lk}$ signifies current in at terminal m, out at n, and voltage measured between terminals l and k. We shall use equation (9.17) to explain two experiments which show how transport in ballistic electron structures can produce results which are quite different from our classical explanations.

9.3.3.1 The Negative Bend Resistance

An immediate consequence of equation (9.17) is that the resistance measured with a 4-terminal configuration is not necessarily positive. This was first shown by Takagaki et al., (1988), who measured the voltage drop across a cross-shaped junction when the current was forced to flow around a corner. Consider the geometry shown in Fig. 9.15(a), in which current is forced to flow from contact

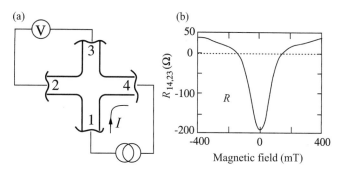

Figure 9.15. (a) The geometry used to measure the negative bend resistance. The dimensions of the island between the probes are less than the elastic mean-free path. (b) The negative bend resistance is destroyed by a weak magnetic field because more electrons are guided into the 'correct' probe (Roukes *et al.*, 1990). (Courtesy T. J. Thornton)

1 to contact 4 with the voltage drop measured across probes 2 and 3. For a large structure in which the transport is diffusive we would expect to measure only a very small voltage drop associated with the finite size of the junction in between the voltage probes and the corresponding resistance $R_{14,23}$ would be positive. However, when the same measurement is made using a sub-micron structure, probe 3 is at a higher potential than probe 2 and the voltage drop $\delta V = V_2 - V_3$ is negative. This result can be explained using equation (9.17) written explicitly for the configuration in Fig. 9.15(a):

$$R_{14,23} = (h/e^2)(T_{21}T_{34} - T_{24}T_{31})/D \tag{9.18}$$

By inspection, the coefficients T_{24} and T_{31} refer to transmission directly across the junction while T_{21} and T_{34} correspond to transmission into adjacent probes. Ballistic electrons injected into the junction with large forward momenta are much more likely to reach the opposite probe than to be scattered through almost $90°$ into an adjacent probe. Consequently, $T_{24}T_{31} > T_{21}T_{34}$ and the resistance in this case is negative. When a magnetic field is applied, the electron trajectories are curved and the probability of an electron reaching the 'correct' probe increases so that at high magnetic fields a positive value for the bend resistance is recovered. A typical result from Roukes *et al.*, (1990) is reproduced in Fig. 9.15(b).

9.3.3.2 Quenching of the Hall Effect

The enhanced probability of forward transmission also has an important effect on the Hall voltage developed across a junction between mesoscopic wires. At low magnetic fields, the Hall resistance can be substantially suppressed or even quenched (Roukes *et al.*, 1988; Ford *et al.*, 1988) when compared with the classical value (Fig. 9.16). To explain this we again evaluate equation (9.17) to

(a)   (b)

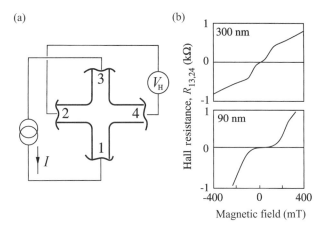

Figure 9.16.   (a) A measurement of the Hall effect using a mesoscopic cross shaped geometry will yield a vanishingly small voltage if transport through the cross is ballistic. (b) Typical results for 300 nm and 90 nm wide quantum wires showing that the Hall voltage is suppressed for small magnetic fields. (Courtesy M. L. Roukes)

find $R_{13,42}$. Now we need to determine the explicit value for D and the result is

$$R_H = R_{13,42} = \frac{T_{41}T_{23} - T_{43}T_{21}}{(T_{21} + T_{41})[(T_{21} + T_{31})^2 + (T_{41} + T_{31})^2]} \, (h/2e^2) \qquad (9.19)$$

To make this result more transparent, we can write the forward transmission T_{31} as T_F and assume that $T_{21} = T_{43} = T_L$ and that $T_{41} = T_{23} = T_R$ where T_L and T_R are the probabilities of transmission into an adjacent probe on the left and right, respectively. As in the case of the negative bend resistance the forward transmission can be much larger than those into the side arms, i.e. $T_F \gg T_L, T_R$, and we get the approximate result that $R_H \sim (h/2e^2)(T_L - T_R)/2T_F^2$ (Roukes et al., 1988). The Hall resistance is strongly suppressed until the magnetic field is large enough for the Lorentz force to guide the electrons into the adjacent Hall probes at which point the Hall resistance is quickly recovered (Beenakker and van Houten, 1989). As in the bend resistance, an enhanced probability of forward transmission leads to results which are markedly different from the classical results.

9.3.4   Possible Applications of Ballistic Electron Devices

Ballistic electron devices have always intrigued device scientists primarily because of their high-speed potential. In fact modern-day transistors already operate in a quasi-ballistic mode with velocity overshoot occurring in the channel of a short-gate FET while electrons in a bipolar device suffer very few collisions as they propagate across the short base region. For this reason, devices which rely on ballistic transport in low-dimensional structures are unlikely to compete on speed grounds alone. If they are to offer any advantage it will have to be in terms of enhanced functionality, i.e. one device performing a function that would normally require several devices. Enhanced functionality will become more important as the difficulty in downscaling conventional devices increases. For this reason it is of interest to discuss new concepts which may offer enhanced device functionality even if no real applications yet exist.

Because of the reasonable accuracy of the resistance quantization, it has been suggested that the output characteristics of a quantum point contact (QPC) could be used to perform decimal logic or as elements within an analogue to digital convertor (ADC). Taking the last case as a specific example, an analogue signal applied to the gate electrodes of the QPC would result in a 'digitized' output voltage if a constant current flowed through the QPC. The output signal could then be analysed using digital signal processing techniques, provided of course that the processing can deal with multiple voltage levels. The QPC analogue to digital converter would unfortunately only operate at low temperatures and is prone to the same irreproducibility as interference devices because the presence of an impurity atom within the vicinity of the constriction will alter the threshold voltage of the QPC and will disrupt the quantization. Any advantage would come from the very small capacitance of the constriction, but silicon-based ADCs already operate at room temperature at frequencies in excess of 1 GHz (Harame et al., 1995) and this application of QPCs may be confined to academic interest.

Another interesting idea is the possibility of steering beams of ballistic electrons as they propagate through a 2DEG. One of the first demonstrations involved electron focussing between two point contacts by means of a magnetic field (van Houten et al., 1988). By tuning the cyclotron radius such that $2r_c = d$, the distance between the point contacts, the electrons can be steered along a semi-circular trajectory from one QPC to the other.

Instead of using a magnetic field to steer electrons it is possible to alter their trajectory by a mechanism analogous to refraction. Consider a 2DEG which has a gate deposited over half of the surface while the rest is left ungated as shown in Fig. 9.17(a). If the gate is reverse biassed, the electron concentration under the gate will be smaller than in the rest of the 2DEG. The conduction band will bend in a step-like fashion and there will be a force acting normal to the interface between gated and ungated regions. The component of the momentum parallel to this interface between the two regions will therefore remain unchanged when an electron crosses the interface, i.e. $p_1 \sin \theta_1 = p_2 \sin \theta_2$. The electrons which carry the current are within kT of the Fermi energy and we can write the momentum as $p = \hbar k_F$. For a 2DEG, the Fermi wavevector is given by $k_F = (2\pi n)^{1/2}$, where n is the electron concentration (per unit area) and we obtain the result

$$\frac{\sin \theta_1}{\sin \theta_2} = (n_2/n_1)^{1/2} \qquad (9.20)$$

(a)

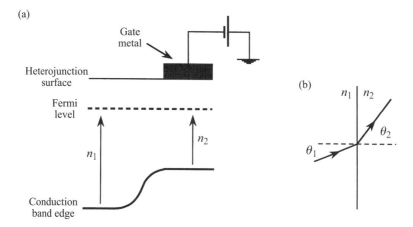

Gate
metal

Heterojunction
surface

Fermi
level

n_1

n_2

Conduction
band edge

(b)

n_1 n_2

θ_2

θ_1

Figure 9.17.   (a) At the interface between gated and ungated regions of a 2DEG there will be a discontinuity in the electron density. If $V_g < 0$, then $n_2 < n_1$. (b) An electron crossing an interface from a region of high electron density to a region of lower density will be bent away from the normal.

The above equation resembles Snells' law for the refraction of light at the interface between materials of different refractive indices. For the example of the partially gated 2DEG, the trajectory of an electron crossing the interface will be deflected away from the normal when it moves from a region of high to low electron density (Fig. 9.17(b)).

A variety of structures has been designed to demonstrate the switching potential of electron refraction in a 2DEG. The surface gate geometries for prismatic and biconcave refractive switches are shown in Fig. 9.18. For the case of the prismatic switch, the ballistic electron beam is swept by the electron collectors as the reverse bias to the prism-shaped gate is increased. For the biconcave lens a divergent electron beam can be brought into focus at the collector by applying a suitable reverse bias. The experimental results (Spector *et al.*, 1990a,b) demonstrate the possibility of switching current by electron refraction but we should ask the question – how does such a device compare with a conventional FET? The gate area of the structure shown in Fig. 9.18 is quite large and the associated RC charging time would certainly make it slower than, for example, a HEMT. But, unlike a HEMT, the refraction FET can switch between multiple outputs and even make use of multiple inputs because crossed beams of ballistic electrons seem to have negligible interaction. Refractive electron devices could therefore be configured to perform fairly sophisticated operations such as switching elements for parallel signal processing. However, an important outstanding problem is the very small voltage generated at the output of the device. To switch the device, the gate voltage has to be changed by a few 100 mV whereas the output voltage is only of the order of 10 μV. This large mismatch makes it almost impossible for the output

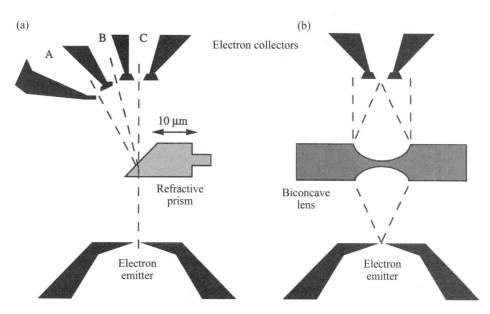

Figure 9.18. Surface gate geometries which can be used to define (a) a refractive prism and (b) a biconcave refractive lens FET.

of one device to drive the input of the next and is a problem which plagues many of the proposed mesoscopic devices.

9.3.5   Boundary Scattering in Ballistic Structures

Before leaving the subject of ballistic devices we should consider the important issue of boundary scattering. Up to now we have implicitly assumed that when an electron interacts with a hard wall boundary it is scattered specularly just as if it were reflected from a mirror. As mentioned in Section 9.2.4.3 it is impossible to make, for example, a quantum wire or ring, with perfectly smooth edges and as a result *diffuse* scattering at the boundaries, rather than specular scattering, becomes important. In the limit of completely diffuse boundary scattering the mean free path of the electrons is limited by the width of the quantum wire. In narrow wires diffuse boundary scattering can dominate the resistivity and might limit the applications of ballistic transport discussed above.

The scattering from a rough edge is often described in terms of a single parameter, p, which varies between zero and unity corresponding to the extremes of completely diffuse scattering and completely specular scattering. In the spirit of Matthiessen's rule we can sum the rates of diffuse boundary scattering and bulk scattering to define an effective mean free path given by $1/\ell_{\mathrm{eff}} = 1/\ell_0 + 1/\ell_{\mathrm{b}}$ where ℓ_0 is the mean free path in a sample with no boundaries and ℓ_{b} is the boundary scattering length defined as the average distance an electron will travel along the wire before suffering a diffuse scattering event at the boundaries. Over a distance ℓ_b an electron will collide approximately $\ell_b/W \sim 1/(1-p)$ times with the edges of a wire of width W before scattering diffusely so that $\ell_b \sim W/(1-p)$. We can therefore write the effective resistivity of the wire ϱ_{eff} as

$$\varrho_{\text{eff}} = \varrho_0 \ell_0 \left(\frac{1}{\ell_0} + \frac{1-p}{W} \right) \tag{9.21}$$

where ϱ_0 is the resistivity of a large sample of the material and for a 2DEG $\varrho_0 \ell_0 = \hbar k_F/ne^2$ (Exercise 2). From equation (9.21) it is clear that the diffuse boundary scattering will dominate the resistance when $1-p > W/\ell_0$ and for a $0.1\,\mu\text{m}$ wide wire with a bulk mean free path $\ell_0 = 10\,\mu\text{m}$ this will be the case for $p < 0.99$.

The contribution that boundary scattering makes to the total resistivity of a narrow wire can be varied by the application of a perpendicular magnetic field. At low fields the resistivity is initially increased but can be reduced to the bulk value ϱ_0 by a sufficiently large magnetic field. The reasons for this behaviour are illustrated in Fig. 9.19. At zero and very small magnetic fields (Fig. 9.19(a)) the electrons with a large component of momentum parallel to the wire interact infrequently with the edges and can contribute significantly to the conductivity. However, as the field is increased these electrons are forced to collide with the

(a)　^1eff

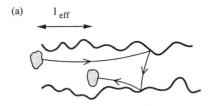

(b)　^1eff

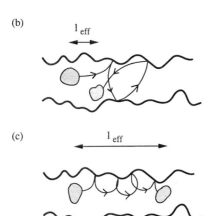

(c)　^1eff

Figure 9.19. Boundary scattering in the presence of a perpendicular magnetic field: (a) zero field, (b) intermediate fields, and (c) high fields.

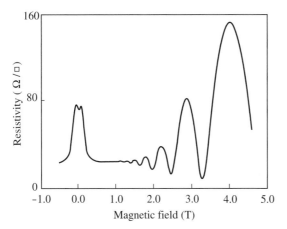

Figure 9.20.   Resistivity as a function of magnetic field for a $1\,\mu$m wide GaAs quantum wire (Thornton *et al.*, 1990). (Courtesy T. J. Thornton)

edges (Fig. 9.19(b)) and for $p < 1$ the extra diffuse scattering will increase the resistivity. The low field positive magnetoresistance saturates and a maximum in the resistance occurs at a magnetic field such that $W/L_c = 0.55$, where $L_c = \hbar k_F/eB$ is the cyclotron radius. This value for the ratio W/L_c is derived from a calculation based on classical electron trajectories (Pippard, 1989) but a fully quantum mechanical calculation gives a very similar result (Akera and Ando, 1990). At still higher magnetic fields the cyclotron diameter will be smaller than the wire width and the electrons will be confined to the edges (Fig. 9.19(c)). At this stage diffuse backscattering is suppressed over distances of order ℓ_0 and we recover the bulk resistivity ϱ_0. This behaviour is clearly observed in a $1\,\mu$m wide GaAs quantum wire defined by ion beam damage as shown in Fig. 9.20 (Thornton *et al.*, 1990).

According to equation (9.21) diffuse boundary scattering increases the wire resistivity by an amount $\delta\varrho = (1 - p)\varrho_0\ell_0/W$. In Fig. 9.21 the extra resistivity is plotted as function of wire width and a best fit of the data to equation (9.21) gives a value for p of 0.85. Even in this example, where only 15% of the boundary collisions are diffuse, their total effect on the wire resistivity is significant.

9.4   Quantum Dot Resonant Tunnelling Devices

Unlike quantum interference transistors, devices based on resonant tunnelling do exist and have important applications as high frequency oscillators and mixers. The Esaki tunnel diode (Esaki, 1958) was the first device to exploit electron tunnelling. Since then the underlying physics has changed very little but the device structures have progressed from the original *p*–*n* junction diode to multi-junction quantum well devices which exploit resonant tunnelling. All of these structures

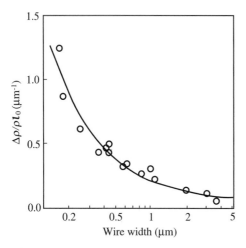

Figure 9.21. Fractional increase in resistivity with reduction in quantum wire width due to diffuse boundary scattering. (Courtesy T. J . Thornton)

have a very non-linear *I–V* characteristic which can yield a negative differential resistance as described below. These and related structures are also discussed in Section 10.3.

9.4.1   Resonant Tunnelling through Quantum Wells

Consider the GaAs/AlGaAs quantum well shown in Fig. 9.22(a). The GaAs layer is assumed to be thin enough for the bottom of the conduction band to be raised by quantum confinement, so that one or more 2D energy levels are formed in the well. In addition the AlGaAs barriers are thin enough that, under certain conditions, electrons can tunnel through the quantum well and appreciable current will flow between the heavily doped contacts.

At zero, or vanishingly small, applied bias the conduction band energy levels across the quantum well are shown in Fig. 9.22(b). Some band bending occurs because of electron diffusion from the contacts into the nominally undoped quantum well. The quantum well is thin enough for the lowest subband to be higher than the Fermi energy in the left-hand contact. As a result no current flows because there are no available states in the quantum well for electrons to tunnel into (Schiff, 1968). Of course, some current can flow by thermal activation over the barriers and inelastic processes, but we shall ignore these effects for the time being. As the applied bias is increased, most of the potential is dropped across the resistive regions (i.e. the undoped barriers and quantum well), the energy bands are tilted and at some point the lowest subband in the quantum well will drop below the Fermi energy in the contact (Fig. 9.22(c)). There are now states in the well which electrons can tunnel into while conserving energy and momentum and,

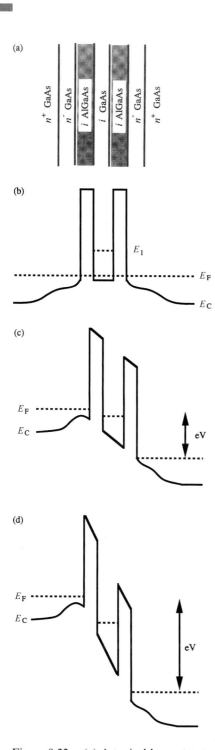

Figure 9.22.   (a) A typical layer structure for a resonant tunnelling quantum-well device. (b), (c) and (d) Energy band diagrams of the same layer under various bias conditions.

as a result, current will begin to flow. The current will increase with increasing bias until the subband drops below the bottom of the conduction band in the contact as shown in Fig. 9.22(d). Once again there are no states for electrons to tunnel into and the current drops sharply. The reduction in current with increasing bias is called a negative differential resistance (NDR). If more than one subband is present in the quantum well it is possible to observe a series of NDR peaks as each subband is swept past the occupied states in the contact. The I–V characteristic of an ideal resonant tunnelling device is shown in Fig. 9.23. The NDR peak is superposed on a slowly rising background which is due to the activation of hot electrons over the barriers and perhaps also to inelastic effects (i.e. electron-phonon scattering) which allow tunnelling without conservation of electron energy and momentum.

Resonant tunnelling devices have been used in oscillator circuits working beyond $400\,\mathrm{GHz}$ (Brown *et al.*, 1989; Sollner *et al.*, 1990). Although this is an extremely high frequency for an electronic device there would probably be significantly more applications if it could be designed as a 3-terminal device. At present the current flows between two terminals and the position of the NDR peak is determined by the layer structure and to some extent by the external circuit. A third terminal which could be used to switch the NDR 'on' and 'off' would be highly desirable. One way of making such a resonant tunnelling FET would be to use a quantum dot of variable area. Before describing a variable area device we shall first look at resonant tunnelling through quantum dots.

9.4.2 Resonant Tunnelling through Quantum Dots

The layer material shown in Fig. 9.22 could be made into a quantum dot simply by patterning a mask on the surface and etching back the unmasked material to form a pillar structure with height and width both of something less than a

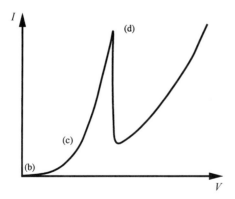

Figure 9.23. Schematic illustration of the current–voltage characteristics of a large area resonant tunnelling device. The biasses marked as (b), (c) and (d) correspond to the energy band diagrams of Fig. 9.22.

micron. In the simplest picture, the extra lateral confinement will split each 2D subband into a number of zero-dimensional (0D) levels. The new energy levels will be at an energy E_n above the bottom of the conduction band of bulk GaAs, where E_n is given approximately by (see Sections 2.3.5 and 2.3.6 and Exercise 4 of Chapter 2)

$$E_n \approx \frac{\hbar^2 \pi^2}{2m} \left[\left(\frac{1}{a} \right)^2 + 2 \left(\frac{n}{w} \right)^2 \right] \tag{9.22}$$

where a is the thickness of the quantum well and w is the edge of the quantum dot pillar, which is assumed to have a square cross-section. The result of the additional quantization is that there are now several levels to sweep by the occupied states in the emitter contact and the single NDR peak of the large-area quantum-well device will split into a number of peaks at higher applied bias. The position in voltage of the peaks is now a function of the width of the quantum dot and if this can be varied by an external bias then we have the basis of the resonant tunnelling FET. Various approaches have been adopted to do just this (Reed *et al.*, 1988; Tewordt *et al.*, 1990; Goodings *et al.*, 1992) but we will concentrate on a technique which employs surface gates to control the conducting area (Dellow *et al.*, 1991; Guéret *et al.*, 1992) in much the same way as split-gate devices have been used to define quantum wires.

9.4.3   Gated Resonant Tunnelling through Quantum Dots

A cross-section through a gated quantum dot double tunnel barrier is shown in Fig. 9.24(a). The layer material could well be the same as that shown in Fig. 9.22 but a small-diameter ohmic contact is first deposited on the surface. As well as providing an ohmic contact the metal layer acts as a resist mask for a reactive ion etching stage during which a shallow pillar is etched into the material. A ring gate electrode can then be evaporated around the contact and by taking special precautions it will not short to the ohmic contact. Current can be made to flow between the top and bottom contacts and, if the ring gate is left unbiassed, the I–V characteristics will be much the same as in a large area device. However, when the gate is reverse biassed, it will deplete any free charge from the contacts below and the current will be confined to flow in a restricted path centred below the top contact. By increasing the reverse bias, the fringing field from the gate will further confine the current flow and as a result the conducting cross-section can be varied electrically. Surface gated structures have been made by Dellow *et al.* (1991) and Guéret *et al.* (1992) while Goodings *et al.* (1992) have used ion implantation to form local *p*–*n* junctions which confine the dot. In all cases the dot diameter can be varied in the range ≈ 0.1–$1 \, \mu m$.

In Fig. 9.24(b) the I–V curves of a surface gated quantum dot are reproduced from Dellow *et al.* (1992). The data seem to fit the simple picture described above in that the NDR peak moves to higher bias as the width of the dot is reduced.

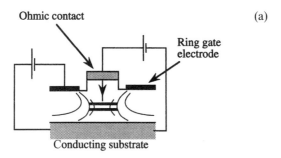

(a)

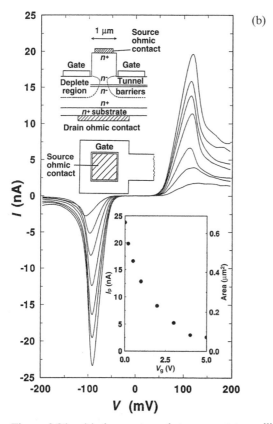

(b)

Figure 9.24.   (a) A quantum dot resonant tunnelling structure defined by electrostatic depletion from a surface gate electrode. (b) Current I versus voltage V obtained from such a structure by Dellow *et al.* (1992). (Courtesy L. Eaves)

However, there are two important effects which urge caution when attempting to interpret such curves. In Section 9.2.3 we saw how impurity atoms can influence the conductance of nominally identical samples and a very similar effect occurs with the quantum dot devices. Even in the very best semiconductor material grown by molecular-beam epitaxy there are likely to be between 1 and 10 impurity atoms located close to the barriers of a sub-micron quantum dot. The fields associated with the impurity atoms will alter the potential profile of the double barrier well in a random fashion. If the change in the local potential is large enough, extra energy levels will be formed in the barriers or the well and current will now be able to flow by resonant tunnelling through these impurity levels (Dellow *et al.*, 1992). The *I–V* curves will show NDR peaks but they will be of little practical use because the energy levels of the impurity states are essentially random and therefore the bias required for a current peak will vary from device to device.

The interpretation of *I–V* curves from sub-micron resonant tunnelling diode (RTD) devices is complicated by the fact that in a real quantum dot structure the confining potential is more likely to resemble an hour glass than the abrupt 3D–0D–3D profile that the simple picture assumes. Close to the barriers, the emitter and collector contacts will be small enough for quantized levels to form. The energy of the levels in the contacts will also vary as the dot diameter is reduced and current will flow due to resonant transmission between 1D levels in the contacts and 0D levels in the dot. This is quite a complicated problem and, although the *I–V* characteristics of a real quantum dot can be calculated from a microscopic picture of the confining potential (Mizuta *et al.*, 1992), this will always be difficult to obtain because of the random distribution of impurity atoms.

9.5   Coulomb Blockade and Single-electron Transistors

So far in this discussion of mesoscopic devices we have ignored the fact that the electronic charge is quantized. The charge on the electron has been almost incidental except for the fact that it carries current. Devices based on the Coulomb blockade, however, depend on the fact that the electron is charged and that an energy penalty has to be paid for charging up an isolated conductor with a single electron. The energy penalty is just the charging energy, $e^2/2C$, where C is the capacitance of the body being charged. The capacitance of a macroscopic conductor is large enough so that this energy penalty is negligible compared with the thermal energy at room temperature, $kT \approx \frac{1}{40}\text{eV}$, so for this situation the Coulomb blockade is unobservable. However, for very small conductors at low temperature it is possible for the charging energy to exceed the thermal energy. As a result it is energetically unfavourable for an electron to charge the conductor until the external driving force is sufficient to supply the extra energy. This is the regime of the Coulomb blockade, in which no current can flow through the conductor.

Many systems are capable of displaying a Coulomb blockade at low temperatures. Some of the earliest work dates back to the 1960s where zero-bias anomalies in the current flowing through a large array of small tin particles were explained in terms of the charging energy of the particles (Zellar and Giaever, 1969). In these original experiments the current flowed through a large number of tin islands and only the average or dominant properties could be observed. Nanofabrication technology has now developed to the stage where we can look for the Coulomb blockade in the current flowing through single, double or multiple tunnel junctions. We shall first consider the Coulomb blockade in a single tunnel junction and then consider the gated double junction which forms the basic element of a single electron transistor.

9.5.1   Coulomb Blockade in the Current-biassed Single Junction

The case of a single junction is the classic textbook tunnelling problem of two conducting electrodes separated by a thin tunnel barrier (Eisberg and Resnick, 1974). This type of tunnel junction is readily made in the $Al/Al_2O_3/Al$ system which exploits the natural oxide formed on aluminium when exposed to air. The oxide barrier can be made thin enough so that for large area junctions, where the charging energy is negligible, transport is ohmic with a reasonably small resistance.

In a small junction of low capacitance we must consider the charging energy. The circuit of Fig. 9.25 shows a current-biassed single junction of capacitance C and conductance G_T shunted by a conductance G_S (Likharev, 1990). The shunt conductance represents the conductivity between the leads of the junction, which is non-zero because the environment around the junction (i.e. the substrate) can never be completely insulating. For very small junctions it turns out that $G_S \gg G_T$ and most of the current flows through the shunt. The voltage across the junction can be written as $V = I_S/G_S$ and the corresponding stored charge is

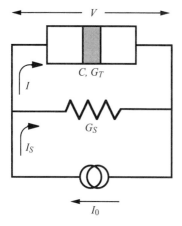

Figure 9.25.   A current-biassed single tunnel junction. The shaded region represents a tunnel barrier.

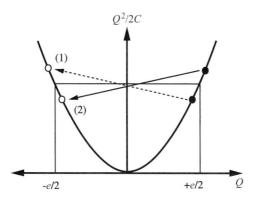

Figure 9.26.   Energy of a charged capacitor as a function of the stored charge.

given by $Q = CV = CI_S/G_S$. It is important to realize that this stored charge does not contribute to the DC current flowing through the capacitor. It is merely a polarization charge brought about by the applied bias and can be a small fraction of the electronic charge, e.

The Coulomb blockade of the current can be explained following the procedure of Averin and Likharev (1986), which is summarized in Fig. 9.26. The energy stored in the junction is given by $Q^2/2C$ and increases quadratically as the bias current, I_0, and therefore as Q increases. If any current is to flow through the junction, the stored charge must change in discrete units of e as the result of a tunnelling event which moves an electron from one side of the junction to the other. If the initial polarization charge, Q, is less than $\frac{1}{2}e$, a drop in charge of e leads to an increase in energy. As a result the transition marked as (1) in Fig. 9.26 is energetically unfavourable. Only when the stored charge exceeds $\frac{1}{2}e$ can an electron tunnel with either no change or a reduction in the total energy – see the transition marked as (2) in Fig. 9.26. This argument suggests that no current can flow unless $|Q| > \frac{1}{2}e$, which corresponds to a threshold voltage of $V_T = e/2C$. The I–V curve for the ideal single junction at zero temperature is shown in Fig. 9.27, where a region of Coulomb blockade of the current exists over a voltage range $-e/2C < V < e/2C$. At higher temperatures the thermal energy, kT, can exceed the charging energy and any evidence of Coulomb blockade will vanish.

The single junction case is very appealing but is almost impossible to achieve in practice because the environment, in the form of the wire leads and the substrate, couples strongly to the junction and enormously increases its capacitance. The charging energy becomes vanishingly small and the Coulomb blockade disappears. To observe the Coulomb blockade it is much easier to use a double tunnel junction for which efficient decoupling from the environment can be achieved (Likharev, 1990, 1991).

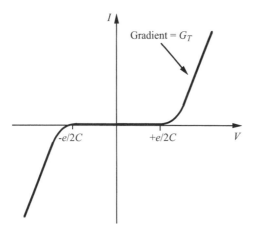

Figure 9.27.   Coulomb blockade of current in an isolated single junction.

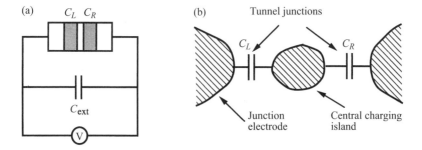

Figure 9.28.   The voltage biassed double junctions: (a) an equivalent circuit representation and (b) a schematic picture of the physical arrangement of such a device.

9.5.2   Coulomb Blockade in Double Junctions

A double junction device consists simply of a small conducting island weakly coupled to external electrodes by means of two tunnel junctions each of which has a small capacitance. A schematic picture of a voltage biassed double junction is shown in Fig. 9.28. The total capacitance of the environment, including that due to the electrical connections, is marked as C_{ext}. The capacitance of the central charging island is just equal to the sum $C_L + C_R = C_T$ and, as mentioned above, the double junction is effectively decoupled from the environment. The charging energy is of the order e^2/C_T and, as a result, the I–V curves will show a region of Coulomb blockade for which $I = 0$ over a voltage range $\Delta V = e/C_T$. The capacitance of the charging island varies with its linear dimension and the smaller the island, the larger and more robust the region of Coulomb blockade. By adding a gate electrode we now have the beginnings of a single-electron transistor (Likharev, 1987). This will be discussed further in Section 9.5.5.

9.5.3   Necessary Conditions for Efficient Coulomb Blockade

If thermal fluctuations are not to mask the Coulomb blockade charging energy the temperature has to be low enough to ensure that the inequality $kT \ll e^2/2C_T$ is satisfied. This condition represents the greatest challenge to the manufacture of single-electron devices which would operate at room temperature. At $T = 300\,\mathrm{K}$ the thermal energy is $25.8\,\mathrm{meV}$, corresponding to a total capacitance $C_T \approx 3\,\mathrm{aF}$ ($1\mathrm{a} = 10^{-18}$). For robust operation the capacitance should be 10–100 times smaller than this value, leading to total device capacitances of the order of $10^{-20}\,\mathrm{F}$. At present most single-electron structures have values of $C_T > 1$ aF and only operate at cryogenic temperatures. Room temperature operation will require devices in which the charging islands are less than $100\,\text{\AA}$ in size and, although this represents a significant challenge to microfabrication techniques, recent progress has been quite encouraging.

In addition to a small total capacitance, the charging island of a single-electron device must be connected to the outside world via leads whose resistance exceeds $26\,\mathrm{k\Omega}$; otherwise, the Coulomb blockade will be masked by quantum fluctuations (Averin and Likharev, 1992). To see why, consider the Heisenberg uncertainty relation in the form $\Delta E \Delta t > \frac{1}{2}h$. Quantum fluctuations will destroy the Coulomb blockade if the uncertainty in the energy ΔE exceeds the charging energy. To ensure this is not the case an electron must stay on the charging island for a time $\Delta t > hC/e^2$. We can equate the charge/discharge time of the island to its RC time constant, i.e. $\Delta t \approx \tau \approx RC$, where R is the total resistance through which the island is charged. This leads to the result that the resistance of any junction in the single electron device must exceed $R_{\mathrm{min}} = h/e^2 = 25.8\,\mathrm{k\Omega}$.

9.5.4   Single-electron Transistors

In a conventional device, such as a field effect transistor, the current is carried by a large number of electrons. We can estimate this number by considering a state-of-the-art transistor such as the HEMT mentioned in Section 9.1. These devices make use of expensive technology to achieve channel lengths (i.e. gate lengths) as small as $0.1\,\mathrm{\mu m}$. An electron travelling at $10^5\,\mathrm{m/s}$, which is close to the saturation velocity in GaAs, will take $10^{-12}\,\mathrm{s}$ to travel from source to drain. If the current flowing is $1\,\mathrm{mA}$ then a charge of $10^{-15}\,\mathrm{C}$ will pass between source and drain in 10^{-12} s corresponding to more than 6000 electrons. In the 'off' state, almost zero current will be flowing so that, if a $1\,\mathrm{mA}$ current pulse was used to represent a single digital 'bit', the transfer of information will depend on the properties of over 6000 electrons in the channel. For the case of a single-electron transistor (SET) the situation is markedly different. When the SET is in the 'on' state, current can flow because one, and only one, electron at a time can tunnel onto and off the conducting island. As soon as the electron has tunnelled off the

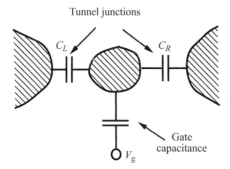

Tunnel junctions

Figure 9.29.   A schematic picture of the single-electron transistor.

island another can tunnel on and in this way the current is maintained, but at any given instant only a single electron is passing through the device.

The SET is perhaps the ultimate logic device where digital '1' and '0' are represented by the presence and absence of a single-electron. Device structures which exhibit single electron behaviour have been demonstrated in a variety of material systems including Al:Al$_2$O$_3$ tunnel junctions (Fulton and Dolan, 1987; Geerlings et al., 1990; Zimmerli et al., 1992) gated heterojunctions (Meirav et al., 1990) and delta-doped layers (Nakazato et al., 1992, 1993). In essence all of these structures are a form of the gated double junction shown schematically in Fig. 9.29. The gate electrode is fundamentally different from the other two contacts in the sense that although it is capacitively coupled to the central island it is not a tunnel junction and therefore no electrons can pass from the gate to the island. The purpose of the gate is to induce additional charge on the island, $\Delta Q = C_g V_g$, where V_g is the voltage applied to the gate. Unlike the current-carrying charge flowing between the two tunnel junctions, the gate induced charge can be varied continuously, i.e. in units smaller than the electronic charge, e, because it is essentially a polarization charge brought about by a change in the local chemical potential due to the applied gate voltage. The charging energy of the gated double junction now includes a term from the quasi-charge ΔQ, i.e.

$$U(N, V_g) = \frac{Q^2}{2C_T} = \frac{(-Ne + C_g V_g)^2}{2C_T} \tag{9.23}$$

where C_T is the total capacitance of the island, $C_T = C_L + C_R + C_g$, and N is the number of electrons on the island.

As the gate voltage is increased, the energy of the SET grows quadratically in a similar fashion to that shown in Fig. 9.26. If the voltage applied across the source and drain is very small (i.e. much less than V_T), the SET is in the regime of Coulomb blockade, its conductivity is small and the device is in the 'off' state. If we increase the gate voltage by an amount ΔV_g, we will reach a condition such that $U(N, V_g) = U(N + 1, V_g + \Delta V_g)$. At this point an extra electron can tunnel onto the island and a current will flow. The conductivity of the SET is no longer negligible and the device is in the 'on' state. Substituting the expression for $U(N, V_g)$ in the above condition shows that $\Delta V_g = e/C_g$. An extra electron can tunnel onto the island each time the gate voltage is increased by ΔV_g and the conductivity of the SET will oscillate with a period given by e/C_g.

A variety of different approaches have been used to demonstrate Coulomb blockade phenomena. Meirav et al., (1990) have used a combination of back gates and surface gates to create SETs which show clear oscillations in the conductance. Ford et al. (1993) have also used surface gates (marked as g_1, g_2 and g_4 in the inset to Fig. 9.30) to define an isolated island of charge. The gating action is derived from another surface gate, g_3, which can independently control the number of electrons on the island. Below a certain gate threshold voltage (i.e. $V_{g_3} < -0.48$ V in this case) the conductance of the device is close to zero but as V_{g_3} increases, well-defined periodic oscillations are observed, as shown in Fig. 9.30. The period of the oscillations is $\delta V_{g_3} = 7$ mV, and equating this to e/C_g gives a value of $C_g \approx 23$ aF. The other gates are likely to have a similar capacitance and we might reasonably expect a total capacitance of $C_T \approx 100$ aF. Knowing the total capacitance we can estimate an upper limit to the temperature at which we shall still be able to observe the Coulomb blockade by equating the thermal energy to the charging energy, i.e. $kT_{max} = e^2/2C_T$. If we assume that $C_T = 10^{-16}$ F, the temperature must be less than $T_{max} \approx 9$ K and the results shown in Fig. 9.30 were taken at a temperature less than 100 mK using a dilution refrigerator.

Practical applications of SETs will be much more likely to develop if they can be made to operate at temperatures higher than the mK regime. To achieve room temperature operation a very considerable decrease in the capacitance and, therefore, a corresponding reduction in the size of the SET will have to be achieved. One approach that lends itself to further size reduction is side gating (Wieck and

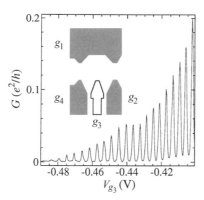

Figure 9.30.   Periodic conductance (G) oscillations for negative gate voltages (V_{g_3}) produced by surface gates as in inset (Reprinted from Ford et al., 1993 with permission from Elsevier Science). (Courtesy C. J. B. Ford)

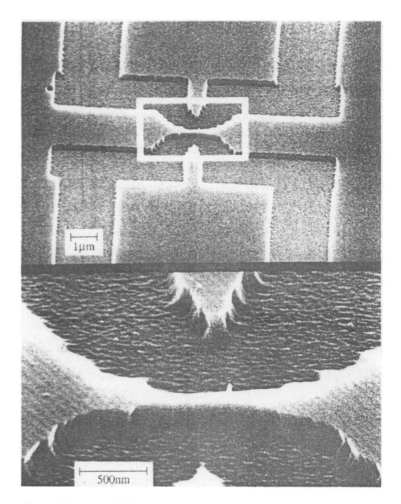

Figure 9.31.   An SEM (scanning electron microscopy) image of the side gated SET used by Nakazato *et al.* (1992). (Courtesy T. J. Thornton)

Ploog, 1990) in delta-doped layers (Feng *et al.*, 1992). Nakazato *et al.* (1992) have fabricated SETs using this approach and a photograph of the device taken with a scanning electron microscope is given in Fig. 9.31. The 2DEG in the narrow constriction breaks up into a number of islands and electrons can propagate from source to drain by tunnelling between the islands. The quasi-charge on the islands can be influenced by capacitive coupling from the finger electrodes at the side of the channel (the so-called side gating technique). The I–V characteristic of the side gated constriction are reproduced in Fig. 9.32. For $V_g = 0$ there is a clear region of Coulomb blockade at low bias. As the side gates are reverse biassed the transistor oscillates between regions of Coulomb blockade and finite conductance when electrons can tunnel through adjacent islands. For the case

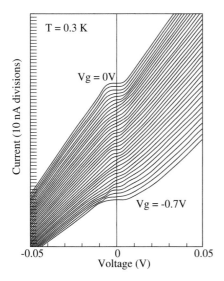

Figure 9.32.   Current–voltage characteristics of a side gated single-electron transistor. The curves are offset by increments of 10 nA. The gate voltage is decreased from 0 to −0.7 V in 20 mV steps (Nakazato *et al.*, 1992). (Courtesy T. J. Thornton)

$V_g = 0$ the threshold voltage, V_T, is approximately 5 mV, from which we can estimate a capacitance $C = e/2V_T$ of approximately 10^{-17} F.

9.5.5   Co-tunnelling and Multiple Tunnel Junctions

In the above discussion of the SET it has been implicitly assumed that the tunnelling process through the device proceeds in a sequential fashion, i.e. an electron first tunnels onto an island and remains there for some time before tunnelling off again. The tunnelling through a single junction is not instantaneous but occurs at a rate given by (Averin and Likharev, 1992; Grabert, 1991)

$$\Gamma = \frac{1}{e^2 R_T} \frac{\Delta E}{1 - \exp(-\Delta E/kT)} \tag{9.24}$$

where R_T is the resistance of the tunnel junction and ΔE is the change in energy of the system after tunnelling, i.e. $\Delta E = e^2/C_T$. At zero temperature this result reduces to $\Gamma = \Delta E/e^2 R_T$ and, substituting ΔE, shows that the charging rate for a single junction device is governed by the charging time of the capacitor, i.e. $\Gamma \sim 1/R_T C_T$.

It turns out that for the case of a double junction device an electron can tunnel through both junctions simultaneously via a process known as macroscopic quantum tunnelling or co-tunnelling. Any single-electron circuit relies on the fact that an electron stays on a particular transistor until it is allowed to move on to the next and any unwanted tunnelling events will reduce the reliability of the circuit.

If an electron can co-tunnel through two or more junctions it can bypass one SET completely and upset the timing of the circuit. For a double junction SET with high-resistance barriers, the co-tunnelling rate can be significant. However, if the electron has to tunnel simultaneously through a multiple array of junctions connected in series the co-tunnelling rate falls dramatically. For this reason any crucial charging nodes in a single-electron circuit should be coupled to the environment via multiple tunnel junctions whenever possible.

9.5.6 Possible Applications of Single-electron Transistors

As mentioned above, the possibility of using SETs to control the flow of individual electrons make this a very attractive device for use in digital logic circuits. In general, digital and analogue circuits are easier to design if the voltage gain of the individual devices is greater than unity so that the output of one SET can be used to switch the input of another. The research SET structures used by many groups would typically have low voltage gains, although values approaching 3 have been measured (Zimmerli et al., 1992). There is probably room for considerable improvement in the properties of individual SETs and this has encouraged a number of groups to consider the implementation of a range of circuits using SETs.

Averin and Likharev have considered a variety of logic elements built from SETs including NOT and OR gates (Averin and Likharev, 1992). Compared with the present complementary metal-oxide semiconductor (CMOS) implementation of these circuits, single-electron logic has the potential for extremely low power dissipation and ultra-large-scale integration perhaps extending into the molecular regime.

Low power dissipation and dense integration are important features of computer memory chips such as DRAMs (dynamic random access memory). DRAM manufacture is a very expensive technology which is close to reaching a mature level beyond which a further increase in the integration level will be almost impossible. SETs, however, may offer new device architectures with integration levels allowing for 1 TBit (10^{12} Bit) chips dissipating acceptable levels of power. Nakazato et al. (1993, 1994) have demonstrated a single-electron memory cell using the side gating approach described above.

One application of single-electron circuits which may be developed in the near future is the so-called single-electron turnstile. Turnstile devices use SETs to accurately clock electrons around a circuit and may provide an accurate current standard for metrological applications. Compared with the ohm and the volt, which can both be defined with reference to physical phenomena (the quantum Hall and Josephson effects, respectively), the ampere is based on a less accurate standard. Turnstile devices are designed to clock single electrons around a circuit at an accurately defined frequency giving a current $I = ef$. If one and only one electron moves through the circuit each cycle then the current will be as accurately

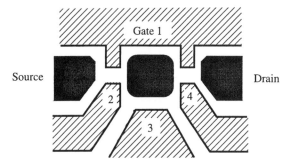

Figure 9.33.   The surface gate arrangement of a single-electron turnstile.

defined as the clock frequency and, since timing can be measured extremely precisely, single-electron turnstiles are expected to lead to significant improvement in the accuracy of the current standard.

A single-electron turnstile using oscillating tunnel barriers has been proposed by Odintsov (1991) and experimentally realized by Kouwenhoven *et al.* (1991) who used a split-gate structure to define the tunnel barriers. A schematic picture of the surface gate geometry is shown in Fig. 9.33. The gates are reverse biassed to define the SET in the underlying 2DEG. The geometry is similar to the generic double junction shown in Fig. 9.28(b) with the exception that now the transmissivity (i.e. barrier resistance) of the tunnel junctions can be varied by means of the gate electrodes marked as 2 and 4. These electrodes are independently biassed at microwave frequencies and modulate the barrier heights in anti-phase.

An energy profile across the device from source to drain is shown in Fig. 9.34. A sufficient voltage, V_{SD}, is applied to ensure that one, and only one, electron can tunnel onto the island during each cycle. As we have seen, the tunnelling is not instantaneous and for the zero temperature approximation we can write a tunnelling rate $\Gamma = 1/R_T C_T$ where R_T and C_T are the resistance and capacitance of the tunnel junction. A high barrier will have a large resistance and the tunnelling rate will be small compared with the oscillating barrier frequency. When the barrier is lowered (Fig. 9.34(b)) the tunnelling probability increases and an electron can rapidly charge the island before the barrier is raised again (Fig. 9.34(c)). Half the period later, the right-hand barrier is lowered allowing the electron to leave the turnstile and the process is ready to repeat giving an average current $I = ef$.

The gate modulation frequency and the number of electrons flowing through the turnstile determine the current flowing. More than one electron could be clocked through each cycle and, provided the number of electrons, N, remained constant, the current would be quantized at a value of $I = Nef$. However, if electrons can pass through the turnstile randomly by, for instance, co-tunnelling, then the accuracy of the current quantization will be reduced. As we saw earlier, multiple tunnel junctions reduce the co-tunnelling rate and will have to be used in

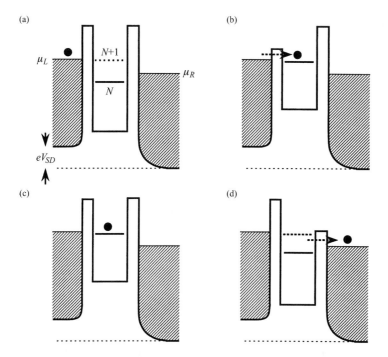

Figure 9.34.   (a)–(d) The clock cycle required to drive a single-electron through the turn-stile.

turnstile devices if they are to provide current standards with an accuracy of better than 1 part in 10^7.

9.6   The Future of Mesoscopic Devices

As mentioned in Section 9.1, much of the driving force behind the study of low-dimensional structures is the hope that new devices will eventually be developed. However, despite generating a great deal of interesting physics, it is clear that mesoscopic devices still have a long way to go before real applications can be considered. In fact there are compelling arguments which suggest that the diffi-culties in developing a new technology may be so overwhelming as to make any applications extremely unlikely (Landauer 1989, 1990). This may well be the case, but for the time being it is reasonable to explore ways in which the new physics might be exploited.

For the case of the quantum interference transistors the future is quite clear; it seems unlikely that any device based on quantum interference will emerge and their acronym QUIT seems rather appropriate. Some of the many difficulties associated with quantum interference transistors have been described in Section 9.2.4.3. In principle, all the individual problems could be overcome but taken together they justify our rather pessimistic conclusion. The same is probably true for the ballistic electron devices for the reasons discussed in Section 9.3.4.

Negative differential devices based on electron tunnelling have a long history dating back to the Esaki tunnel diode. These types of device generated a lot of excitement but never developed into a dominant technology. The new generation of double barrier resonant tunnelling devices will oscillate at frequencies beyond 400 GHz and have been used as mixers in high frequency circuits. This technology is just in its infancy and may develop to satisfy certain niche markets. The next step would be the miniaturization and integration of resonant tunnelling devices (RTDs) and the future of quantum dot RTDs is probably not as bleak as that of the quantum interference transistor.

Of the three types of device described in this chapter the single-electron transistor has the most promising future. This is because the performance of the SET improves as it is made smaller and it therefore lends itself to extreme integration levels in contrast to conventional devices such as the MOSFET which will cease to function when it is made so small that, for instance, electrons can tunnel through the gate oxide. The DRAM example used above is a case in point. The present technology may ultimately produce chips with an integration level of more than 1 GBit but the power dissipated will make it very hard to go beyond this to the 1 TBit level. In other words, even if technologies are invented which allow integrated circuit fabrication using design rules of 100 Å it will probably be impossible to use scaled down versions of present-day devices and there will be a real need for new devices based on fundamentally different physical principles. The single-electron transistor may be such a device.

EXERCISES

1. By considering the number of electrons travelling at the Fermi velocity, v_F, that cross unit area per second, show that the coefficient, D, for a degenerate 2DEG is given by

$$D = \tfrac{1}{2}v_F^2 \tau_e$$

2. Starting from the Drude semiclassical expression for conductivity, $\sigma = (ne^2\tau_e)/m_e$, show that, if the electrons which scatter after length ℓ_0 are those moving with the Fermi momentum $\hbar k_F$, then

$$\ell_0 \varrho_0 = \frac{\hbar k_F}{ne^2}$$

3. A GaAs sample is patterned into a gated ring as shown in the figure below and is to be used as an electron interferometer. The electron concentration per unit length under the gate is $n = 10^8 \, \text{m}^{-1}$ for $V_g = 0$ volts. The electron mass is given by $m^* = 0.067m_e$.

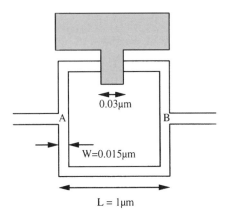

Figure 9.35.   Diagram for Exercise 3.

(a) Assuming a square well confining potential with hard walls, calculate the energy difference between the first and second subbands.

(b) If the Fermi velocity of an electron in a 1D subband is given by $v_F = \pi h n/m^*$, show that only one subband is occupied in the arms of the ring.

(c) A gate voltage is now applied such that the electron concentration under the gate is reduced to $5 \times 10^7\,\mathrm{m}^{-1}$. Estimate the phase difference between two electrons travelling through different arms of the ring when they emerge at point B. You may assume that the electrons are in phase when they enter the ring at point A and that they suffer no phase breaking events as they propagate through the ring.

(d) Calculate the magnetic field that would give the same change in phase for electrons propagating through an ungated ring of the same geometry.

(e) Comment briefly on the effect an increase in wire width would have on the phase of the electron waves.

4. Describe in some detail the essential features of a single-electron transistor (SET). Explain why the conductance (resistance^{-1}) of an SET oscillates as a function of the applied gate bias. Show that the period of the oscillations is given by $\Delta V_g = e/C_g$, where C_g is the gate capacitance.

5. The graph in Fig. 9.32 shows the current–voltage characteristics of a SET for a number of different gate biasses ranging from 0 to $-0.7\,\mathrm{V}$. The difference in gate voltage between successive traces is 20 mV and they are offset vertically for clarity. Using the features of the graph, calculate the gate capacitance and total capacitance of the SET. Estimate the temperature above which this SET would stop operating.

6. Figure 9.20 shows the magnetoresistance of a GaAs quantum wire of nominal width 1 μm. By extracting data from the figure calculate:

(a) the sheet electron density of the 2DEG;
(b) the mobility of the bulk sample, μ_0;
(c) the electron mean-free path in the bulk sample, ℓ_0;
(d) the wire width, W;
(e) the boundary scattering coefficient, p.

References

E. Abrahams and T. V. Ramakrishnan, *J. Non. Cryst. Solids* **35 & 36**, 15 (1980).

E. Abrahams, P. W. Anderson, D. C. Licciardello and T. V. Ramakrishnan, *Phys. Rev. Lett.* **42**, 718 (1979).

Y. Aharonov and D. Bohm, *Phys. Rev.* **115**, 485 (1959).

H. Akera and T. Ando, *Phys. Rev. B* (1990).

P. W. Anderson, E. Abrahams and T. V. Ramakrishnan, *Phys. Rev. Lett.* **43**, 718 (1979).

B. L. Altshuler, A. G. Aronov and D. E. Khmelnitskii, *J. Phys. C* **15**, 7367 (1982).

B. L. Altshuler and D. E. Khmelnitskii, *JETP Lett.* **42**, 359 (1986).

N. W Ashcroft and N. D. Mermin, *Solid State Physics* (Holt, Reinhart and Winston, New York, 1976).

D. V. Averin and K. K. Likharev, *J. Low Temp. Phys.* **62**, 345 (1986).

D. V. Averin and K. K. Likharev, in *Single Charge Tunneling*, H. Grabert and M. H. Devoret, eds. (Plenum, New York, 1992).

H. U. Baranger and A. D. Stone, *Phys. Rev. Lett.* **63**, 414 (1989).

C. W. J. Beenakker and H. van Houten, *Phys. Rev. Lett.* **63**, 1857 (1989).

G. Bergmann, *Phys. Rev. Lett.* **43**, 1357 (1979).

G. Bergmann, *Phys. Rev. B* **28**, 2914 (1983).

E. R. Brown, T. C. L. G. Sollherm, C. D. Parker, W. D. Goodhue and C. L. Chen, *Appl. Phys. Lett.* **55**, 1777 (1989).

M. Büttiker, *Phys. Rev. B* **38**, 12724 (1986).

S. Datta, *Superlatt. Microstructures* **6**, 83 (1989).

M. W. Dellow, P. H. Beton, M. Henini, P. C. Main, L. Eaves, S. P. Beaumont and C. D. W. Beaumont, *Electron. Lett.* **27**, 134 (1991).

M. W. Dellow, P. H. Beton, C. J. G. M. Langerak, T. J. Foster, P. C. Main, L. Eaves, M. Henini, S. P. Beaumont and C. D. W. Wilkinson, *Phys. Rev. Lett.* **68**, 1754 (1992).

R. Dingle, H. L. Stormer, A. C. Gossard and W. Weigmann, *Appl. Phys. Lett.* **33**, 665 (1978).

R. Eisberg and R. Resnick, *Quantum Physics* (Wiley, New York, 1974).

L. Esaki, *Phys. Rev.* **109**, 603 (1958).

Y. Feng, T. J. Thornton, J. J. Harris and D. Williams, *Appl. Phys. Lett.* **60**, 94 (1992).

D. K. Ferry, *Semiconductors* (Macmillan, New York, 1991).

C. J. B. Ford, T. J. Thornton, R. Newbury, M. Pepper, H. Ahmed, D. D. Peacock, R. A. Ritchie and J. E. F. Frost, *Phys. Rev. B* **38**, 8518 (1988).

C. J. B. Ford, T. J. Thornton, R. Newbury, M. Pepper, H. Ahmed, D. D. Peacock, R. A. Ritchie and J. E. F. Frost, *Appl. Phys. Lett.* **54**, 21 (1989).

C. J. B. Ford, P. J. Simpson, M. Pepper, D. Kern, J. E. F. Frost, D. A. Ritchie and G. A. C. Jones, *Nanostructured Materials* **3**, 283 (1993).

A. B. Fowler, Semiconductor Interferometer, U.S. Patent 4, 550, 330 (1984).

T. A. Fulton and D. J. Dolan, *Phys. Rev. Lett.* **59**, 109 (1987).

L. J. Geerligs, V. F. Anderegg, P. A. M. Holweg, J. E. Mooij, H. Pothier, D. Esteve, C. Urbina and M. H. Devoret, *Phys. Rev. Lett.* **64**, 2691 (1990).

C. J. Goodings, J. R. A. Cleaver and H. Ahmed, *Electron. Lett.* **28**, 1535 (1992).

L. P. Gorkov, A. I. Larkin and D. E. Khmelnitskii, *JETP Lett.* **30**, 228 (1979).

H. Grabert, *Z. Phys. B* **85**, 319 (1991).

P. Guéret, N. Blanc, R. Germann and H. Rothuizen, *Phys. Rev. Lett.* **62**, 1896 (1992).

D. L. Harame, J. H. Comfort, J. D. Cressler, E. F. Crabbé, J. Y. Sun, B. S. Meyerson and T. Tice, *IEEE Trans. Electron. Devices* **42**, 469 (1995).

S. Hikami, A. I. Larkin and Y. Nagaoka, *Prog. Theor. Phys.* **63**, 707 (1980).

H. van Houten, B. J. van Wees, J. E. Mooij, C. W. J. Beenakker, J. G. Williamson and C. T. Foxon, *Europhys. Lett.* **5**, 721 (1988).

T. Ikoma, T. Odagiri and K. Hirakawa in *Quantum Effect Physics, Electronics and Applications (IOP Conf. Ser. 127)*, K. Ismail, T. Ikoma and H. I. Smith, eds. (IOP, London, 1992).

C. Jacobini and D. K. Ferry, in *Quantum Transport in Ultrasmall Devices*, D. K. Ferry, H. L. Grubin, C. Jacobini and A.-P. Jauho, eds. (Plenum, London, 1995).

Y. Kawaguchi, H. Hitahari and S. Kawaji, *Surf. Sci.* **73**, 520 (1978).

L. P. Kouwenhoven, A. T. Johnson, N. C. van der Vaart, C. J. P. M. Harmans and C. T. Foxon, *Phys. Rev. Lett.* **67**, 1626 (1991).

R. Landauer, *IBM J. Res. Dev.* **1**, 223 (1957).

R. Landauer, *Physics Today* **42**, 119 (1989).

R. Landauer, *Physica A.* **168**, 75 (1990).

P. A. Lee, A. D. Stone and H. Fukuyama, *Phys. Rev. B* **35**, 1039 (1987).

K. K. Likharev, *IEEE Trans. Mag.* **23**, 1142 (1987).

K. K. Likharev, in *Granular Nanoelectronics*, D. Ferry, J. R. Barker and C. Jacaboni, eds. (Plenum, New York, 1990).

K. K. Likharev, in *Mesoscopic Phenomena in Solids*, B. L. Altshuler, P. A. Lee and R. A. Webb, eds. (North Holland, Amsterdam, 1991).

U. Meirav, M. A. Kastner and S. J. Wind, *Phys. Rev. Lett.* **65**, 771 (1990).

T. Mimura, S. Hiyamizu, T. Fujii and K. Nambu, *Jpn. J. Appl. Phys.* **19**, L225 (1980).

H. Mizuta, C. J. Goodings, M. Wagner and S. Ho, *J. Phys. C* **4**, 8783 (1992).

K. Nakazato, T. J. Thornton, J. White and H. Ahmed, *Appl. Phys. Lett.* **61**, 3145 (1992).

K. Nakazato, R. J. Blaikie, J. R. A. Cleaver and H. Ahmed, *Electronics Lett.* **29**, 384 (1993).

K. Nakazato, R. J. Blaikie and H. Ahmed, *Appl. Phys. Lett.* **75**, 5123 (1994).

A. A. Odintsov, *Appl. Phys. Lett.* **58**, 2695 (1991).

A. B. Pippard, *Magnetoresistance in Metals* (Cambridge University Press, Cambridge, 1989), Ch. 6.

R. Prasad, T. J. Thornton, A. Matsumura, J. M. Fernández and D. Williams, *Semicond. Sci. Technol.* **10**, 1084 (1995).

M. A. Reed, J. N. Randall, R. J. Aggarwal, R. J. Matyi, T. M. Moore and A. E. Wetsel, *Phys. Rev. Lett.* **60**, 535 (1988).

M. L. Roukes, A. Scherer, S. J. Allen Jr, H. G. Graighead, R. M. Ruthen, E. D. Beebe and J. P. Harbison, *Phys. Rev. Lett.* **59**, 3011 (1988).

M. L. Roukes *et al.*, in *Electronic Properties of Multilayers and Low Dimensional Semiconductor Structures*, J. M. Chamberlin, L. Eaves and J. C. Portel, eds. (New York, Plenum, 1990).

L. I. Schiff, *Quantum Mechanics* (McGraw-Hill, New York, 1968).

W. J. Skocpol, P. M. Mankiewich, R. E. Howard, L. D. Jackel, D. M. Tennant and A. D. Stone, *Phys. Rev. Lett.* **56**, 2865 (1986).

T. Sollner *et al.*, in *Physics of Quantum Electron Devices*, F. Capasso, ed. (Springer, Berlin, 1990).

F. Sols, M. Macucci, U. Ravaioli and K. Hess, *Appl. Phys. Lett.* **54**, 350 (1989).

J. Spector, H. L. Störmer, K. W. Baldwin, L. N. Pfeiffer and K. W. West, *Appl. Phys. Lett.* **56**, 1290 (1990a)

J. Spector, H. L. Störmer, K. W. Baldwin, L. N. Pfeiffer and K. W. West, *Appl. Phys. Lett.* **56**, 2433 (1990b).

A. D. Stone, *Phys. Rev. Lett.* **54**, 2692 (1985).

A. Szafer and A. D. Stone, *Phys. Rev. Lett.* **62**, 300 (1989).

Y. Takagaki, K. Gamo, S. Namba, S. Ishida, K. Ishibashi and K. Murase, *Solid State Comm.* **68**, 1051 (1988).

M. Tewordt, V. J. Law, M. J. Kelly, R. Newbury, M. Pepper, D. C. Peacock, J. E. F. Frost, D. A. Ritchie and G. A. C. Jones, *J. Phys.: Condens. Matter* **2**, 8969 (1990).

T. J. Thornton, M. Pepper, H. Ahmed, G. J. Davies and D. Andrews, *Phys. Rev. Lett.* **56**, 1198 (1986).

T. J. Thornton, M. Pepper, H. Ahmed, G. J. Davies and D. Andrews, *Phys. Rev. B* **36**, 4514 (1987).

T. J. Thornton, M. L. Roukes, A. Scherer and B. P. Van der Gaag, *Phys. Rev. Lett.* **63**, 2128 (1989).

T. J. Thornton, M. L. Roukes, A. Scherer and B. P. Van der Gaag, in *Granular Nanoelectronics*, D. K. Ferry, J. R. Barker and C. Jacoboni, eds. (New York, Plenum, 1990).

G. Timp, A. M. Chang, J. E. Cunningham, T. Y. Chang, P. Mankiewich, R. Beringer and R. E. Howard, *Phys. Rev. Lett.* **58**, 2814 (1987).

C. P. Umbach, S. Washburn, R. B. Laibowitz and R. A. Webb, *Phys. Rev. Lett.* **30**, 4048 (1984).

P. G. N. de Vegvar, G. Timp, P. M. Mankiewich, R. Behringer and J. Cunningham, *Phys. Rev. Lett.* **40**, 3491 (1989).

S. Washburn and R. A. Webb, *Adv. Phys.* **35**, 375 (1986).

S. Washburn, H. Schmid, D. Kern and R. A. Webb, *Phys. Rev. Lett.* **59**, 1791 (1987).

R. Webb, S. Washburn, C. P. Umbach and R. B. Laibowitz, *Phys. Rev. Lett.* **54**, 2696 (1985).

B. J. van Wees, H. van Houten, C. W. J. Beenakker, J. G. Williamson, L. P. Kouwenhoven, D. van der Marel and C. T. Foxon, *Phys. Rev. Lett.* **60**, 848 (1988).

D. A. Wharam, T. J. Thornton, R. Newbury, M. Pepper, H. Ahmed, J. E. F. Frost, D. G. Hasko, D. D. Peacock, D. A. Ritchie and G. A. C. Jones, *J. Phys. C* **21**, L209 (1988).

A. D. Wieck and K. Ploog, *Surf. Sci.* **229**, 252 (1990).

G. Zellar and I. Giaever, *Phys. Rev.* **181**, 798 (1969).

G. Zimmerli, R. L. Kantz and J. M. Martinis, *Appl. Phys. Lett.* **61**, 2616 (1992).

10 High-speed Heterostructure Devices

J. J. Harris

10.1 Introduction

High-speed semiconductor devices are key elements in the development of electronic systems for data processing or analogue signal handling at ever-higher frequencies (the state-of-the-art is currently in excess of 100 GHz; a detailed discussion of a wide range of devices is given by Sze (1990)). Most of these systems use circuits based on Si devices prepared by implantation or diffusion, but these fabrication processes are limited in their ability to produce the small-scale device structures required for high-speed operation. However, the advent of highly controllable epitaxial deposition processes such as molecular-beam epitaxy (MBE) and metal-organic vapour-phase epitaxy (MOVPE) has enabled semiconductor structures to be prepared with compositional or dopant properties defined in layers with thicknesses down to the atomic scale. In many cases, this results in the mobile charge carriers being confined in a quasi-two-dimensional sheet, giving rise to a wide range of quantum confinement effects, as described in the earlier chapters of this book. Historically, these heterostructures have been prepared in III–V compound semiconductors, principally GaAs and (Al,Ga)As, although recently much effort is being put into investigating epitaxial Si and SiGe layers. This chapter will review two classes of electronic device, namely those involving transport of charge along the two-dimensional sheets, based on variations on the field-effect transistor (FET) principle, and those using 'vertical' transport, i.e. perpendicular to the sheets. It is interesting to note that, with the exception of resonant tunnelling and superlattice structures, the performance characteristics of these devices are not primarily due to quantum confinement effects associated with their two-dimensional nature, although these may play a secondary role in some instances. Nevertheless, as the following sections will demonstrate, the ability to prepare semiconductor structures with control on the nanometre scale has resulted in the development of a wide range of new or improved devices.

10.2 Field-effect Transistors

The basic principle of the FET is illustrated in Fig. 10.1. The flow of charge carriers between two ohmic contacts, the source and drain, is controlled by the

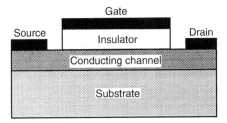

Figure 10.1.   Schematic structure of a basic FET.

voltage applied to a third electrode, the gate, isolated from the conducting region either by a layer of insulator, as in a metal-insulator-semiconductor-FET (MISFET, or if the insulator is an oxide, MOSFET), or by the depletion region of a Schottky barrier, as in the metal-semiconductor-FET (MESFET). The gate, insulator/depletion region and conducting channel form a parallel-plate capacitor structure, so that application of a voltage across this structure will determine the number of charge carriers in the conducting channel, and hence the magnitude of the current flowing between source and drain. Four implementations of this device are considered below: the silicon MOSFET, the high-electron-mobility transistor (HEMT) in GaAs/AlGaAs and in GaAs/InGaAs/AlGaAs, and the delta-doped FET (δ-FET).

10.2.1 The Si MOSFET

Although epitaxial techniques are currently not employed in the fabrication of this device (the insulating region is generally formed by oxidation of the Si surface), conduction nevertheless takes place through a quasi-two-dimensional channel, i.e. the accumulation or (more usually) the inversion layer formed at the Si-SiO$_2$ interface when an appropriate gate voltage is applied. Indeed, as mentioned in Section 5.5.1, the quantum Hall effect was first observed in a Si MOSFET cooled to low temperatures (von Klitzing *et al.*, 1980). Figure 10.2(a) shows the structure of an *n*-channel MOSFET formed in *p*-type Si; a positive gate bias creates an inversion layer at the interface, allowing current to flow between the *n*-type contact regions. Figure 10.2(b) shows the DC characteristics of such a device. The high-frequency performance of Si MOSFETs will be compared with that of various HEMT structures in Sections 10.2.2 and 10.2.3.

  The operating characteristics of the Si MOSFET are well understood, and are treated in most textbooks on semiconductor devices (see, e.g., Sze, 1985; Bar-Lev, 1984). There are strong similarities in basic operation between these devices and the heterojunction transistors described below, and expressions will be derived here for the current–voltage behaviour of an ideal device operating in the low and high field limits. Consider the situation depicted in Fig. 10.3, of an *n*-channel MOSFET formed in *p*-type substrate material. In equilibrium (Fig. 10.3(a)), the

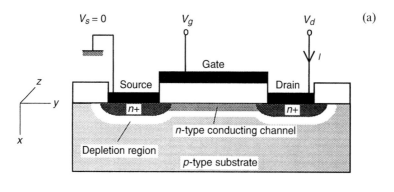

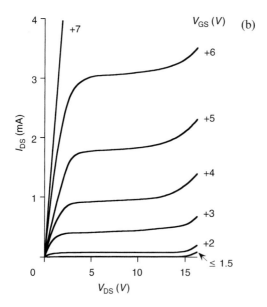

Figure 10.2.   *n*-type inversion channel MOSFET: (a) structure; (b) DC characteristics of a typical device.

contact potential difference, V_{bi}, is seen to be composed of a voltage drop, $V_{ox}(0)$, across the oxide layer, and the depletion potential, $\phi_s(0)$, at the semiconductor surface. Application of a voltage V_{app} to the gate further depletes, and eventually inverts, the surface. A convenient criterion for inversion is as shown in Fig. 10.3(b), where the Fermi level at the surface lies as far above the intrinsic level as it does below it in the bulk, i.e. $|\psi_s| = |\psi_b|$. With a higher applied voltage V_{app}, further surface carriers are induced, but this requires only a relatively small change in ψ_s, which is therefore usually taken as constant. Figure 10.3(c) shows this situation, together with the result of applying a voltage V_d to the drain. At an arbitrary point, y, along the channel, the difference between bulk and surface Fermi levels is $V(y)$. The total potential difference between the bulk of

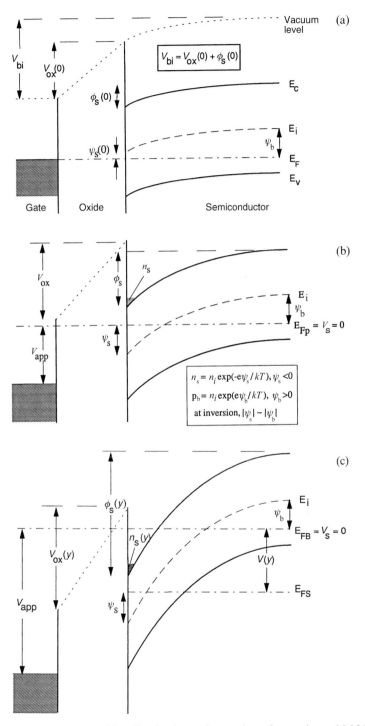

Figure 10.3.   Band-bending in the surface region of an n-channel MOSFET: (a) in equilibrium; (b) at inversion; (c) with an applied source-drain voltage.

the semiconductor and the gate, V_g, is given by

$$V_g = V_{bi} + V_{app} = V_{ox}(y) + \phi_s(y) \tag{10.1}$$

where V_{ox}, ϕ_s and the surface carrier density, n_s, all vary with position along the channel. Treating the gate-oxide-semiconductor combination as a parallel plate capacitor, the total charge per unit area stored in this capacitor due to a voltage V_g between gate and source is given by

$$Q_{ox} = \frac{\epsilon_i}{d} V_{ox}(y) = \frac{\epsilon_i}{d}[V_g - \phi_s(y)] \tag{10.2}$$

where ϵ_i and d are the dielectric constant and thickness of the insulator, and $\phi_s(y)$ is the surface potential at a distance y from the source contact (taken as the zero of potential). For positive gate voltages in excess of the threshold for inversion, and in the absence of interface states, this charge is distributed between free electrons in the inversion layer and charged acceptors in the depletion region of the substrate. The depletion potential, $\phi_s(y)$, is

$$V(y) + \psi_b + \psi_s \approx V(y) + 2\psi_b \tag{10.3}$$

so that, for an acceptor density N_A, the depletion charge Q_{dep} is

$$Q_{dep} = \{2\epsilon_s N_A e[V(y) + 2\psi_b]\}^{1/2} \tag{10.4}$$

where ϵ_s is the dielectric constant of the semiconductor. The sheet electron density in the inversion layer is $n_s = (Q_{ox} - Q_{dep})/e$. Writing dR for the resistance of an element of the 2DEG conducting channel of length dy and transverse dimension z, then the voltage drop, dV, across this element is given by

$$dV = I\, dR = \frac{I\, dy}{zn_s e\mu} \tag{10.5}$$

where μ is the electron mobility. It follows that, for a gate of length L and a source-drain voltage of V_d,

$$\int_0^L I\, dy = \int_0^{V_d} z\mu \left[\frac{\epsilon_i}{d}(V_g - 2\psi_b - V) - \sqrt{2\epsilon_s e N_A(V + 2\psi_b)}\right] dV \tag{10.6}$$

and hence, upon integration,

$$I = \frac{z\mu}{L}\left\{\frac{\epsilon_i}{d}[(V_g - 2\psi_b)V_d - \tfrac{1}{2}V_d^2] - \tfrac{2}{3}\sqrt{2\epsilon_s e N_A}[(V_d + 2\psi_b)^{3/2} - 2\psi_b^{3/2}]\right\} \tag{10.7}$$

In the limit of small V_d,

$$\lim_{V_d \to 0}(V_d + 2\psi_b)^{3/2} = (2\psi_b)^{3/2}\left[1 + \frac{3V_d}{4\psi_b} + \cdots\right] \tag{10.8}$$

so, retaining the first two terms in this expansion, (10.7) becomes

$$I \approx \frac{z\mu\,\epsilon_i}{L\,d}\left\{V_g - \left[2\psi_b + \frac{d}{\epsilon_i}\sqrt{2\epsilon_s e N_A(2\psi_b)}\right]\right\}V_d \tag{10.7a}$$

This equation predicts that, for fixed V_g, there is a linear increase of I with source-drain voltage, V_d, at small V_d. The threshold gate voltage, V_{th}, at which current just starts to flow is given by

$$V_{th} = 2\psi_b + \frac{d}{\epsilon_i}\sqrt{2\epsilon_s e N_A (2\psi_b)} \tag{10.9}$$

Equation (10.7) shows that at higher drain voltages the current increase becomes sublinear, and eventually $\partial I / \partial V_d$ becomes zero. The drain voltage at which this occurs is the 'saturation voltage', and corresponds to the situation in which the conducting channel at the drain end of the gate is completely pinched-off, i.e. depleted of mobile charge. Beyond this voltage, the current remains roughly constant, although this is not predicted by equation (10.7), which is invalid beyond saturation. Higher values of positive gate bias produce larger saturation currents and voltages, and thus equation (10.7) provides a good description of the behaviour shown in Fig. 10.2(b). For some short gate-length FETs, however, the experimental saturation currents fall below those predicted by equation (10.7), which describes the case of devices with a large source-drain separation, for which the low-field mobility, μ, can be treated as independent of V_d. For a short-gate device, the electric field in the channel can be high enough for velocity saturation to occur (Fig. 10.4), and it is this, rather than channel pinch-off, which is responsible for the levelling-off of the I–V_d characteristics in these structures. The velocity-field characteristic, expressed in terms of a field-dependent mobility, can be approximated as

$$\mu = \mu_0 \left[1 + \frac{1}{E_c}\frac{\partial V(y)}{\partial y}\right]^{-1} \tag{10.10}$$

where the critical field E_c is defined by $\mu_0 E_c = v_{sat}$, the saturation drift velocity. Substituting (10.10) into (10.6) and integrating, the following expression is

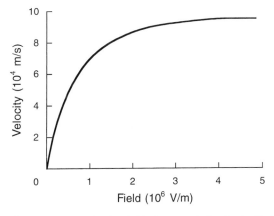

Figure 10.4.   Velocity-field characteristics for bulk Si. (Replotted on linear axes from the data of Canali *et al.*, 1975).

obtained for the drain current for negligible depletion charge, i.e. in the limit that $N_A \to 0$:

$$I = \frac{z\epsilon_i \mu_0}{Ld} \frac{V_g' V_d - \frac{1}{2}V_d^2}{1 + (V_d/V_c)} \tag{10.11}$$

where $V_g' = V_g - V_{th}$, $V_{th} = 2\psi_b$ for $N_A \to 0$, and $V_c = LE_c$. This equation reduces to the first term of equation (10.7) in the limit of large L, i.e. $V_c \gg V_d$. The saturation drain voltage, V_{dsat}, and current, I_{sat}, are found by differentiating (10.11) and setting $\partial I/\partial V_d = 0$, giving

$$V_{dsat} = V_c \left[\sqrt{1 + (2V_g'/V_c)} - 1 \right] \tag{10.12}$$

and

$$I_{sat} = V_c^2 \frac{z\epsilon_i \mu_0}{Ld} \left[\sqrt{1 + (2V_g'/V_c)} - 1 \right]^2 \tag{10.13}$$

In the long-gate, low-field limit (i.e. $V_c \gg V_g'$), these reduce to

$$V_{dsat} \approx V_g' \quad \text{and} \quad I_{sat} \approx \frac{z\epsilon_i \mu_0}{2Ld} V_g'^2 \tag{10.14}$$

while in the opposite limit (i.e. $V_c \ll V_g'$) we have

$$I_{sat} \approx \frac{z\epsilon_i \mu_0}{Ld} V_c V_g' \tag{10.15}$$

i.e. the transfer function, relating channel current to gate voltage, becomes linear. An important parameter of FET performance is the transconductance, g_m, defined as the differential transfer function relating changes in source-drain current to changes in gate voltage, i.e.

$$g_m = \frac{\partial I}{\partial V_g}\bigg|_{V_d=\text{constant}} \tag{10.16}$$

This is usually evaluated in the saturation region, where it is a maximum. Differentiating (10.11) and setting $V_d = V_{dsat}$ from (10.12) gives

$$g_m = \frac{\partial I}{\partial V_g}\bigg|_{V_d=V_{dsat}} = V_c \frac{z\epsilon_i \mu_0}{Ld} \left[1 - \frac{1}{\sqrt{1 + (2V_g'/V_c)}} \right] \tag{10.17}$$

In the long-gate limit, this reduces to

$$g_m = \frac{z\epsilon_i \mu_0}{Ld} V_g' \tag{10.18}$$

while for the high-field case,

$$g_m = V_c \frac{z\epsilon_i \mu_0}{Ld} = \frac{z\epsilon_i v_{sat}}{d} \tag{10.19}$$

which is independent of V_g' and L.

Note that, in the above discussion, quantization of the electron states in the potential well of the inversion layer has not been considered. As mentioned earlier, the existence of such quantization is demonstrated by the observation of the quantum Hall effect in Si MOSFETs at low temperatures, but the relatively small confinement energies and subband spacing means that for room temperature operation there is sufficient thermal smearing for confinement effects to be generally neglected.

10.2.2 GaAs/AlGaAs High-electron-mobility Transistor

The higher low-field mobility and saturation drift velocity of carriers in GaAs compared with Si make it an attractive material for high frequency devices. Unfortunately, formation of the native oxide on GaAs creates a high density of interface states (unlike SiO_2 on Si), making a MOSFET structure impractical, so that devices are either GaAs MESFETs, usually with a refractory metal such as tungsten or titanium as the Schottky gate material, or heterojunction structures such as HEMTs, based on a combination of GaAs and the ternary alloy $Al_xGa_{1-x}As$. MESFET structures will not be considered further here (for a discussion, see, e.g., Shur, 1987), but Fig. 10.5(a) illustrates a typical sequence of layers employed in the GaAs/$Al_xGa_{1-x}As$ HEMT, while Figs. 10.5(b, c) show the corresponding conduction band bending for two types of device, known as 'enhancement mode' and 'depletion mode' transistors. Since the only regions of this basic structure into which dopant atoms (usually Si donors) are deliberately introduced are part of the $Al_xGa_{1-x}As$ layer and the top GaAs layer, it is often referred to as a 'modulation-doped' heterojunction, leading to an alternative name for the device, the MODFET. This device has been discussed in Section 3.2.3, where approximate expressions for the energies and wavefunctions of the two-dimensional electron gas (2DEG) at the interface have been derived. Because of the spatial separation between electrons and ionized donors, impurity scattering is reduced and this is the origin of the 'high electron mobility' of the device name (for a recent review of these effects, see Harris, 1997). As shown in Fig. 10.5(a), it is usual to have a relatively thick doped region, both in the $Al_xGa_{1-x}As$ layer and the GaAs capping layer (used to improve the quality of the ohmic source and drain contacts), in order to reduce the resistance of the device between the ohmic contacts and the gated region. This, together with the contact resistance, constitutes the 'access' resistance of the device; if this is too high, the performance is degraded. The active region of the device is then created by recessing the gate so that it modulates only the conductivity of the 2DEG. With the shallower recess shown in Fig. 10.5(b), a current path exists at zero gate bias (the device is 'normally on'), and negative bias is required to deplete the channel, whereas the deeper recess of Fig. 10.5(c) gives a 'normally-off' device for which positive bias is used to create a conducting channel. This latter variant is very

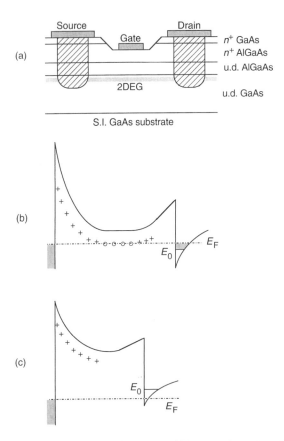

Figure 10.5.   High-electron-mobility transistor (HEMT): (a) basic structure; (b) conduction band profile of depletion mode device; (c) as (b), for enhancement mode.

similar to the n-channel MOSFET described earlier; hence, in what follows the discussion will be confined to the depletion mode device.

By treating the depleted $Al_xGa_{1-x}As$ region as the insulator in the gate capacitor structure, it is possible to derive analytic expressions for the device characteristics, very similar to those for the MOSFET. For a normally-on device, it can be shown (Das and Reszak, 1985) that

$$I_{sat} \approx \frac{z\epsilon_i v_{sat}}{d}\left(V_g - V_{off} - R_sI_{sat} - V_c\right) \tag{10.20}$$

where V_{off} is the pinch-off voltage required to deplete the channel, and R_s is the source access resistance (ignored in the discussion of the MOSFET). This formula is appropriate to a short gate-length device (for high frequency operation this is typically less than 0.5 μm), where the source-drain field is $\sim 10^6$ V/m. As shown in Fig. 10.6, at these fields the carrier velocity has saturated at a value slightly lower than that of lightly-doped bulk GaAs (Masselink, 1993), so that much of the advantage expected from the high low-field mobility is not in fact realized in

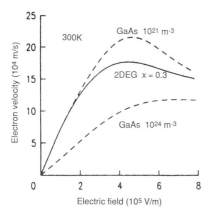

Figure 10.6. Velocity-field characteristic of GaAs and $Al_xGa_{1-x}As/GaAs$ HEMT structures: bulk GaAs, $n = 10^{21}\,m^{-3}$ and $n = 10^{24}\,m^{-3}$ (dashed lines); HEMT, $x = 0.3$ (solid line). (Reprinted from W. T. Masselink, 'Transport of quasi-two dimensional electrons in heterojunction field effect transistors', *Thin Solid Films* **231**, 86 (1993), with kind permission of Elsevier Science SA, P.O. Box 564, 1001 Lausanne, Switzerland.)

practice. Nevertheless, compared with GaAs of comparable electron density ($\sim 10^{24}\,m^{-3}$), there is an increase of $\sim 30\%$. A contributory factor to the low value of v_{sat} may be real-space transfer (Keever *et al.*, 1981), in which carriers gain enough energy to surmount the heterojunction barrier, and suffer increased scattering in the $Al_xGa_{1-x}As$. A Monte-Carlo simulation of this process in a quantum well is shown in Fig. 10.7. Although this effect is generally detrimental, it has been suggested that the negative differential resistance region could be exploited to generate microwave power in a negative resistance FET (NERFET) (Luryi *et al.*, 1984).

Typical DC characteristics of a HEMT are illustrated in Fig. 10.8(a), and the corresponding transconductance data are shown in Fig. 10.8(b). The long linear region (i.e. constant transconductance) in the latter results (Laviron *et al.*, 1981) is an important advantage of this device over the MESFET, for which g_m falls with increasing reversed bias. Figure 10.9 shows that, compared with the Si MOSFET and GaAs MESFET, HEMTs have high-speed performance with low power requirements (Solomon and Morkoç, 1984). They are also particularly notable for their low noise figure, which make them attractive for use as amplifiers in satellite TV receivers. Other applications include digital and millimetre wave integrated circuits.

A limitation of this device for some applications is that the output power is relatively low, because the two-dimensional carrier density cannot be increased much above $10^{16}\,m^{-2}$. To do so would require a larger conduction band offset, i.e. a higher Al content in the $Al_xGa_{1-x}As$ layer, but an undesirable consequence of this is an increase in the population of metastable Si donor states referred to as

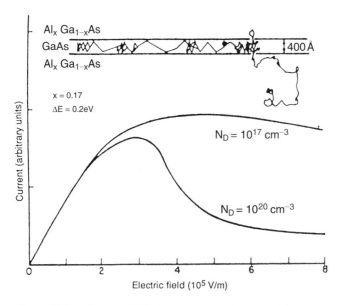

Figure 10.7. Current-field characteristics for a modulation-doped $Al_{0.17}Ga_{0.83}As/GaAs/$ $Al_{0.17}Ga_{0.83}As$ quantum well sample as a function of doping level in the $Al_{0.17}Ga_{0.83}As$. Inset is a Monte Carlo simulation of high field transport, showing real-space transfer. (From Keever *et al.* (1981), with permission of the authors.)

DX centres (Bourgoin, 1989). These states are a source of noise and instability, so that in practice the maximum Al content is restricted to about 25%. Alternative approaches to solving this problem are (a) the use of $In_xGa_{1-x}As$ in place of GaAs, or (b) replacing modulation doping by delta-doping. These concepts are discussed below.

10.2.3 InGaAs HEMTs

As shown in Fig. 10.10, addition of In to GaAs reduces the energy gap and increases the conduction band offset with $Al_xGa_{1-x}As$. Thus the amount of charge transferred can be increased, or alternatively, for the same band offset as in a $GaAs/Al_xGa_{1-x}As$ HEMT, the Al content can be lowered, thereby reducing the DX problem. $In_xGa_{1-x}As$ has several advantages over GaAs: a lower effective mass and higher mobility, a larger energy separation between the Γ and L conduction band minima, giving better high-field characteristics (a higher v_{sat}), and reduced real-space transfer because of the large band offset. Unfortunately, the lattice constant of $In_xGa_{1-x}As$ is greater than that of GaAs so that, when the layer is grown pseudomorphically on GaAs, there is an elastic strain in the layer resulting from a tetragonal distortion of the crystal lattice (Fig. 10.11(a)). This produces an increase in the energy gap, and lifts the degeneracy of the valence

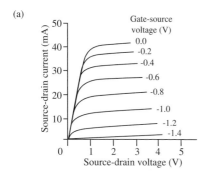

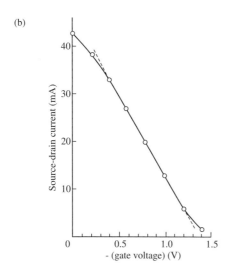

Figure 10.8.   (a) DC characteristics of an $Al_{0.3}Ga_{0.7}As/GaAs$ HEMT with a gate length of 0.3 μm; (b) I_{ds}–V_g curve showing linear transfer characteristic (constant g_m). (From Laviron *et al.* (1981), with permission of the IEE Publication Department.)

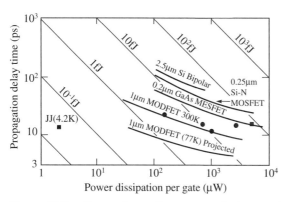

Figure 10.9.   Comparison of power-speed characteristics for Si bipolar and MOSFET devices, GaAs MESFETs and MODFETs. (From Solomon and Morkoç, © 1984 IEEE.)

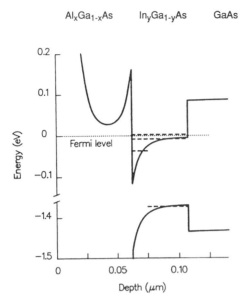

Figure 10.10.   Band-edge profile of a pseudomorphic $Al_{0.25}Ga_{0.75}As/In_{0.15}Ga_{0.85}As/GaAs$ HEMT.

bands (Fig. 10.11(b)); the dependence of band structure on strain must therefore be taken into account in device design, and can in fact be used to advantage in some circumstances. However, as the $In_xGa_{1-x}As$ thickness increases, the strain energy stored in the film eventually exceeds a critical value, and misfit dislocations are formed at the $GaAs/In_xGa_{1-x}As$ interface. These dislocations relieve the strain in the film, but degrade the device performance. The critical thickness depends on In content, so that although a higher In fraction gives a deeper well, the lattice mismatch is greater and the critical thickness is smaller; thus the well must be narrow ($< 200\,\text{Å}$ for 15% In), giving higher confined state energies and hence somewhat reduced charge transfer. A typical structure with 15% In in the well and 25% Al in the doped region will have a sheet electron density of $\sim 2 \times 10^{16}\,\text{m}^{-2}$.

An alternative to the use of strained layers on GaAs is to deposit $In_{0.53}Ga_{0.47}As$ on an InP substrate, since the two materials have the same lattice constant. The modulation-doped layer is then either InP or $Al_{0.48}In_{0.52}As$ (which is also lattice matched), and carrier densities up to $3 \times 10^{16}\,\text{m}^{-2}$ have been achieved in this system. As well as the increased output power resulting from the higher carrier densities, there is also a small but significant improvement in high frequency performance for both types of $In_xGa_{1-x}As$ HEMT. This can been seen in Fig. 10.12, which compares the gate-length dependences of the cutoff frequency, f_T (defined by $f_T = g_m/2\pi C_g$, where $C_g = \epsilon_i zL/d$ is the gate capacitance) for Si MOSFETs, AlGaAs/GaAs HEMTs and pseudomorphic and lattice-matched

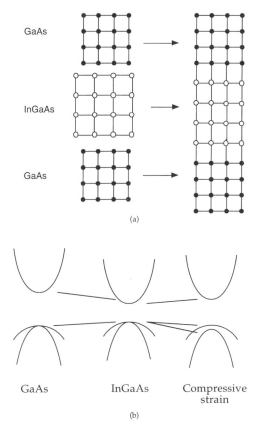

Figure 10.11.   Elastically strained GaAs/InGaAs/GaAs quantum well: (a) tetragonal distortion of the InGaAs lattice; (b) effect of strain on the InGaAs band structure.

InGaAs HEMTs (Weisbuch and Vinter, 1991). Using equation (10.19) gives $f_T = v_{sat}/2\pi L$, showing that the speed improvement is largely due to the increased saturation drift velocity in these materials.

10.2.4 Delta-doped FETs

It is possible to achieve very high quasi-2D sheet carrier densities by using the technique of 'delta-doping', in which the growth of an undoped semiconductor film is interrupted while impurity atoms are deposited on the surface, and growth is then recommenced (Schubert, 1990; Harris, 1993). For Si donors in GaAs (the most-studied system), a limit of $\sim 10^{17}\,\mathrm{m}^{-2}$ has been found, nearly an order of magnitude greater than the GaAs/Al$_x$Ga$_{1-x}$As HEMT. As discussed in Section 3.3.2.1, the electrostatic interaction between the electrons and ionized donors results in a V-shaped potential well, and for the high carrier densities used in devices, several subbands (typically ~ 5) are occupied, compared with 1 or 2 in a

HEMT. A disadvantage of this structure is that the mobility is lower than in a HEMT, because the ionized scattering centres are in the quantum well and not remote. The shape of the wavefunction envelopes means that electrons in the lowest subband are strongly scattered, and hence have a mobility comparable to doped GaAs, but higher-lying bands are more extended spatially, and thus can have significantly higher mobilities (up to a factor of ∼5 greater). The low-field mobility affects primarily the source and drain access resistances, which will therefore be rather higher than in a HEMT of the same geometry, but the high frequency performance is largely determined by the saturation drift velocity, and this is very similar in GaAs MESFETs, HEMTs, and δ-FETs. Other claimed advantages for the device are that (a) because the delta-doped layer can be positioned closer to the surface than the 2DEG in a HEMT, a higher g_m can be obtained, (b) the absence of space charge below the gate in the δ-FET results in a lower surface field and hence higher reverse-bias breakdown voltages, and (c) the absence of doped AlGaAs in the structure means that DX centres are no longer a problem. (This last point is contentious, since several authors have

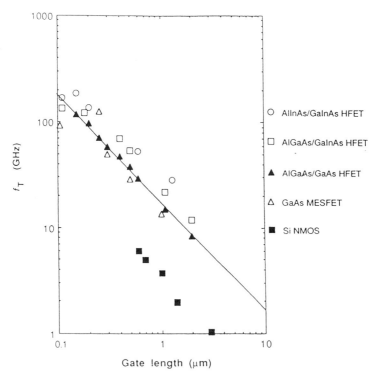

Figure 10.12. Comparison of cutoff frequency, f_T, versus gate length for three types of HEMT: AlGaAs/GaAs, pseudomorphic AlGaAs/InGa-As/GaAs, and lattice-matched AlInAs/InGaAs/InP. (From Weisbuch and Vinter, 1991, with permission of Academic Press).

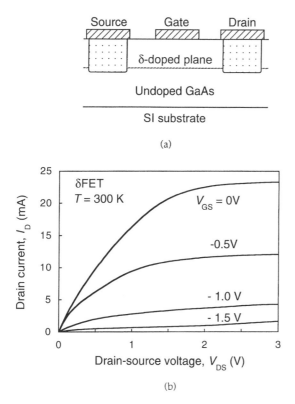

Figure 10.13.   Delta-doped FET: (a) structure of device; (b) DC characteristics. (From Schubert *et al.*, © 1986 IEEE.)

suggested that DX centres *in the GaAs* can become occupied at high doping levels, and this may be responsible for the observed limit to the sheet carrier density (Zrenner and Koch, 1988).) A schematic device structure is shown in Fig. 10.12 and typical DC characteristics are given in Fig. 10.13. The latter results (Schubert *et al.*, 1986) are similar to those of Fig. 10.8(a), and it is perhaps significant that, despite the claimed advantages of the δ-FET structure, research effort on this device appears to be outweighed by that on InGaAs HEMTs.

10.3   Vertical Transport Devices

In this section we will consider mainly unipolar devices, in which potential barriers formed by compositional or doping variations in the growth direction are used to control the flow of one type of carrier between contacts to the top and bottom layers of the device. However, impressive performance characteristics are currently being achieved from a recent variant of the bipolar transistor, the heterojunction bipolar transistor (HBT), and we will conclude with a discussion of this device. The basic concepts of unipolar devices will be introduced by reference to the various forms of unipolar diode which have been studied; the application of this type of structure as a hot-electron injector in the hot-electron transistor and hot-electron Gunn diode will then be described. For thin potential barriers, tunnelling becomes significant, and this is exploited in the tunnelling hot-electron transistor, the resonant tunnelling diode and resonant tunnelling transistor. Some applications of superlattice effects will then be described, before moving on to consider HBTs.

10.3.1 Unipolar Diodes

Much of the early work on this topic was based on the use of a thin, heavily-doped p-type layer, which was placed between two n-type regions in such a way that the p-layer was fully depleted; the resultant space charge created a potential barrier so that electron flow between the n-regions was limited by thermonic emission over this barrier. With the doped regions touching, the device was referred to as a 'camel' diode (because of the hump!) (Shannon, 1979), whereas the form with intermediate undoped regions on either side of the p-layer is known as the planar-doped barrier diode (Malik *et al.*, 1980). Figure 10.14(a) shows the structure and conduction band profile of the latter device; it can be seen that,

(a)

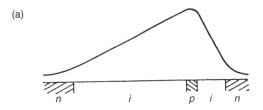

(b)

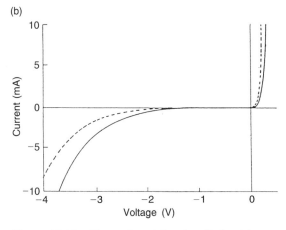

Figure 10.14. Planar-doped barrier diode: (a) structure and conduction band profile; (b) experimental (solid line) and theoretical (broken line) current-voltage characteristic. (From Kearney *et al.*, 1989, with permission of the IEE Publication Department.)

by placing the p^+ layer closer to one n^+ contact, an asymmetric barrier is formed, giving the rectifying I–V characteristic displayed in Fig. 10.14(b). Other variations on this principle have included the modification of Schottky barrier heights by using adjacent thin, heavily-doped layers (Woodcock and Harris, 1983) (a p-layer raises the effective barrier height, and an n-layer reduces it), and the creation of abrupt potential steps by depositing a closely spaced pair of delta-doped planes of opposite type (Shen *et al.*, 1992). The potential profile and I–V behaviour of this 'delta-dipole' diode are shown in Fig. 10.15. Alternatively, a compositional variation of, say, Al in $Al_xGa_{1-x}As$ can be used to create a graded potential profile (Allyn *et al.*, 1980), although for this example the maximum barrier height is limited to about 370 meV at $x = 0.45$, since for higher x the Γ conduction band minimum lies above the X minimum, i.e. the band gap becomes indirect.

10.3.2 Hot-electron Devices

In addition to providing a barrier to thermonic emission of carriers, the unipolar structures described above provide a means of injecting carriers with high kinetic energy into regions of a device where the local carriers are in thermal equilibrium with the lattice. The injected 'hot' carriers will eventually scatter and thermalize with the local population but, if the thickness of this region is small compared with the scattering length, it is possible to exploit the properties of these hot carriers in the hot-electron transistor (HET), shown in Fig. 10.16(a). Under the bias conditions shown in Fig. 10.16(b), an emitter current I_e is injected into the base at energy eV_{eb}, and after some of the electrons lose energy due to scattering in the base, a smaller current I_c passes over the second barrier and enters the collector. The rate of loss of energy of hot electrons due to optical phonon emission is ~ 0.16 eV/ps, and for an electron velocity of 10^5 m/s, this implies that the base thickness should be less than 0.1 μm to avoid significant energy loss. The base transport factor, α, is defined as the fraction of the emitter current which enters the collector, i.e. I_c/I_e (or $\partial I_c/\partial I_e$ in the small signal regime); the base current is given by $I_b = (1 - \alpha)I_e$ and thus the current gain is

$$\beta = \frac{\partial I_c}{\partial I_b} = \frac{\alpha}{1 - \alpha} \tag{10.21}$$

Early studies using $GaAs/Al_xGa_{1-x}As$ structures gave poor results which theoretical analysis (Hayes and Levi, 1986) showed was due to the limited injection energy and strong base scattering, but recent results for a device using $In_xGa_{1-x}As/InP$ have demonstrated β values of over 100 at 77 K (Chen *et al.*, 1992). (The performance of this device was also enhanced by using a tunnel barrier in the emitter region, to reduce the energy spread of injected electrons—tunnelling is discussed further in Section 10.3.3.) A major problem with the HET is the conflicting requirements for the base region, i.e. that it should be thin to maximize α, and reduce the

(a)

ohmic contact
0.5 µm GaAs $N_D = 4 \times 10^{18}$ cm^{-3}
1 nm undoped GaAs
5 nm undoped GaAs
1 nm undoped GaAs
1 µm GaAs $N_D = 1 \times 10^{16}$ cm^{-3}
n^+ GaAs substrate
In back contact

$\delta_n = 1 \times 10^{13}$ cm^{-2}

$\delta_p = 1 \times 10^{13}$ cm^{-2}

(b)

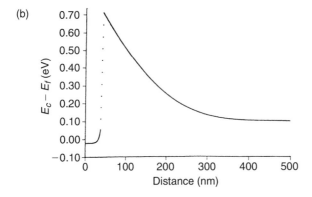

(c)

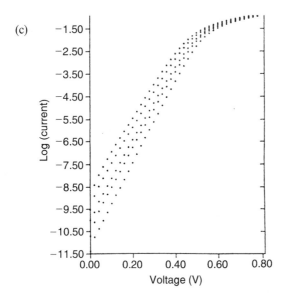

Figure 10.15.   Delta-dipole diode: (a) structure; (b) conduction band profile; (c) current-voltage characteristics. (From Shen *et al.*, 1992, with permission of the authors.)

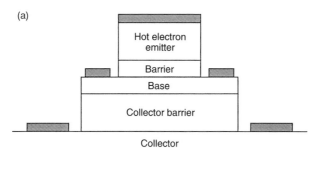

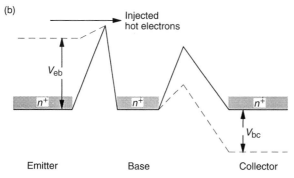

Figure 10.16. Hot-electron transistor: (a) structure; (b) conduction band profile under zero bias (solid lines) and under operating conditions (broken lines).

transit time, but that this gives a large base resistance, R_b, which increases the RC time constant associated with varying the charge on the emitter-base capacitance, C_{eb}, thereby reducing the operating frequency of the device.

Another device which benefits from hot-electron injection is the Gunn diode, which relies for its operation on the transfer of electrons from the lowest-energy, high-mobility Γ conduction band minimum to the higher-energy, low-mobility X and L minima. In the conventional structure, there is a drift region adjacent to the cathode in which the electrons are accelerated to sufficient energy for them to scatter into the subsidiary minima, thereby producing the desired negative differential resistance. By incorporating a potential barrier in the cathode structure, as shown in Fig. 10.17, electrons are injected with high enough energy to scatter immediately into the X and L valleys, thus eliminating this cathode dead space and producing a device with higher output power at high frequencies, greater DC-to-microwave conversion efficiency, lower temperature sensitivity and reduced noise (Couch *et al.*, 1989).

10.3.3 Resonant Tunnelling Structures

Quantum mechanical tunnelling is a non-classical process which arises from the penetration of an electronic wavefunction into a potential barrier such as an

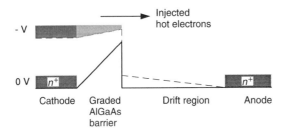

Figure 10.17.   Gunn diode with graded $Al_xGa_{1-x}As$ hot-electron injection barrier in the emitter: for zero bias (solid lines) and for operating conditions (broken lines).

$Al_xGa_{1-x}As$ layer in a GaAs crystal. If the barrier is thin enough, there is a finite probability of the electron emerging on the other side of the barrier (Fig. 10.18(a)). In a double barrier structure (Fig. 10.18(b)), the presence of a confined state in the quantum well region enhances the tunnelling probability of electrons incident at that energy, and hence there is a peak in the transmission probability versus energy curve (Fig. 10.18(c)). This is referred to as 'resonant tunnelling'. In practice, the incident electron energy is varied by applying a bias voltage across the structure. The resultant I–V curve, sketched in Fig. 10.18(d), is the super-position of a peak due to each resonant tunnelling process (only one is shown) on a gradually-increasing background due to non-resonant tunnelling. This type of curve has two important features: firstly, it shows a negative differential resistance (NDR) region on the high-energy side of the peak which can be used as as source of microwave power; and secondly, it allows new devices with novel operating characteristics to be fabricated, as discussed below.

   The simplest device to make use of resonant tunnelling NDR is the resonant tunnelling diode, which consists of a double- (or sometimes triple-) barrier struc-ture sandwiched between n^+ contacts, as shown in Fig. 10.19(a). In order to reduce the applied voltage necessary to bring the electrons in the emitter into resonance with the confined level in the quantum well, it has become usual to make the well region of narrower-gap material than the emitter, as shown. The I–V characteristics of such a device (Broekaert et $al.$, 1989) are displayed in Fig. 10.19(b), where the large ratio of 'peak' to 'valley' currents in the NDR region is close to state-of-the-art. When mounted in a microwave resonator cav-ity, and biassed into the NDR region, the diode becomes unstable and oscillates; frequencies of up to 420 GHz have been measured (Brown et $al.$, 1989), but because of the generally low current densities involved in the tunnelling process, the output power which could be obtained at this frequency was less than 1 µW.

   In the resonant tunnelling transistor, the NDR behaviour is used to add increased functionality to the device characteristics. Several variants have been reported, with the resonant tunnelling structure positioned in the gate of a HEMT device (Sen et $al.$, 1987), in the emitter region of an HET (Yokoyama et $al.$, 1985)

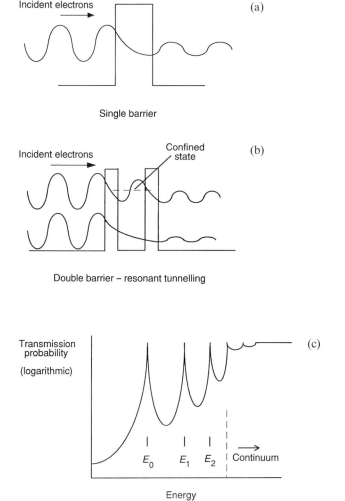

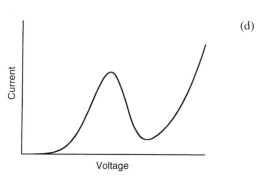

Figure 10.18. (a) Quantum mechanical tunnelling through a thin heterojunction barrier; (b) resonant tunnelling through a double barrier structure when the incident electron energy coincides with the confined-state energy; (c) schematic transmission probability curve as a function of electron energy for a quantum well with three confined states (note that the vertical axis is logarithmic); (d) idealized current–voltage curve showing one resonance.

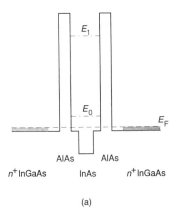

AlAs   AlAs

n^+ InGaAs   InAs n^+ InGaAs

(a)

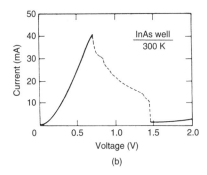

(b)

Figure 10.19. $In_{0.53}Ga_{0.47}As/AlAs/InAs$ resonant tunnelling diode: (a) conduction band profile; (b) room temperature current–voltage characteristic; the broken curve corresponds to the negative differential resistance region. (From Broekaert *et al.*, 1988, with permission of the authors.)

or in the emitter (Futatsugi *et al.*, 1987) or base (Capasso *et al.*, 1986) of an HBT. Figure 10.20 shows a double-barrier structure incorporated into the emitter region of a hot electron transistor, and the corresponding DC characteristics; the presence of three stable operating points for a given current value within the NDR range opens up the possibility of multi-level logic, which gives a reduction in the number of devices required to perform a given logical operation, thereby reducing the complexity of integrated circuits. For example, one resonant tunnelling transistor can perform an exclusive-NOR operation normally requiring eight conventional transistors (Yokoyama *et al.*, 1985).

10.3.4 Superlattice Devices

As discussed in Section 3.5, a superlattice structure is formed by a repeated sequence of wells and barriers in which, as in the resonant tunnelling structure,

(a)

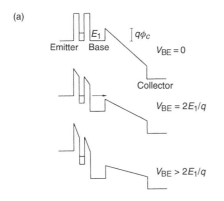

(b)

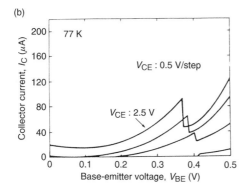

Figure 10.20. Resonant tunnelling hot electron transistor: (a) operating principle; (b) collector current versus base-emitter voltage, as a function of collector-emitter voltage. (From Yokoyama *et al.*, 1985, with permission of the publishers.)

the barrier is sufficiently thin to allow significant coupling of the wavefunctions between adjacent quantum wells. This smears out the sharply defined confined states into a 'superlattice miniband' which extends throughout the structure and allows current to flow in the vertical direction. This structure acts as an energy filter, and has been used, for example, in the superlattice tunnel diode (Davies *et al.*, 1989), which is the unipolar equivalent of the Esaki (p^+n^+) tunnel diode. Its structure, shown in Fig. 10.21(a), consists of two superlattice sections separated by a sightly thicker tunnel barrier. The current-voltage curve for this device, and the corresponding miniband alignments, are shown in Fig. 10.21(b). It can be seen that as the bias level is increased across the tunnel barrier, the current rises initially due to tunnelling of electrons from filled states in one miniband into empty states in the corresponding miniband on the other side of the barrier. Eventually the current will fall when the Fermi level on the left-hand side of the barrier is between the two minibands on the right-hand side of the barrier. When the second miniband on the right-hand side becomes accessible to electrons tunnelling through the barrier, the current rises again.

(a)

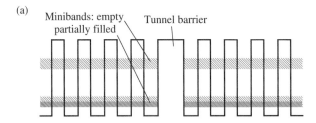

(b)

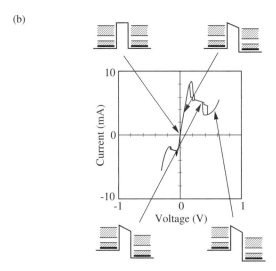

Figure 10.21.   Superlattice tunnel diode: (a) structure of device; (b) current-voltage characteristic showing miniband alignment in the various bias regions. (From Davies *et al.*, 1989, with permission of Institute of Physics Publishing.)

10.3.5 Heterojunction Bipolar Transistors

As with the HET (Section 10.3.2), two of the factors which limit the high frequency performance of a bipolar transistor are (a) the transit time for a carrier, injected from the emitter, to cross the base region and enter the collector, and (b) the RC time constant associated with varying the charge on the emitter-base junction capacitance, C_{eb}, through the base resistance, R_b. The transit time can be reduced by making the base thinner, but since the base current flows into this region from the edges, this also has the effect of increasing R_b and, hence, the time constant, unless the base doping is correspondingly increased. However, since the base-emitter leakage current (and hence the total base current, I_b) increases with base doping, the current gain, β, will consequently fall (β is defined as I_c/I_b, where I_c is the collector current).

A solution was proposed as early as 1957 (Kroemer, 1957), namely the use of a wider band gap material for the emitter, so that the band offset at the emitter-base

(a)

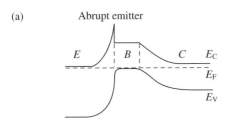

(b)

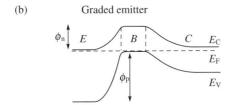

(c)

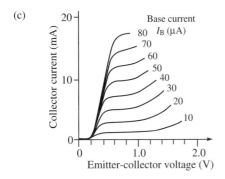

Figure 10.22.   AlInAs/InGaAs heterojunction bipolar transistor: (a) band structure of abrupt-junction device; (b) as (a), for graded junction; (c) DC characteristics of graded-junction device. (From Malik *et al.*, © 1983 IEEE.)

junction would create an enhanced barrier to base-emitter current flow; however, it took more than two decades before such devices were realized in practice. The principle is illustrated in Fig. 10.22(a) for an *n-p-n* device with an abrupt hetero-junction. It can be seen that the valence band barrier to hole flow into the emitter is increased, but this also introduces a narrow triangular barrier in the conduction band. However, compositional grading gives a gradual transition in energy gap between emitter and base, and results in the band structure of Fig. 10.22(b); the conduction band peak has been removed, further enhancing the emitter efficiency.

A simple estimate of this enhancement can be obtained as follows: for negligible electron–hole recombination in the base, the collector current under an applied forward bias, V_a, will be approximately equal to the electron current injected from the emitter, i.e.

$$I_c \approx n_e e v_n \exp\left[-\frac{\phi_n - V_a}{k_B T}\right] \tag{10.22}$$

where n_e is the electron density in the emitter, v_n is the diffusion velocity of electrons across the base and ϕ_n is the equilibrium barrier to electron diffusion from emitter to base. Similarly, the base-emitter hole leakage current is

$$I_{be} \approx p_b e v_p \exp\left[-\frac{\phi_p - V_a}{k_B T}\right] \tag{10.23}$$

where p_b is the base doping level, v_p the hole diffusion velocity and ϕ_p is the barrier for hole flow from base to emitter. Hence

$$\beta \approx \frac{n_e v_n}{p_b v_p} \exp\left[-\frac{\phi_n - \phi_p}{k_B T}\right] \tag{10.24}$$

Inspection of Fig. 10.22(b) shows that

$$\phi_n \approx E_G(\text{base}) \qquad \text{and} \qquad \phi_p \approx E_G(\text{emitter}) \tag{10.25}$$

In a conventional device, these are equal and the exponential factor is unity. However, this is not the case for a heterojunction, so that the enhancement factor of β for the HBT compared with a conventional bipolar device is given by

$$\frac{\beta(\text{HBT})}{\beta(\text{bipolar})} = \exp(\Delta E_G / k_B T) \tag{10.26}$$

For an $Al_{0.25}Ga_{0.75}As/GaAs$ device, $\Delta E_G = 0.31 = 12 k_B T$ at room temperature, giving an enhancement factor of over 10^5. The base doping can therefore be increased by several orders of magnitude compared with the conventional device, to reduce the base resistance, while still maintaining a low base-emitter leakage current and consequent high gain.

Other materials combinations which have been used for HBTs are $InP/In_{0.53}Ga_{0.47}As$, $Al_{0.52}Ga_{0.48}As/In_{0.53}Ga_{0.47}As$ and Si/Si_xGe_{1-x}. This final pairing has the complication that the lattice constant of the narrower-gap SiGe base region is larger than that of Si, producing a strained, pseudomorphic layer, similar to that for the InGaAs pseudomorphic HEMT structures described in Section 10.2.3. Indeed, the Si/SiGe combination is also being investigated for pseudomorphic HEMT applications, but since most of the band-gap discontinuity appears in the valence band (an advantage for n-p-n HBTs), most devices are relatively low mobility p-type HEMTs. However, n-channel structures can also fabricated by taking advantage of the effect of strain on the conduction band energy (People, 1986).

The DC performance characteristics of an HBT are qualitatively similar to that of the conventional bipolar device, as illustrated in Fig. 10.22(c) for an AlInAs/InGaAs structure (Malik *et al.*, 1983). However, the high frequency performance is much enhanced, primarily due to the ability to engineer thin, low resistance base regions; for example, an $InP/In_{0.53}Ga_{0.47}As$ device has been reported with a unity current-gain cutoff frequency, f_T, of 165 GHz (Chen *et al.*, 1989), and IBM is developing 1 GHz analogue ICs using SiGe HBTs.

10.4   Conclusions

The ability of epitaxial growth techniques to deposit quasi-two-dimensional layers of semiconductor materials with controlled thickness on the nanometre scale has allowed the feasibility of a large number of novel electronic device concepts to be demonstrated. Apart from the Si MOSFET, which while being two dimensional is not epitaxial, it is the HEMT, in its various guises, and the HBT, which have made the largest commercial impact, mainly as a result of their high speed and, in the case of the HEMT, its low noise characteristics. Nevertheless the potential for higher-speed operation and increased functionality of many of the other devices discussed here should ensure continued interest from device physicists and engineers for some time to come.

EXERCISES

1.   To compare the performances of a Si MOSFET and a GaAs/AlGaAs HEMT, the two devices are fabricated with the following common dimensions:

gate length, L	1 μm
gate width, z	100 μm
oxide/AlGaAs thickness, d	0.05 μm

The velocity–field characteristics of these devices are shown in Figs. 10.4 and 10.6 (marked 2DEG). From these curves,

(a) estimate the critical field, E_c, and hence the critical voltage, V_c, for each device. Use this information to determine whether the transistors are operating in the low-field or high-field regime at an effective gate voltage, V_g', of 0.1 V and 5 V.

(b) calculate the source-drain voltage, V_{dsat}, at which current saturation occurs for each effective gate voltage, and by converting this into an electric field along the channel, comment on whether the velocity–field data shown above support your assignment of the operating regimes.

(c) Using the appropriate limiting form, calculate the value of the transconductance, g_m, in the current saturation regime for both devices at each effective gate voltage. ($\epsilon_i = 3.4 \times 10^{-11}$ F/m for SiO$_2$ and $\epsilon_i = 1.15 \times 10^{-10}$ F/m for AlGaAs). Also calculate the cut-off frequency, $f_T = g_m/2\pi C_g$, of each device in the high-field limit, and comment on the difference.

2.   Using equation (10.20), derive an expression for the transconductance, g_m', of a HEMT in the presence of a contact resistance, R_C, and relate this to the ideal value, g_m, given in equation (10.19). Calculate g'_m for a device with a g_m of 100 mS and a contact resistance of 10 Ω.

3. From Fig. 10.12, deduce the saturation drift velocity in each type of FET device shown.

4. A delta-doping dipole, grown in GaAs, consists of an n-type plane of 10^{17} m^{-2} Si atoms separated by d nm from a Be-doped p-type plane of the same density.

(a) Calculate the height of the potential step for $d = 5$ nm using $\epsilon_i = 1.15 \times 10^{-10}$ F/m for GaAs. How does this result compare with Fig. 10.15(b)?

(b) What is the step height for $d = 0.3$ nm (i.e. ~1 monolayer)? Comment on the relevance of this to the ultimate accuracy of controlling such potential barriers.

5. The WKB approximation (Section 3.3.1) states that the probability P of an electron of energy E tunnelling through an arbitrarily shaped potential barrier of thickness d, whose height $V(x)$ varies with position x, is given by:

$$P = \exp\left[-2\int_0^d k(x)\,dx\right]$$

where

$$\frac{\hbar^2[k(x)]^2}{2m^*} = V(x) - E$$

(a) Apply the WKB formula to the GaAs/AlGaAs/GaAs tunnel barrier shown in Fig. 10.18(a) for an incident electron of energy E above the GaAs conduction band edge. What is the value of the probability for a barrier height of 0.3 V, a barrier thickness of 10 nm and $E = 0.05$ eV? Use $m^* = 0.067\,m_0$, where $m_0 = 9.1 \times 10^{-31}$ kg.

(b) A potential difference of V_a is applied across the insulating AlGaAs barrier. Sketch the resultant shape of the conduction band edge, and calculate the new tunnelling probability for an electron incident on the barrier from the negatively biassed side at an energy E above the GaAs band edge. Using the values from (a) above, evaluate P when $V_a = 0.1$ V.

(c) The single AlGaAs barrier is replaced by a double-barrier structure with a narrow GaAs quantum well, width a, in the centre (Fig. 10.18(b)). Using the formula for the confined state energies of an infinitely deep well, i.e.

$$E_n = \frac{n^2\pi^2\hbar^2}{2m^*a^2}$$

calculate the applied voltages at which the first two resonances will occur if $a = 2$ nm.

6. A Si bipolar transistor has the following properties:

cross-section, $L \times W$	$100\,\mu m \times 20\,\mu m$
base thickness, t	$1\,\mu m$
base doping, p_b	$10^{24}\,\mu m^{-3}$
hole mobility, μ_h	$0.1\,m^2/Vs$
emitter-base junction thickness, d	$0.05\,\mu m$
dielectric constant, ϵ_i	$1.06 \times 10^{-10}\,F/m$

If the base contact is made to both of the $100\,\mu m$-long edges, estimate the cutoff frequency, f_T, set by the base emitter RC time constant.

7. An HBT of identical dimensions as the bipolar transistor in Exercise 6 has a base region of $Si_{0.9}Ge_{0.1}$ alloy. If the strained $Si_{1-x}Ge_x$ bandgap reduces by 8 meV for each per cent of Ge, calculate the increased base doping level which can be accommodated for the same current gain as the all-Si device. Hence, find the new cut-off frequency.

References

C. L. Allyn, A. C. Gossard and W. Weigmann, *Appl. Phys. Lett.* **36**, 373 (1980).

A. Bar-Lev, *Semiconductors and Electronic Devices* (Prentice Hall International, 1984).

J. C. Bourgoin, ed., *Physics of DX Centers in GaAs Alloys*, Solid State Phenomena **10** (Sci-Tech Publications, Brookfield, USA, 1989).

T. P. E. Broekaert, W. Lee and C. G. Fonstad, *Appl. Phys. Lett.* **53**, 1545 (1989).

E. R. Brown, T. C. L. G. Sollner, C. D. Parker, W. D. Goodhue and C. L. Chen, *Appl. Phys. Lett.* **55**, 1777 (1989).

C. Canali, C. Jacoboni, F. Nava, G. Ottaviana and A. Alberigi-Quaranta, *Phys. Rev. B* **12**, 2268 (1975).

F. Capasso, S. Sen, A. C. Gossard, A. L. Hutchinson and J. H. English, *IEEE Electron. Dev. Lett.* **EDL-7**, 573 (1986).

W. L. Chen, J. P. Sun, G. I. Haddad, M. E. Sherwin, G. O. Munns, J. R. East and R. K. Mains, *Appl. Phys. Lett.* **61**, 189 (1992).

Y. K. Chen, R. N. Nottenburg, M. B. Panish, R. A. Hamm and D. A. Humphrey,*IEEE Electron. Dev. Lett.* **EDL-10**, 267 (1989).

N. R. Couch, H. Spooner, P. H. Beton, M. J. Kelly, M. E. Lee, P. K. Rees and T. M. Kerr, *IEEE Electron. Dev. Lett.* **EDL-10**, 288 (1989).

M. B. Das and M. L. Reszak, *Solid State Electr.* **28**, 997 (1985).

R. A. Davies, E. G. Bithell, A. Chew, P. G. Harris, C. Dineen, M. J. Kelly, W. M. Stobbs, D. E. Sykes and T. M. Kerr, *Semicond. Sci. Technol.* **4**, 35 (1989).

T. Futatsugi, Y. Yamaguchi, K. Imamura, S. Muto, N. Yokoyama and A. Shibatomi, *Japan. J. Appl. Phys.* **26**, L131 (1987).

J. J. Harris, *J. Mater. Sci.: Materials in Electronics* **4**, 93 (1993).

J. J. Harris, in *Properties of GaAs*, M. Brozel and G. E. Stillman, eds., EMIS Dataview No. 16 (INSPEC, IEE, Stevenage, UK, 1997), p. 61.

J. R. Hayes and A. F. J. Levi, *IEEE J. Quant. Electr.* **QE-22**, 1744 (1986).

M. J. Kearney, M. J. Kelly, R. A. Davies, T. M. Kerr, P. K. Rees, A. Condie and I. Dale, *Electron. Lett.* **25**, 1454 (1989).

M. Keever, H. Shichijo, K. Hess, S. Banerjee, L. Witkowski, H. Morkoç and B. G. Streetman, *Appl. Phys. Lett.* **38**, 36 (1981).

K. von Klitzing, G. Dorda and M Pepper, *Phys. Rev. Lett.* **45**, 494 (1980).

H. Kroemer, *Proc. IRE.* **45**, 1535 (1957).

M. Laviron, D. Delagebeaudeuf, P. Delescluse, J. Chaplart and N. T. Linh, *Electron. Lett.* **17**, 536 (1981).

S. Luryi, A. Kastalsky, A. C. Gossard and R. H. Hendel, *Appl. Phys. Lett.* **45**, 1294 (1984).

R. J. Malik, T. R. Aucoin, R. L. Ross, K. Board, C. E. C. Wood and L. F. Eastman, *Electron. Lett.* **16**, 836 (1980).

R. J. Malik, J. R. Hayes, F. Capasso, K. Alavi and A. Y. Cho, *IEEE Electron. Dev. Lett.* **EDL-4**, 383 (1983).

W. T. Masselink, *Thin Solid Films* **231**, 86 (1993).

R. People, *IEEE J. Quantum Electr.* **QE-22**, 1696 (1986).

E. F. Schubert, *J. Vac. Sci. Technol. A* **8**, 2980 (1990).

E. F. Schubert, A. Fischer and K. Ploog, *IEEE Trans. Electron. Devices* **ED-33**, 625 (1986).

S. Sen, F. Capasso, F. Beltram and A. Y. Cho, *IEEE Trans. Electron. Dev.* **ED-34**, 1768 (1987).

J. M. Shannon, *Appl. Phys. Lett.* **35**, 63 (1979).

T.-H. Shen, A. C. Ford, M. Elliott, R. H. Williams, D. I. Westwood, D. A. Woolf, J. P. Marlow, J. E. Aubrey and G. Hill, *J. Vac. Sci. Technol. B* **10**, 1757 (1992).

M. Shur, *GaAs Devices and Circuits* (Plenum, New York, 1987).

P. M. Solomon and H. Morkoç, *IEEE Trans Electron Devices* **ED-31**, 1015 (1984).

S. M. Sze, *Semiconductor Devices: Physics and Technology* (Wiley, New York, 1985).

S. M. Sze, *High Speed Semiconductor Devices* (Wiley, New York, 1990).

C. Weisbuch and B. Vinter, *Quantum Semiconductor Structures* (Academic, San Diego, 1991), p. 151.

J. M. Woodcock and J. J. Harris, *Electron. Lett.* **19**, 93 (1983).

N. Yokoyama, K. Imamura, S. Muto, S. Hiyamizu and H. Nishi, *Japan. J. Appl. Phys.* **24**, L853 (1985).

A. Zrenner and F. Koch, *Inst. Phys. Conf. Ser.* **95**, 1 (1988).

Solutions to Selected Exercises

Chapter 1

1. $J = 2.43 \times 10^{22}$ atoms m^{-2} s^{-1}
2. $\lambda = 1.6$ cm (with $d = 2.5$ Å)
4. (a) $J = 0.73$ ML/s
   (b) $\delta J \sim 3\%$.
5. If $J = 0$, then $c(x) = c_{eq}$.
6. $a^2 c_{max} = 3.32 \times 10^{-6}$
11. The L- and ℓ-dependent part of the energy is

$$E - \sum_i C_0 = 2C_m \ln[\tfrac{1}{4}(L^2 - \ell^2)] - C_m \ln(L\ell) + 16C_d \frac{L^2 + \ell^2}{(L^2 - \ell^2)^2} + C_d \frac{L^2 + \ell^2}{L^2 \ell^2}$$

Chapter 2

2. $E_0^A / E_0^B = 4$
3. (a) One subband in well A, three in well B.

   (b) $N_A = \dfrac{\pi}{L^2}$, $N_B = \dfrac{11}{4L^2}$

   (c) For electrons in well A, $v_F = \sqrt{2}\,\dfrac{\hbar\pi}{m^*L}$. For electrons in the nth subband in well B, $v_F(n)$ is given by

   $$v_F(1) = \sqrt{11}\,\frac{\hbar\pi}{2m^*L}, \quad v_F(2) = \sqrt{2}\,\frac{\hbar\pi}{m^*L}, \quad v_F(3) = \sqrt{3}\,\frac{\hbar\pi}{2m^*L}$$

4.

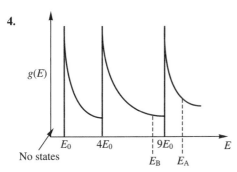

Figure A.1. Answer to Exercise 4: density of states for a quantum wire consisting of a narrow two-dimensional strip with straight parallel sides.

379

5. (a) $j_{total} = 0.35E_0 \dfrac{2e}{h}$

Chapter 3

1. (b)

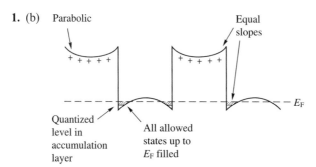

Figure A.2.   Answer to Exercise 1(b).

4. The WKB quantization condition is

$$\int_0^d \left\{ \frac{2m^*}{\hbar^2}[E_n - V(z)] \right\}^{1/2} dz = (n + \tfrac{3}{4})\pi$$

where E_n is related to the total energy of the electron by

$$E_{total} = E_n + \frac{\hbar^2}{2m^*}(k_x^2 + k_y^2)$$

5. (a) The system is charge-neutral.

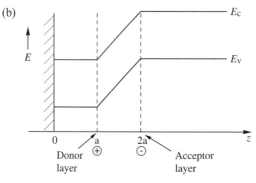

Figure A.3.   Answer to Exercise 5(b).

(c) $\Delta = \dfrac{e^2}{\epsilon_r \epsilon_0} Na$

6. (a) $A = \dfrac{e^2}{2\epsilon_r \epsilon_0} n_D, \quad E_n = (n + \tfrac{1}{2})\hbar\omega$ (c) $n = \dfrac{2.0m^*\omega}{\pi\hbar}$ (d) $n_D = \dfrac{4.0m^*e^2}{\pi^2\hbar^2 W^2 \epsilon_r \epsilon_0}$

$$\omega = \left(\frac{e^2}{2\epsilon_r \epsilon_0 m^*} n_D \right)^{1/2} = \sqrt{\frac{2A}{m^*}}$$

7. (a) 2ℓ

(b)

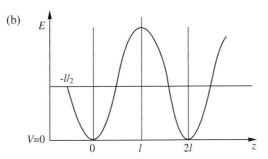

Figure A.4.   Superlattice band-edge potentials for Exercise 7(b).

(c) For $-\frac{1}{2}\ell \le z \le \frac{1}{2}\ell$, $V = \dfrac{e^2}{2\epsilon_r\epsilon_0} n_0 z^2$

For $\frac{1}{2}\ell \le z \le \frac{3}{2}\ell$, $V = \dfrac{e^2}{2\epsilon_r\epsilon_0} n_0 [\frac{1}{2}\ell^2 - (z-\ell)^2]$

8. (a) $V_0 = \dfrac{e^2}{4\epsilon_r\epsilon_0} n_0 \ell^2$

(b) $E_0 = \frac{1}{2}\hbar\omega = \frac{1}{2}\hbar \left(\dfrac{e^2 n_0}{\epsilon_r\epsilon_0 m^*} \right)^{1/2}$

(c) $\Delta E = \dfrac{\hbar\omega}{\pi} [2\cos(\frac{1}{10}\pi)]e^{-\tau}$, where ω is given in (b) and τ is given by equation (3.59) with appropriate limits of integration.

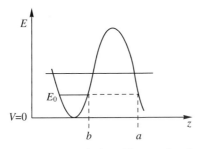

Figure A.5.   Limits of integration b, a for Exercise 8.

9. (a) $E_0 \rightarrow \sqrt{2}E_0$

(b) $V_0 \rightarrow 2V_0$

(c) Bandwidth: $\omega \rightarrow \sqrt{2}\omega$ (this would increase the miniband width ΔE); $\tau \rightarrow$ much bigger (this decreases ΔE strongly because of its exponential dependence on τ [see solution to Exercise 8(c)].

10. (b)

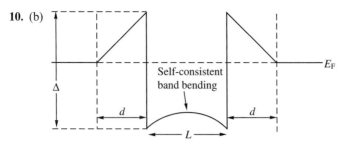

Figure A.6.   Answer to Exercise 10(b).

(d) Using $m^*/m_e = 0.067$, $N_s = 1.57 \times 10^{12}\,\mathrm{cm}^{-2}$, $E_F = 112\,\mathrm{meV}$

(f) $E_F = \Delta - \dfrac{e^2 N_\delta}{\epsilon_r \epsilon_0}d$

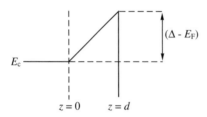

Figure A.7.   Band-edge potential for Exercise 10(f).

Chapter 4

1. (a)

$$\omega^2 = \frac{\kappa}{m_1 m_2}\left(m_1 + m_2 \pm \{m_1^2 + m_2^2 + 2m_1 m_2[\cosh(\beta a)\cos(\alpha a) - i\sinh(\beta a)\sin(\alpha a)]\}^{1/2}\right)$$

4. The derivation of the quantum wire temperature dependence in the EP regime follows closely the derivation for the quantum well [see part (b) of Exercise 3]. In the BG regime, $\overline{\tau_m^{-1}} \propto \exp(-2\hbar s k_F / k_B T)$.

Chapter 5

9. (b) $\varrho_{xy} = -\dfrac{1}{\sigma_{xy}}$ **(c)** $\varrho_{xx} \to 0$

11. $n_s = 1.9 \times 10^{11}\,\mathrm{cm}^{-2}$, $\mu = 2.4 \times 10^6\,\mathrm{cm}^2\,\mathrm{V}^{-1}\,\mathrm{s}^{-1}$

Chapter 6

1. $g_{QW}(E) = \dfrac{m_e^*}{\pi \hbar^2}\displaystyle\sum_{n=1}^{\infty}\theta(E - E_n)$, where $E_n = \dfrac{\hbar^2 \pi^2 n^2}{2m_e^* L^2}$.

2.   (a) $g(E)$

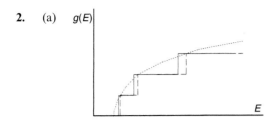

(b) $g(E)$

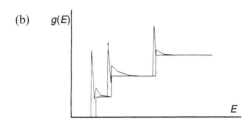

(c) $g(E)$

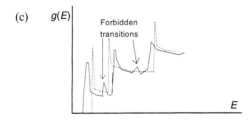

Figure A.8.   Figures for solution to Exercise 2(a,b,c).

4. The electron energy level must be more than $0.198\,\text{eV}$ above the bottom of the QW for the system to become Type II. The system will become Type II when the well width is less than $53\,\text{Å}$.

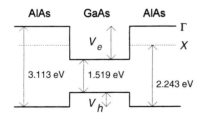

Figure A.9.   Figure for solution to Exercise 4.

7. (a) The electron effective mass is replaced by the *reduced* effective mass μ, where

$$\frac{1}{\mu} = \frac{1}{m_e^*} + \frac{1}{m_h^*}$$

8. We expect to see only two electron states.

(i) $\Delta L/L = 5\%$ (ii) $\Delta m/m = 11\%$ (iii) $\Delta V/V = 10\%$

Chapter 7

3. (a) $k_{opt} = 1.99 \times 10^8$ m^{-1}
   (b) $n = 2.66 \times 10^{17}$ cm^{-3}
   (c) Energy density $= 1.8 \times 10^{-3}$ J m^{-2}
   (d) $I = 1.8$ MW m^{-2}
   (e) 28 µs
4. (a) $N_{2D} = 7.368 \times 10^{16}$ m^{-2}
   (b) $L_z = 18.33$ nm
   (c) Electron and heavy hole confinement energies are 154 meV and 84.7 meV, respectively.
   (d) $L_z = 27.7$ nm
   (e) $L_z^{crit} \sim 308$ nm

Chapter 8

1. (a) $L = \dfrac{\hbar\pi}{\sqrt{2}}\left[\dfrac{m_e^* + m_h^*}{m_e^* m_h^*}\left(\dfrac{1}{E - E_g}\right)\right]^{1/2}$, where m_e^* and m_h^* are the effective masses of the lowest electron and hole states, respectively, E_g is the band gap, and E is the lasing energy.
   (b) For $\lambda = 1.55$ µm, $L = 128.6$ Å. For $\lambda = 1.3$ µm, $L = 63.7$ Å.
2. (b) $L = 94$ Å
3. $L = 500$ µm. $\Delta_{gth} = 20.4$ cm^{-1}

4. $P_0 = \eta_{inj} N_c \left[\dfrac{\ln(1/R)}{\alpha L + \ln(1/R)}\right]$

Chapter 9

3. (a) $\Delta E = 74.4$ meV
   (c) $\Delta\phi = \frac{3}{2}\pi$
   (d) $B = 3.3$ mT
5. $C_g = 1.6$ aF, $C_T = 16$ aF, $T_{max} = 58$ K
6. (a) $n_0 = 4.6 \times 10^{11}$ cm^{-2}
   (b) $\mu_0 = 545\,000$ cm^2/V s
   (c) $\ell_0 = 6.1$ µm
   (d) $W = 1.02$ µm
   (e) $p = 0.68$

Chapter 10

1. (a)

	MOSFET	HEMT*	
$E_c = v_{sat}/\mu_0$	7.0×10^5	2.1–2.5×10^5	(V/m)
$V_c = LE_c$	0.70	0.21–0.25	(V)

* Here and below, the range of values for the HEMT corresponds to the range of velocities in the saturation regime. Any answer in this range is acceptable.

For $V'_g = 0.1\,\text{V}$, this is less than V_c, which thus corresponds to the low-field regime in both devices, while for $V'_g = 5\,\text{V}$, this is greater than V_c, which corresponds to the high-field regime.

(b)

	MOSFET	HEMT*	
$V_{\text{dsat}}(0.1\ \text{V})$	9.4×10^{-2}	$\sim 8.5 \times 10^{-2}$	(V)
E	9.4×10^{4}	$\sim 8.5 \times 10^{4}$	(V/m)

These fields are in the linear region of the $v - F$ curves, i.e the low-field is confirmed.

	MOSFET	HEMT*	
$V_{\text{dsat}}(5\ \text{V})$	2.0	1.35	(V)
E	2.0×10^{6}	1.35×10^{6}	(V/m)

For Si, the field is close to the 'knee' of the $v - F$ curve, i.e. just below the high-field regime. However, for the HEMT, the result is far into the velocity saturation region.

(c)

	MOSFET	HEMT*	
g_m (0.1 V)	9.7×10^{-4}	1.6×10^{-2}	(A/V)
g_m (5 V)	6.8×10^{-3}	$3.5 - 4.1 \times 10^{-2}$	(A/V)

Thus,

	MOSFET	HEMT*	
f_T	15.9	23.9–28.6	(GHz)

2. $g'_m = 50\,\text{mS}$

3. Taking the values at 1 μm gate length:

Device	MOSFET	MESFET	HEMT[1]	HEMT[2]	HEMT[3]	
f_T	3.8	13.0	17.5	23.3	32.8	(GHz)
v_{sat}	2.4×10^{4}	8.2×10^{4}	1.1×10^{5}	1.5×10^{5}	2.1×10^{5}	(m/s)

HEMT[1] is GaAs/AlGaAs, HEMT[2] is InGaAs/GaAs, and HEMT[3] is InGaAs/AlInAs.

4. (a) $V = 0.697\,\text{V}$
(b) $V = 0.042\,\text{V}$

5. (a) $P = 1.76 \times 10^{-6}$
(b) $P = 7.38 \times 10^{-6}$

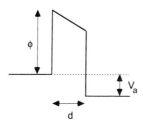

Figure A.10.   Figure for solution to Exercise 5(b).

(c) $V_a = 2.6\,\text{V}$ and $V_a = 10.4\,\text{V}$ for the first and second resonances, respectively.

6. $f_T = 12\,\text{GHz}$

7. $p_b = 2.3 \times 10^{25}\,\text{cm}^{-3}$, $f_T = 273\,\text{GHz}$

Index

holes (*cont.*)
  and strain 281–4
  transport 373, 374
hopping conductivity
  variable range 154
hot-electron transistor (HET) 365, 367, 368, 372
hut clusters 40–2, 50

impurities 94, 97, 112, 113, 114, 131, 133, 135,
    141, 142, 163, 166, 235, 313, 355, 361
InAs/GaAs 255–256
InAs/GaAs(001) 10, 38–40, 45
InAs/GaSb 33, 34, 116
InGaAs 34, 40, 267, 289, 363, 370
InGaAsP 293
injection 262, 267, 278, 279
  current 270, 277
  efficiency 269–70
  of hot carriers 365, 367
  levels 263, 266, 267, 271, 292
InP 269, 360
InP/InGaAs 217, 219, 365, 374
interface states 352, 355
inversion (*see* population inversion)
inversion
  criterion for 350–1
  threshold for 352
  inversion layer 85, 89, 90, 93, 95, 96, 298
inversion symmetry (in second-order
    susceptibility) 230
irreversible aggregation 25

Kane model 184–5, 187, 192, 194
  optical spectra 207, 211, 212
k.p. theory (*see* effective mass)
kinetic roughening 28
Knudsen cell 4, 7, 11, 47, 48
Kramers–Kronig relation 232

Landau gauge 177
Landau levels 166, 177
  fractional quantum Hall effect 174
  quantum Hall effect 171–2
  Shubnikov–de Haas effect 167
  two-dimensional electron gas 167
Landau–Baber scattering 303
Landauer–Büttiker formula 315–16, 318
lasers 2, 7, 23, 32, 42, 46, 215
  distributed feedback laser 271
  double heterostructure Fabry–Perot in-plane
    laser 269
  in-plane laser (IPL) 263–4, 266–7, 269, 271–1,
    286–91
  surface emitting laser (SEL) 286–8
  vertical cavity surface emitting laser (VCSEL)
    288–92
lattice constants
  AlAs 25
  GaAs 25
  of binary semiconductors 182
lattice misfit/mismatch 9, 10, 29, 30, 33, 34, 38,
    40, 42, 43, 117, 180, 222, 360, 374

localization 152, 171, 176
  and disorder 151
  length 157, 158
  phonon 131, 133
  in quantum Hall effect 172
  and quantum interference 158–9
  strong 298
  of water waves 152, 153
  weak 157–8, 163, 297, 298, 300, 301

Mach–Zender laser interferometer 310
magnetoconductance 308
magnetoresistance 164, 306, 325, 344
  (*see also* negative magnetoresistance)
matrix elements
  dipole 184, 185, 206, 207, 208, 222
  for electron–phonon scattering 136, 138, 144
  interband 276, 284–6
  Kane model 185, 192, 207
mean free path 158, 296, 297, 307, 323, 324, 345
  elastic 298, 314
  inelastic 97, 98
  semi-classical 154
mesoscopic structures/devices 32, 33, 57, 296,
    297, 309, 315, 318, 319, 323, 331, 342–3
metal-insulator-semiconductor field-effect
    transistor (MISFET) 84, 349
metal–insulator transition 155–7, 176
metalorganic molecular-beam epitaxy
    (MOMBE) 4, 7
metalorganic vapour-phase epitaxy (MOVPE) 4,
    6–7, 348
metal–oxide-semiconductor field-effect transistor
    (MOSFET) 84, 149, 158, 164, 167, 176,
    307, 308, 315, 343, 349–55, 356, 357, 359,
    360, 375
metal–semiconductor field-effect transistor
    (MESFET) 349, 355, 357, 362
minibands 108, 109–12, 114, 118, 120, 371
minimum metallic conductivity 154–5, 176
misfit dislocations (*see* dislocations)
mobility
  adatom 17–18
  edge 151–3, 176
  electron–phonon interactions 123, 139, 140,
    141, 143, 144, 145
  exciton 199
  of injected carriers 269
  in low-dimensional systems 136
  phonon-limited 139, 140, 141
  in two-dimensional electron gas 85, 93, 94,
    310
modulation-doped field-effect transistor
    (MODFET) 85, 355
modulation doping 82, 97, 144, 242, 257, 301,
    310, 314, 355, 358, 360
modulators
  light 235
  electro-optic 239, 249
molecular-beam epitaxy (MBE) 3–6, 7, 10, 11,
    16, 21, 23, 35, 38, 47, 61, 172, 330
molecular flow 5, 47–48